Periodic Table of the Elements

Symbol → C 6 ← Atomic Number
Name → Carbon
Atomic Weight → 12.011

Metal

Semimetal

Nonmetal

Period \ Group	1	2	3	4	5	6	7	8	9
1	H 1 Hydrogen 1.007								
2	Li 3 Lithium 6.941	Be 4 Beryllium 9.0122							
3	Na 11 Sodium 22.9898	Mg 12 Magnesium 24.305							
4	K 19 Potassium 39.0983	Ca 20 Calcium 40.078	Sc 21 Scandium 44.9559	Ti 22 Titanium 47.867	V 23 Vanadium 50.9415	Cr 24 Chromium 51.9961	Mn 25 Manganese 54.938	Fe 26 Iron 55.845	Co 27 Cobalt 58.9332
5	Rb 37 Rubidium 85.4678	Sr 38 Strontium 87.62	Y 39 Yttrium 88.9059	Zr 40 Zirconium 91.224	Nb 41 Niobium 92.9064	Mo 42 Molybdenum 95.94	Tc 43 Technetium 98.9063	Ru 44 Ruthenium 101.07	Rh 45 Rhodium 102.9055
6	Cs 55 Cesium 132.9055	Ba 56 Barium 137.327		Hf 72 Hafnium 178.49	Ta 73 Tantalum 180.9479	W 74 Tungsten 183.84	Re 75 Rhenium 186.207	Os 76 Osmium 190.23	Ir 77 Iridium 192.217
7	Fr 87 Francium 223.0197	Ra 88 Radium 226.0254		Rf 104 Rutherfordium 261.1088	Db 105 Dubnium 262.1142	Sg 106 Seaborgium 266.1219	Bh 107 Bohrium 264.1247	Hs 108 Hassium 269.13	Mt 109 Meitnerium 268.1388

Lanthanides	6	La 57 Lanthanum 138.9055	Ce 58 Cerium 140.116	Pr 59 Praseodymium 140.9077	Nd 60 Neodymium 144.24	Pm 61 Promethium 146.9151	Sm 62 Samarium 150.36
Actinides	7	Ac 89 Actinium 227.0278	Th 90 Thorium 232.0381	Pa 91 Protactinium 231.0359	U 92 Uranium 238.0289	Np 93 Neptunium 237.0482	Pu 94 Plutonium 244.0642

10	11	12	13	14	15	16	17	18
								He 2 Helium 4.0026
			B 5 Boron 10.811	C 6 Carbon 12.011	N 7 Nitrogen 14.0067	O 8 Oxygen 15.9994	F 9 Fluorine 18.9984	Ne 10 Neon 20.1797
			Al 13 Aluminium 26.9815	Si 14 Silicon 28.0855	P 15 Phosphorus 30.9738	S 16 Sulfur 32.0650	Cl 17 Chlorine 35.4533	Ar 18 Argon 39.948
Ni 28 Nickel 58.6934	Cu 29 Copper 63.546	Zn 30 Zinc 65.409	Ga 31 Gallium 69.723	Ge 32 Germanium 72.64	As 33 Arsenic 74.9216	Se 34 Selenium 78.96	Br 35 Bromine 79.904	Kr 36 Krypton 83.798
Pd 46 Palladium 106.42	Ag 47 Silver 107.8682	Cd 48 Cadmium 112.411	In 49 Indium 114.818	Sn 50 Tin 118.71	Sb 51 Antimony 121.76	Te 52 Tellurium 127.6	I 53 Iodine 126.9045	Xe 54 Xenon 131.293
Pt 78 Platinum 195.078	Au 79 Gold 196.9666	Hg 80 Mercury 200.59	Tl 81 Thallium 204.3833	Pb 82 Lead 207.2	Bi 83 Bismuth 208.9804	Po 84 Polonium 208.9824	At 85 Astatine 209.9871	Rn 86 Radon 222.0176
Ds 110 Darmstadtium 271.15	Rg 111 Roentgenuim 272.1535	Cn 112 Copernicium (277)	Nh 113 Nihonium (286.2)	Fl 114 Flerovium (289.19)	Mc 115 Moscovium (288.19)	Lv 116 Livermorium (293)	Ts 117 Tennessine (294)	Og 118 Oganesson (294)

Eu 63 Europium 51.96 14	Gd 64 Gadolinium 157.25	Tb 65 Terbium 22.9898	Dy 66 Dysprosium 162.5	Ho 67 Holmium 164.93	Er 68 Erbium 167.259	Tm 69 Thulium 168.9342	Yb 70 Ytterbium 173.04	Lu 71 Lutetium 174.967
Am 95 Americium 243.0614	Cm 96 Curium 247.0703	Bk 97 Berkelium 247.0703	Cf 98 Californium 251.0796	Es 99 Einsteinium 252.083	Fm 100 Fermium 257.0951	Md 101 Mendelevium 258.0984	No 102 Nobelium 259.101	Lr 103 Lawrencium 262.1097

4th edition

Electrochemistry

Seung Mo Oh

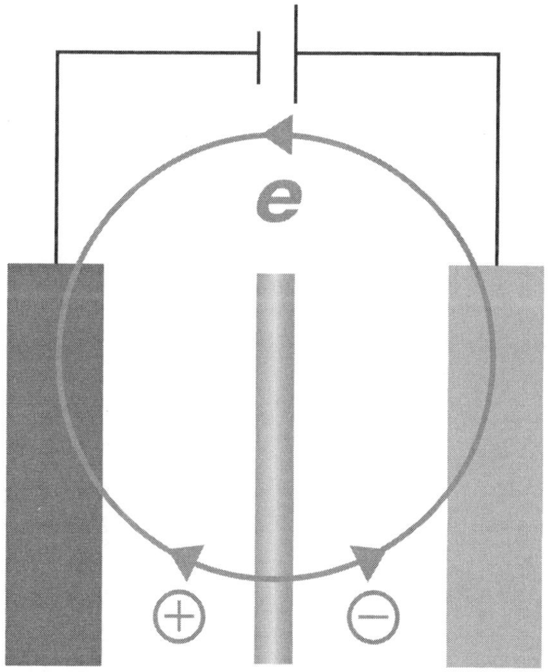

자유아카데미

Preface

In 2010, the book entitled "Electrochemistry" written in Korean was published. The intention of writing the 1st edition was to provide the reader a general overview on the basic concepts of electrochemistry, experimental techniques and electrochemical industries. Due to the limited space, all the concepts and techniques were not included; rather, the explanation was simplified as much as possible. However, an emphasis was laid throughout on the concept of "closed-loop" that should form for current flow in electrochemical cells, resulting in "cell polarization", which eventually has a significant impact on the performance of electrochemical cells.

Through the 2nd and 3rd edition, errors and incomplete explanations have been corrected and supplemented. During the last decades, as battery and fuel cell industries emerged world wide, interests in this field increased significantly. Accordingly, in this 4th edition, the content of Chap. 12 has been increased to supplement this part. This book is directed towards the students in early years of their graduate studies; the difficulty and quantity have been adjusted to be suitable for one semester course. If the contents marked with * are skipped, however, it is likely that one semester lecture at the level of undergraduate junior or senior is also possible.

I am indebted to many peoples who made valuable contributions to this book. I would like to particularly thank my former graduate students in Seoul National University for their drawing of the figures. Also, I would like to express my gratitude to the peoples in FREEDOM ACADEMY Publisher for their hard work for publication.

Please note that any revisions that may occur after publication of the 4th edition will be provided in the data room on the FREEDOM ACADEMY Publisher website (www.freeaca.com).

February 2024

Seung Mo Oh

Contents

CHAPTER

01 Introduction

CHAPTER

02 Electrode potential

CHAPTER

03 Electrolytes and ion conduction

CHAPTER

04 Electrochemical kinetics

CHAPTER

05 Electrochemical instrumentations

CHAPTER

06 Hydrodynamic methods

CHAPTER

07 Potential step methods

CHAPTER

08 Potential sweep methods

CHAPTER

11 Electrochemical industries

CHAPTER

12 Galvanic cells

CHAPTER 13 Photoelectrochemistry and solar cells

CHAPTER

01

Introduction

1-1 Some distinct concepts of electrochemistry

Electrons are involved in electrochemical reactions. In electrochemical oxidation reaction, electrons being emitted from an electroactive chemical species move to an electrode, whereas electrons move from an electrode to an electroactive species in electrochemical reduction reaction. The electrode on which electrochemical oxidation reaction takes place is called an ***anode***, whereas the electrode for an electrochemical reduction is called a ***cathode***.

$$Fe^{2+} \rightarrow Fe^{3+} + e \quad \text{(electrochemical oxidation on anode)}$$
$$Ce^{4+} + e \rightarrow Ce^{3+} \quad \text{(electrochemical reduction on cathode)}$$

For electrochemical reactions, electrons move between an electrode and an electroactive chemical species by tunneling. Note that the electroactive chemical species means a chemical species that can be electrochemically oxidized or reduced; it is also called a redox-active species or redox center. As shown in Fig.1-1, the tunneling rate decreases with an exponential function as the distance increases, manifesting itself that electron transfer is only possible near the electrode surface. Hence, the electroactive species located far from the electrode surface must move to the electrode surface to participate in electrochemical reactions; mass transfer of chemical species is required in solution. Some distinct

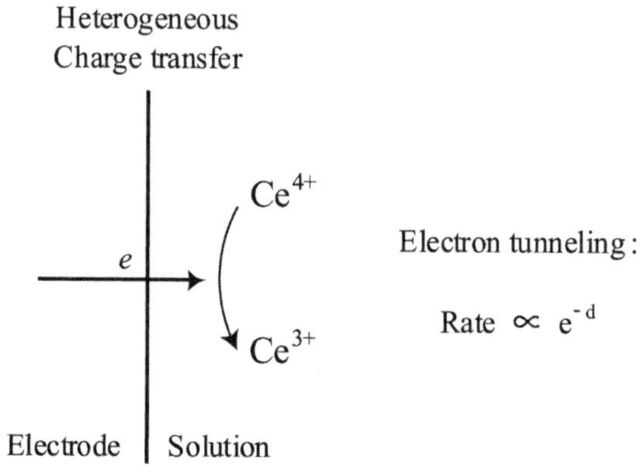

Fig. 1-1 Charge (electron) transfer by tunneling.

features of electrochemical reactions, which differ in many ways from chemical reactions, are described below.

Self learning 1-1

When the redox centers located inside a large-sized biomolecule (e.g. cytochrome *c* or ferredoxins) are electrochemically oxidized or reduced on an inert electrode (e.g. Pt), the rate is very slow. Give an explanation on this.

1-1-1 Electrode potential is an expression of electron energy

In physics, electrical potential is defined as the energy required to bring a unit positive charge from the point at infinity to the interior of a phase. Fig. 1-2 compares the electrical potentials for two phases (in this case, two electrodes) that have either excessive positive charges or excessive negative charges. Since the energy required to bring a positive charge to the electrode possessing an excess of positive charge is higher, the electrical potential of this electrode is higher (has a more positive value) than that for the other. Here, if the electrode has an excessive positive charge, the energy of the electrons within this electrode is, however, lower. Therefore, the higher the electrode potential, the lower the electron energy in the electrode; electrode potential and electron energy in electrode tend to be opposite to each other.

As mentioned above, electron energy within an electrode is determined by the type

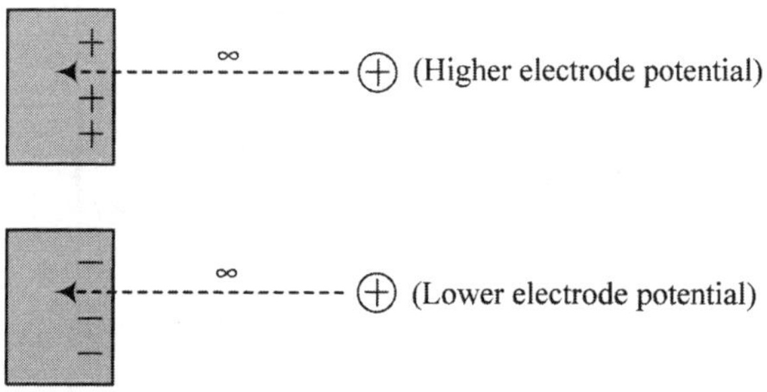

Fig. 1–2 Definition of electrical potential.

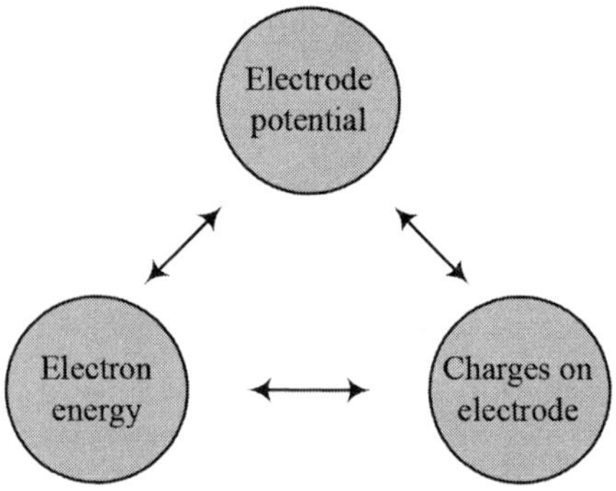

Fig. 1–3 Relationship between the type/amount of electric charges in electrode, electrode potential, and electron energy.

and amount of charges within that electrode; with an increase of negative charges, the electron energy increases but the electrode potential decreases. This implies that electrode potential can be controlled by intentionally adjusting the type and amount of charges in the electrode. In real, this is possible by using a device called ***potentiostat***, which is described in Chapter 5. Fig. 1-3 illustrates the relationship between the type and amount of electric charges in electrode, the electron energy within electrode, and the electrode potential. Again, if an electrode has a larger amount of electrons, the electron energy in this electrode is higher, which appears as a more negative electrode potential.

Fig. 1-4 shows the process of inducing an electrochemical reduction reaction by controlling the electron energy within an electrode. Note here that the electron energy is shown to be higher toward the upper side; the electrode potential has a more negative value going up. Also shown in Fig. 1-4 are the highest electron energy level in the electrode,

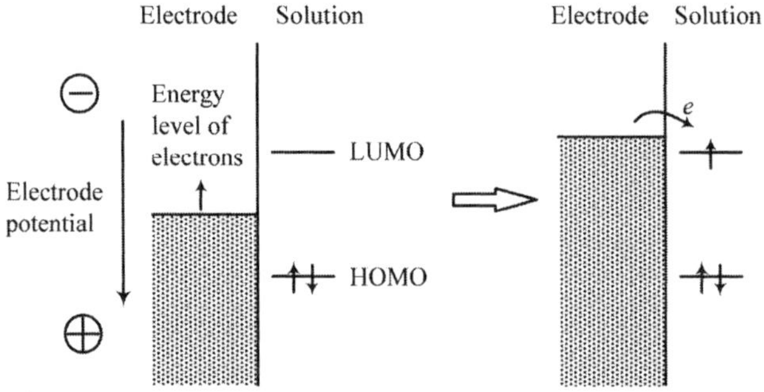

Fig. 1–4 Electrochemical reduction according to a change of electron energy (electrode potential).

which is called ***Fermi level*** (see Chap. 13), and LUMO and HOMO level of a redox-active chemical species. In molecules, electrons fill the molecular orbitals from the lowest one. The highest molecular orbital filled by electrons is called the highest occupied molecular orbital (HOMO), but the empty lowest molecular orbital is called the lowest unoccupied molecular orbital (LUMO). The LUMO and HOMO levels are fixed for a given electro-active species, whereas the Fermi level in the electrode can be arbitrarily adjusted by using a potentiostat as described above. Now, comparing the Fermi level with the LUMO and HOMO levels, it is possible to judge the direction of electron flow since electrons move from a higher energy state to a lower one. In Fig. 1-4, the Fermi level is lower than the LUMO initially. When the Fermi level becomes higher than the LUMO level by electrode potential control, electrons will move from the electrode to the redox-active species; electrochemical reduction is possible. Conversely, if the Fermi level is controlled to be lower than the HOMO level, an electrochemical oxidation reaction, in which electrons move from the redox-active species to the electrode, becomes possible. This discussion, however, does not account for how fast the electrochemical reactions take place, but only the thermodynamic feasibility (see Section 1-2).

1-1-2 Electrochemical reactions are possible only near electrode surface

Let us consider that ferrocenium ($FeCp_2^+$) dissolved in solution is electrochemically reduced to ferrocene ($FeCp_2$) through a potential-step experiment using a three-electrode cell shown in Fig. 1-5.

$$FeCp_2^+ + e \rightarrow FeCp_2$$

In the potential-step experiment, the electrode potential is controlled to be E_1 before the potential step as shown in Fig. 1-6-a. Under this condition, the electrochemical reduction is thermodynamically impossible since the Fermi level is lower than the LUMO of $FeCp_2^+$. Now, a potential step is applied to the electrode from E_1 to E_2. The step potential (E_2) is negative enough to push the Fermi level to be higher than the LUMO; the electro-

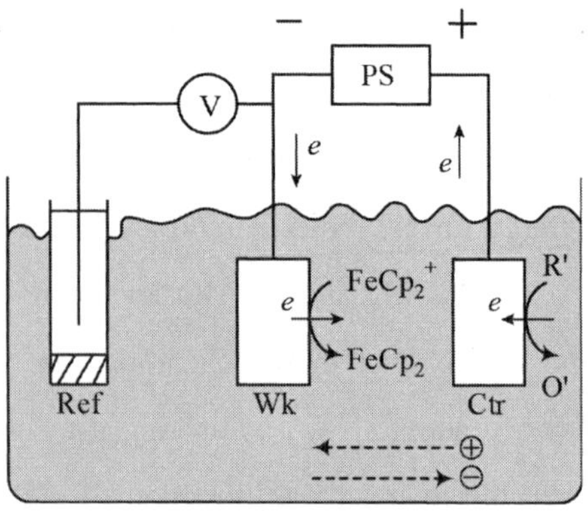

Fig. 1–5 Three–electrode cell consisting of a working electrode (Wk), counter electrode (Ctr) and reference electrode (Ref) for the electrochemical reduction of ferrocenium ($FeCp_2^+$).

chemical reduction of $FeCp_2^+$ is now possible. The concentration profile of $FeCp_2^+$ in the solution after the potential step is sketched as a function of time in Fig. 1-6-b. Before the step (t = 0), the concentration of $FeCp_2^+$ is the same at every place in the cell. It is assumed that there is no stirring in the solution. After the step, $FeCp_2^+$ is reduced to $FeCp_2$ in solution near the electrode surface where electron tunneling rate is appreciable. The concentration of $FeCp_2^+$ is zero on the electrode surface, which is maintained as far as the step potential (E_2) is applied. On the other hand, since the concentration of $FeCp_2^+$ in the solution far from the electrode surface is not zero, diffusion due to the concentration difference occurs; $FeCp_2^+$ diffuses to the electrode surface and continuously reduced. The concentration profile of $FeCp_2^+$ changes with time as shown in

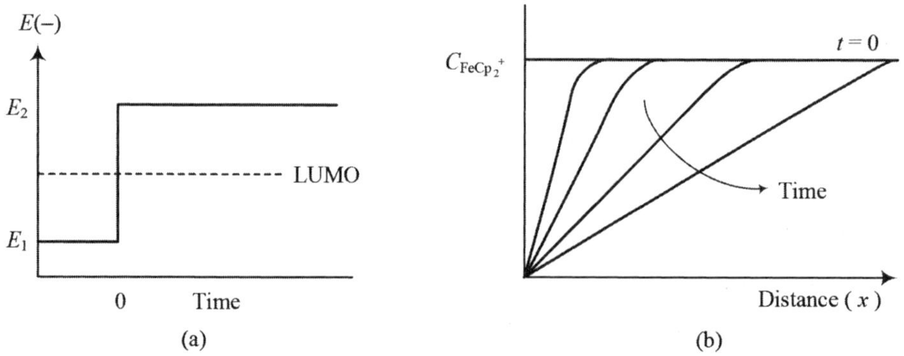

Fig. 1–6 (a); Change of the electron energy in an electrode for the potential step experiment and (b); Evolution of concentration profile of $FeCp_2^+$ after the potential step.

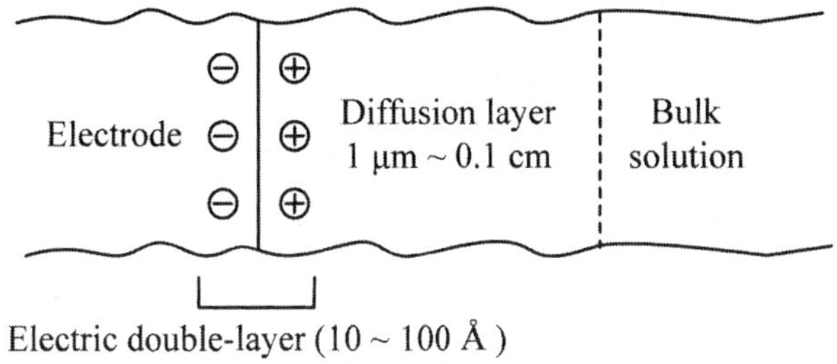

Fig. 1-7 Electric double-layer, diffusion layer and bulk solution in unstirred solution.

Fig. 1-6-b. The concentration gradient of $FeCp_2^+$ is the largest immediately after the step. However, the slope becomes gradually flatter with time, and the diffusion rate of $FeCp_2^+$ gradually decreases; the amount of $FeCp_2^+$ diffused into the electrode becomes smaller and smaller with an increase in the step time. Hence, it can be said that only $FeCp_2^+$ existing within a certain distance from the electrode surface participates in the electrochemical reaction.

In normal chemical reactions, A + B → C, the reaction can take place everywhere in the reactor if the reactant A and B are well mixed. However, as described above, in electrochemical reactions, only the reactants existing near electrode surface can participate in the reaction. The range of the solution, in which the reactants participating in electrochemical reaction are populated, is referred to as a ***diffusion layer***, and the solution far from the electrode surface is called a ***bulk solution***. Generally, the diffusion layer has a thickness of about 1 μm ~ 0.1 cm (Fig. 1-7). Meanwhile, if the electrode has an excess negative charge, extra positive ions arrange in the solution side to maintain electroneutrality. The electrode/solution interface where different charges are arranged is called an ***electric double-layer***.

1-1-3 Many things can happen in series in electrochemical reactions

As exemplified by the reduction of $FeCp_2^+$, an electroactive species located in bulk solution must move to the electrode surface region by mass transfer processes to participate in electrochemical reaction. Mass transfer is possible by three routes; diffusion due to concentration difference, convection by either mechanical stirring or thermal motions of reactants, and migration due to an ion movement within electric field. The electroactive species reaching the electrode surface is reduced by charge transfer (or electron transfer), and the as-formed product continues to move toward the bulk solution (Fig. 1-8). Since the charge transfer takes place at the interface of two different phases (electrode and solution), this is referred to as ***heterogeneous charge transfer***.

Some electrochemical reactions involve adsorption, desorption, migration of electrodeposited atoms on electrode surface, crystallization, and phase change in addition to mass transfer and charge transfer. Multiple charge transfer is also possible. For example, in the *ECE*-type reactions, the R produced by electrochemical reaction (*E*) is converted into Y by a chemical reaction (*C*) in solution, and the Y is further electrochemically reduced to Z (*E*).

$$\mathrm{O} + ne \rightarrow \mathrm{R} \quad (E)$$
$$\mathrm{R} \xrightarrow{k} \mathrm{Y\ in\ solution} \quad (C)$$
$$\mathrm{Y} + n'e \rightarrow \mathrm{Z} \quad (E)$$

The rate of overall reaction is determined by the slowest step among the serially occurring elementary steps, which is referred to as a rate-determining (rate-limiting) step. In a simple electrochemical reaction comprising mass transfer and charge transfer, if the mass transfer is slower than the charge transfer, the mass transfer is the rate-determining step; it is referred to as mass-transfer limited. Also, some electrochemical reactions are charge transfer limited due to slower charge transfer rate. Obviously, mass transfer and charge transfer can show a comparable rate in electrochemical reactions; both steps together determine the overall rate.

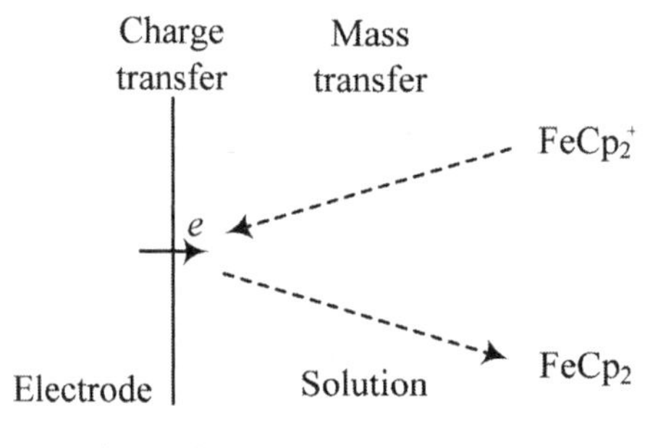

Fig. 1-8 Mass transfer and charge transfer for electrochemical reactions.

1-1-4 Current is an expression of reaction rate

Fig. 1-5 shows the three-electrode electrochemical cell. The electrode where the targeted electrochemical reaction takes place is called a working electrode (Wk). The potential of working electrode is controlled to be V against a reference electrode (Ref) by using a potentiostat.

Let us assume that an electrochemical reaction (O + ne → R) proceeds on a working electrode. The electric charges (Q) consumed for the production of N moles of R are

$$Q = nF \times N$$

where n is the number of electrons participating in the reaction and F is the Faraday constant (9.6485 × 10^4 C/mol). Since the current $i = dQ/dt$, <1-1> and <1-2> are derived. As is seen in <1-2>, the reaction rate is proportional to current (i). Thus, it can be said that "Current is an expression of the rate for electrochemical reactions".

$$\text{Current (amperes)} = \frac{dQ}{dt}\text{ (coulombs/s)} = nF\frac{dN}{dt}$$ <1-1>

$$\text{Reaction rate (mol/s)} = \frac{dN}{dt} = \frac{i}{nF}$$ <1-2>

Self learning 1-2

The electric charge for one electron is 1.60218 × 10^{-19} C. From this, confirm that the Faraday constant, which represents the electric charge of 1 mole (or equivalent) of electrons, is 9.6485 × 10^4 C/mol.

1-1-5 Current flows by forming a closed loop

Fig. 1-9 shows the flow of electrons and ions when an electrochemical reduction occurs at a working electrode. Electrons are required for the reduction reaction at the working electrode, which are provided from the counter electrode (Ctr), on which electrons are

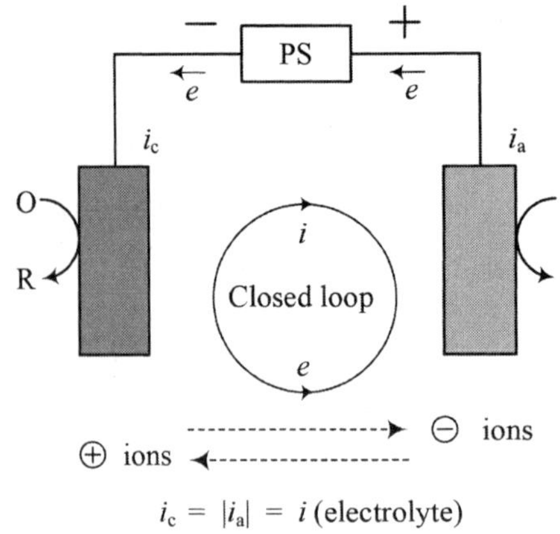

Fig. 1-9 Formation of a closed loop through the anode, external circuit, cathode, and electrolyte solution.

produced by an electrochemical oxidation of R'; the as-produced electrons move to the working electrode through the external circuit and power supply. As indicated, in order for an electrochemical reduction reaction to occur at a working electrode, electrochemical oxidation of another electroactive species must be accompanied at the counter electrode. In opposite, for an oxidation reaction at the working electrode, a reduction should proceed at the counter electrode by accepting electrons being produced at the working electrode. This illustrates that at least two electrodes must be paired in electrochemical cells and different electrochemical reactions must proceed. The counter electrode is also referred to as an ***auxiliary electrode*** because it serves an auxiliary function of receiving electrons generated by an oxidation reaction at the working electrode or providing electrons for the reduction reaction at the working electrode.

Electric charges must also flow in the electrolyte for electrochemical reactions and ionic species existing in the electrolyte solution are responsible for the flow of electric charges. A solution containing ionic species is called an ***electrolyte solution***. In Fig. 1-9, an electric field forms between the working and counter electrode since two electrodes are negatively or positively polarized. Within this electric field, anions migrate to the counter electrode that is positively polarized, whereas cations to the working electrode that is negatively polarized. The movement of ions corresponds to the flow of electric charges and thus appears as a current flow in the electrolyte solution. Here, a distinct feature in the current flow is that the cathodic current (i_c) is the same with the anodic current (i_a); it is obvious because the electron generation rate on an anode (i_a) must be identical with the electron consumption rate on a cathode (i_c). In addition to this, the current flow across the electrolyte solution is also the same with the two; $i_c = |i_a| = i_{electrolyte}$. Now, let us trace the movement of the negatively charged species (electrons and anions); they move in the counterclockwise within a ***closed loop***. Note here that the movement of cations toward the negative electrode has the same effect as the movement of anions toward the positive electrode; it can be assumed that cations also act to carry negative charges in the counterclockwise direction like anions. Obviously, current flows in the clockwise direction since the flow direction for current and electrons is opposite to each other. Here, i_c is expressed a positive value while i_a as a negative one. Hence, i_a is expressed as an absolute value in Fig. 1-9.

Example 1-1

There is a phenomenon called ***crevice corrosion*** among the causes of collapse of steel structures. Explain how crevice corrosion occurs by using the concept of closed loop.

Solution

Fig. 1-10 illustrates the mechanism of crevice corrosion. In order for steel corrosion to proceed, a galvanic cell should be formed combining an anode, cathode, electrolyte, and external circuit. When steel surface is exposed to air and moisture, the surface is corroded and an electron-conductive oxide layer deposits. Oxygen reduction can take place on the surface of oxide layer, which plays

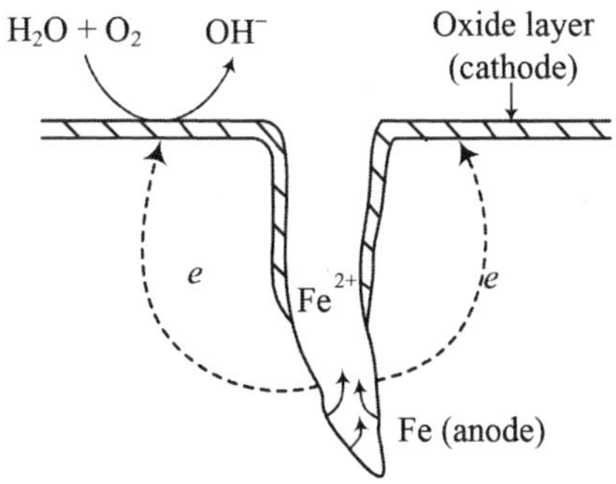

Fig. 1–10 Electrochemical reactions that induce a crevice corrosion and the formation of a closed loop.

as a cathode. If some part of oxide layer is damaged, the steel surface is exposed to air, where Fe oxidation can take place; it is an anode. The combined reaction of Fe oxidation and oxygen reduction is thermodynamically allowed since $\Delta G^0 < 0$.

$$\text{Anode: Fe} \rightarrow \text{Fe}^{2+} + 2e$$

$$\text{Cathode: O}_2 + 2\text{H}_2\text{O} + 4e \rightarrow 4\text{OH}^-$$

$$\text{Overall: 2Fe} + \text{O}_2 + 2\text{H}_2\text{O} \rightarrow 2\text{Fe}^{2+} + 4\text{OH}^- \quad (\Delta G^0 < 0)$$

Meanwhile, the moisture and ions on the oxide surface can play as an electrolyte, and the electron-conducting steel can play as an external circuit. In short, a galvanic cell forms. In galvanic cells, the oxidation and reduction reactions should proceed under the condition of $i_c = |i_a|$ by forming a closed loop. The surface area of oxide is much larger than that for the inner part of crevice, such that the cathode (oxide surface) receives a plentiful supply of oxygen to result in an appreciable cathodic current (i_c). On the other hand, the metal oxidation is localized on such a small surface exposed at the inner part of crevice. To meet the condition for $|i_a| = i_c$ for a closed loop formation, the rate of metal dissolution (i_a) should be comparable with the rate of oxygen reduction on the oxide layer of large surface area and plentiful supply of oxygen; the metal corrosion ($Fe \rightarrow Fe^{2+} + 2e$) is substantial. Consequently, once crevice corrosion has commenced (steel surface is exposed to air by a damage of oxide layer in some reasons), the depth of crevice increases very rapidly to destruct the steel structures.

1-1-6 Current can be controlled by changing the electrode potential under charge transfer limiting condition

In a chemical reaction ($A + B \rightarrow C$), the reaction rate is expressed as: $Rate = k[A][B]$, where k is the rate constant. Since $k \propto \exp(-E_A/RT)$, the reaction rate can be increased by lowering the activation energy (E_A) by using a catalyst or increasing reaction temperature. Furthermore, the reactant mixture is mechanically stirred to increase the collision between reactant A and B. The rate for heterogeneous charge transfer in electrochemical reactions can be expressed by the same way by using a rate constant (k') and the concentration of two reactants; $[FeCp_2^+]_s$ and electrons.

$$\text{Charge transfer rate} = k' [FeCp_2^+]_s[e] = k [FeCp_2^+]_s$$

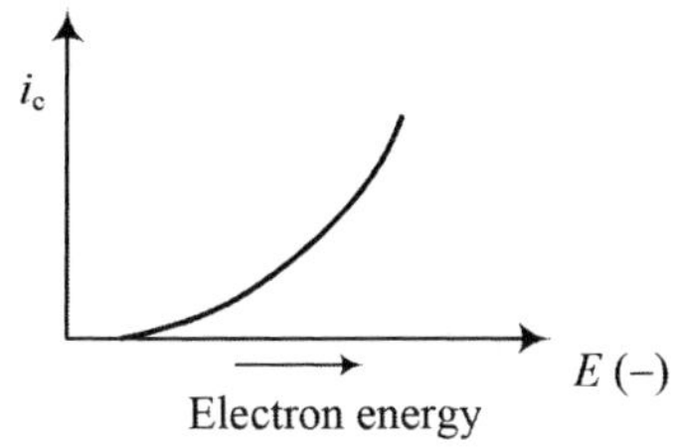

Fig. 1–11 The current–potential profile observed under charge–transfer limiting condition.

Note that the reactants in the charge transfer step are the $FeCp_2^+$ on the electrode surface region and electrons placed in the electrode. Hence, the subscript ($_s$) in $[FeCp_2^+]_s$ indicates the $FeCp_2^+$ concentration on the electrode surface. Here, since the amount of electrons in the electrode is sufficiently large, even if the reaction progresses, it can be assumed to be constant; therefore, the relation $k'[e] = k$ is valid.

The rate constant for charge transfer reaction has the relationship of <1-3> and thus the current (reaction rate) has the relationship of <1-4>.

$$k \propto e^{\eta} \propto e^{|E|} \quad \text{<1-3>}$$

$$i \propto \text{charge transfer rate} \propto k \propto e^{\eta} \propto e^{|E|} \quad \text{<1-4>}$$

Here, $\eta = |E - E_{eq}|$ is called ***overpotential***. In particular, it is ***charge transfer overpotential*** (or ***activation overpotential***) because it is related with charge transfer reaction. E is the potential (or voltage) applied to the electrode and E_{eq} is the equilibrium potential. As is seen in <1-3>, the rate constant varies with an exponential function of applied electrode potential (E). As shown in Fig. 1-11, the reduction current (i_c) increases exponentially with the applied electrode potential being pushed to more negative values; the current can be controlled by changing the electrode potential. Note that this explanation is valid only when charge transfer is slower than mass transfer (under charge transfer limiting condition).

1-1-7 Current and electrode potential can not be controlled simultaneously

As shown in <1-4>, the current is determined by the applied electrode potential. In opposite, electrode potential is determined when a current is given. Two variables can not be controlled simultaneously. The potentiostat monitors the current while electrode potential being controlled. The galvanostat traces the electrode potential by applying a current. In practical electrolytic cells, electrolysis is performed by controlling either the applied electrode potential or applied current. In the latter case, the applied current is predetermined, which means that the reaction rate is predetermined. This is impossible in normal chemical reactions.

1-1-8 Current is determined by the mass transfer rate under mass transfer limiting condition

Let us assume that, in an electrochemical reaction, $O + ne \rightarrow R$, the overall rate is determined by the mass transfer of O, and assume that the convection and migration are absent. Then, diffusion is the only way for mass transfer. In this case, the diffusion rate of O is the reaction rate (current); the current is decided by the diffusion rate of O near the electrode surface (diffusion layer as will be mentioned later).

$$i_d \propto \text{mass transfer rate} \propto D_O \left.\frac{dC_O(x,t)}{dx}\right|_{x=0} \qquad \text{<1-5>}$$

Here, D_O is the diffusion coefficient of O and $C_O(x,t)$ is the concentration of O at a position x and time; x is the distance from the electrode surface to the bulk solution. $\frac{dC_O(x,t)}{dx}$ is the concentration gradient of O. In <1-5>, $x = 0$ is explicitly noticed to emphasize the fact that the diffusion rate near the electrode surface ($x = 0$) determines the current; the diffusion rate at the position far from the electrode surface does not affect the current values.

Fig. 1-12 shows the current change in the potential-step experiment, which is estimated by using <1-5> and the change of concentration gradient of $FeCp_2^+$ near electrode surface shown in Fig. 1-6-b. If the mass transfer of $FeCp_2^+$ determines the overall reaction rate and further diffusion is the only way for mass transfer, the current decreases as time elapses after the potential step because the concentration gradient of $FeCp_2^+$ near the electrode surface decreases with time. Here, i_d is called a diffusion current, implying that the current is decided by the diffusion rate.

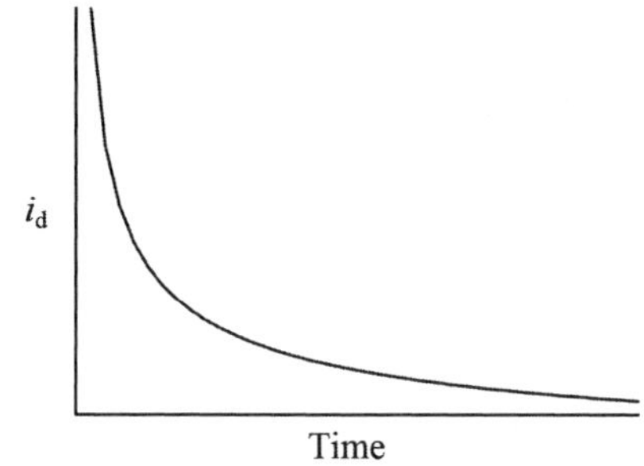

Fig. 1–12 The current–time profile observed in the potential step–experiment under mass transfer (diffusion) limiting condition, which is estimated from the change of concentration gradient in Fig. 1–6–b.

1-2 Thermodynamics and kinetics of electrochemical reactions

As discussed in Fig. 1-4, if the electron energy in an electrode is higher than the LUMO of an electroactive species present in solution, an electrochemical reduction reaction is possible. This is a prediction from a thermodynamic point of view. However, thermodynamics only tell us the feasibility of reactions. The issue of how fast the reaction goes is explained in terms of kinetics.

Fig. 1-13 illustrates the change of electron energy before and after the potential step,

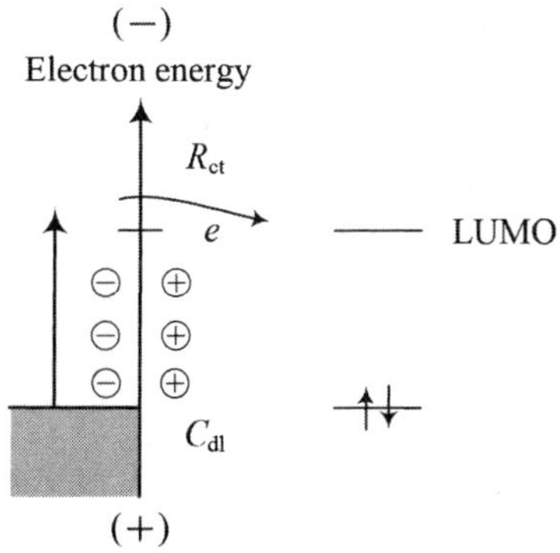

Fig. 1–13 Electric double–layer capacitance (C_{dl}) and charge transfer resistance (R_{ct}).

which is explained in Fig. 1-6-a. The electron energy in the electrode, which is lower than the LUMO of the electroactive species before the potential step, is raised over the LUMO immediately after applying potential step. Now, the electrochemical reduction reaction is thermodynamically possible. Before the electron energy surpasses the LUMO, electrons are not consumed for the reduction reaction but accumulated on the electrode surface. Excess amount of cations are placed in the solution side for electroneutrality. Such an electrode/solution interface where electrons and cations are accumulated is called ***electric double-layer***, which behaves like a capacitor since electric charges are stored. This means that a current should flow to fill the capacitor prior to the reduction current flow. The current to fill the capacitor is called ***electric double-layer charging current***, whereas the current for the reduction reaction is called ***faradaic current***. In general, the current for electrochemical reactions (O + $ne \rightarrow$ R) is the faradaic current, whereas the current for other processes such as filling capacitors is a ***non-faradaic current***.

The sequence of current flow just mentioned can be explained by using the equivalent circuit shown in Fig. 1-14. Note that this is a simulation for the potential step experiment. When the circuit is closed, a direct voltage (V) is imposed across the parallel-connected resistor and capacitor. Then, current flows to the capacitor in advance and then to the resistor, which is identical with the sequence of current flow explained in Fig. 1-13. The C_{dl} refers to the ***electric double-layer capacitance***, whereas the R_{ct} is ***charge transfer resistance*** since it is related with charge transfer reaction. As mentioned above, the electrochemical reduction is thermodynamically allowed when the electron energy is higher than the LUMO level. Here, R_{ct} is the kinetic parameter; it represents how fast the charge transfer reaction takes place; a larger R_{ct} gives a smaller faradaic current.

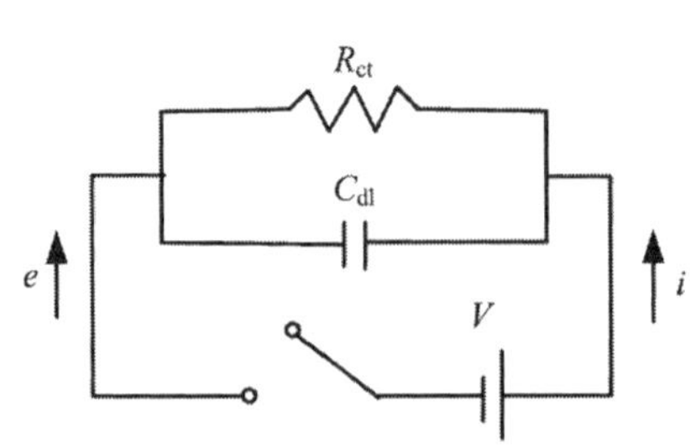

Fig. 1–14 The equivalent circuit illustrating the sequence of electron (current) flow in Fig. 1–13.

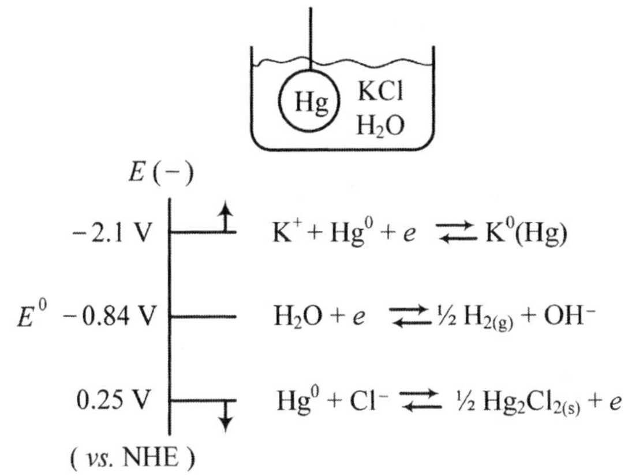

Fig. 1-15 Half-cells that involve in an electrochemical system comprising Hg electrode and aqueous KCl solution.

Fig. 1-15 shows the standard electrode potential (E^0) of three half-cells that are involved in an electrochemical system comprising Hg electrode and aqueous KCl solution. The E^0 of $H_2O + e = ½ H_2 + OH^-$ half-cell is -0.84 V (*vs.* NHE), where NHE (normal hydrogen electrode) is a reference electrode, against which the potential of Hg electrode is controlled by using a potentiostat. Here, E^0 is an equilibrium potential when the activity and fugacity is unity for all the chemical species involved in electrochemical reactions (Chap. 2). Thermodynamics dictates that a net reduction current for H_2 evolution can flow when the potential of Hg electrode is more negative than the equilibrium potential (E_{eq}) of hydrogen half-cell (-0.84 V *vs.* NHE); here, $E_{eq} = E^0$ because a standard state is assumed. In real, however, the reduction current is negligibly small until the $K^0(Hg)$ amalgam forms at -2.1 V. This illustrates that the reduction at $E < -0.84$ V is thermodynamically feasible but kinetically slow to give a negligible hydrogen evolution. Thermodynamics also dictates that the hydrogen oxidation ($½ H_2 + OH^- \rightarrow H_2O + e$) is possible when the applied potential is more positive than the equilibrium potential ($E_{eq} = -0.84$ V). The absence of H_2 in this electrochemical cell can discard this reaction. Even if H_2 is present, this oxidation reaction is negligible since it is kinetically very slow. When the electrode potential is driven to 0.25 V, the oxidation reaction to generate Hg_2Cl_2 takes place. In summary, the thermodynamically feasible electrochemical reduction and oxidation at 0.25 V ~ -2.1 V is kinetically too slow to give faradaic currents. Note, however, that the double-layer charging current flows because the electrode potential changes. The current appeared at 0.25 V ~ −2.1 V is the non-faradaic double-layer charging current (Fig. 1-16).

The kinetic slowness can be explained by using the concept of resistance in Fig. 1-13. Once again, the charge transfer reaction is thermodynamically allowed when the Fermi level surpasses the LUMO level; electrons can move from the electrode to the electroactive species by a charge transfer process. Here, one can envisage a resistor that impedes the electron flow, and the negligible current (i) after potential step (V) can be accounted for by a large resistance (R) from $i = V/R$. Here, since this resistor controls the charge transfer rate, it is called ***charge transfer resistance*** (R_{ct}). If $R_{ct} = \infty$, the electron flow from the electrode to the electroactive species in the charge transfer process is negligible; the faradaic current is also negligible. The electrochemical systems comprising electrode/electrochemical reactions with $R_{ct} = \infty$ are called ***ideally polarizable electrode***. In this case, electrons can not cross the interface because of extremely large resistance; then electrons are accumulated in the electrode even if the Fermi level is higher than the LUMO.

As discussed in 1-1-1, the type and amount of excess charges can be controlled by adjusting electrode potential. One can thus assume a certain electrode potential, at which the electrode does not have either excess negative or positive charges. This is called the ***potential of zero charge*** (E_z) (Chap. 3). When electrode is driven to a negative or positive potential from E_z, the electrode is polarized due to the accumulation of negative or positive charges within the electrode. Here, the "polarization" means the charge accumulation

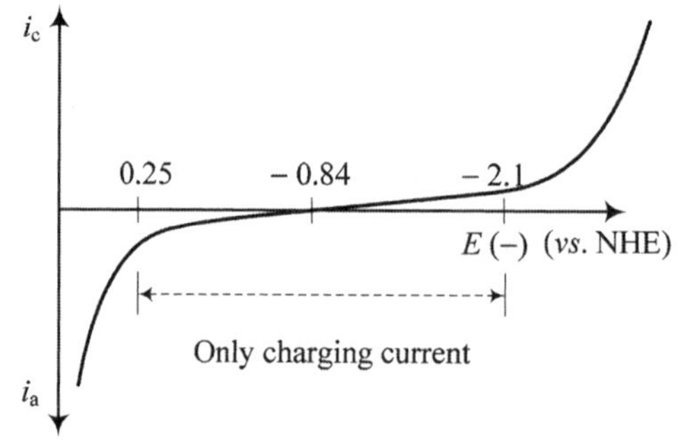

Fig. 1-16 The electric double-layer charging current observed at 0.25 V ~ −2.1 V.

in the electrode and the "ideal" means the polarization without any charge loss. In opposite, the electrochemical systems with R_{ct} = 0 is called ***ideally non-polarizable electrode***. In this case, electrons can not be accumulated because all of them are consumed for charge transfer reaction. Here, "ideally non-polarizable" means no accumulation of electrons within the electrode at the condition of Fermi level > LUMO. In practice, there are no electrochemical systems with $R_{ct} = 0$ or $R_{ct} = \infty$. But some electrochemical systems are close to these two extremes; $R_{ct} \rightarrow \infty$ or $R_{ct} \rightarrow 0$. The electrochemical system comprising Hg electrode/aqueous KCl solution (Fig. 1-15) can be considered to be "close" to the ideally polarized electrode at the potential range of 0.25 V ~ −2.1 V with $R_{ct} \rightarrow \infty$ for H_2 evolution and oxidation.

If any electrochemical systems are ideally polarized in a wide potential range, electric charges can be stored in the electrode without any loss. Ions are arranged at the solution side for electroneutrality; electric double-layer capacitor forms at the electrode/solution interface. The device that stores electric charges in ideally polarized electrodes is called ***supercapacitor*** (ultracapacitor or electric double-layer capacitor). Electric double-layer forms only at electrode/solution interface, such that the electrode materials of large surface area such as activated carbons are used as the electrode.

1-3 Components of electrochemical cells

1-3-1 Configuration of electrochemical cells

The reactor where electrochemical reaction takes place is called electrochemical cell, which is classified into two; electrolytic cell and galvanic cell. In both cases, when oxidation (or reduction) occurs at one electrode, reduction (or oxidation) must be accompanied at the other electrode. Therefore, at least two electrodes are required to construct an electrochemical cell, which is called a two-electrode cell. When a reference electrode is added, it becomes a three-electrode cell.

As electrodes, the substance having a high electronic conductivity is used. Metals (e.g. Pt or Au), carbons, metal oxides (e.g. PbO_2) or semiconductors can be employed. In addition, the materials having both electronic conductivity and ionic conductivity, which are called ***mixed conductors***, are also used as electrodes. For example, graphite (Li_xC_6) and tungsten oxide bronze (H_xWO_3) have good ionic conductivity as well as electron conductivity. Graphite is used as the negative electrode of lithium secondary batteries. Meanwhile, $La_{1-x}Sr_xCoO_3$ with a perovskite structure is a mixed conductor having electron conductivity and O^{2-} conductivity at high temperature (800~1000°C), and thus used as the cathode of a solid-oxide fuel cell.

In electrochemical cells, the electrolyte is another key component for the closed loop formation. Various types of ionic conductors can be used. For example, the salt-dissolved water or organic solvents, and the salt-dissolved polymers (e.g. poly (ethylene oxide), PEO) are used as an electrolyte. When the Li_2CO_3/K_2CO_3 mixture is heated above 600°C, it becomes a molten salt; it has ionic conductivity and thus used as the electrolyte for molten carbonate fuel cell. Yttria-stabilized zirconia (Y_2O_3/ZrO_2, YSZ) is also used as an electrolyte for solid-oxide fuel cell because of its high O^{2-} conductivity at elevated temperature (800 ~ 1000°C).

In electrochemical cells, separators may be required to prevent mixing of anolyte and catholyte; in an electrochemical cell, the electrolyte solution is divided into two by the presence of an intervening separator; the electrolyte solution in anode side is called an anolyte, while that for cathode side is called catholyte. For the closed loop formation, the separator must have appreciable ionic conductivity. Porous materials (e.g. fritted glass, filter paper, and porous poly (ethylene)) are used as the separator, wherein ion conduction is possible by the ions dissolved in the solution that fills the pores. These porous materials do not have ion-selective conduction property, but only provide a function of preventing mixing of solutions. A material having ion-selectivity in addition to prevention of solution mixing can be used as a separator. Typical example is the polymer membrane called ***Nafion***, which has a sulfonic acid ($-SO_3H$) functional group inside it. When the sulfonic acid moiety is hydrated, it dissociates into $-SO_3^-$ and H^+. The anion ($-SO_3^-$) is immobilized on the polymer chain, but the cation (H^+) is free to move. It has a proton (H^+) conduction property. Obviously, it also has a function of preventing mixing of solutions when formulated into a membrane. Further, protons can be replaced by other cations (e.g. Na^+), so it is also called as a cation-exchange membrane.

Self learning 1-3

Make sure that the electrode materials used for batteries have either electronic conductivity or mixed conductivity. The materials for electrolyte, either in solid or solution, should have ionic conductivity but no electronic conductivity. Explain why.

1-3-2 Electrolytic cell

The brine electrolysis is performed in an electrolytic cell. The electric energy from a power source is supplied, and chlorine (Cl_2) and NaOH are obtained as a product (Fig. 1-17).

The half-cell reactions occurring at the two electrodes and the overall reaction are as follows.

$$\text{Anode: } 2Cl^- \rightarrow Cl_2(g) + 2e \qquad E^0 = 1.36 \text{ V } (vs. \text{ NHE})$$

$$\text{Cathode: } 2H_2O + 2e \rightarrow H_2(g) + 2OH^- \qquad E^0 = -0.84 \text{ V } (vs. \text{ NHE})$$

$$\text{Overall: } 2H_2O + 2Cl^- \rightarrow Cl_2(g) + H_2(g) + 2OH^- \quad (\Delta G^0 > 0)$$

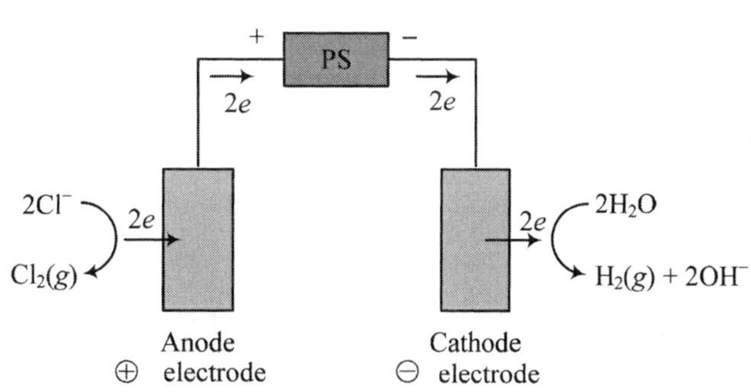

Fig. 1–17 Electrode reactions in the brine electrolysis.

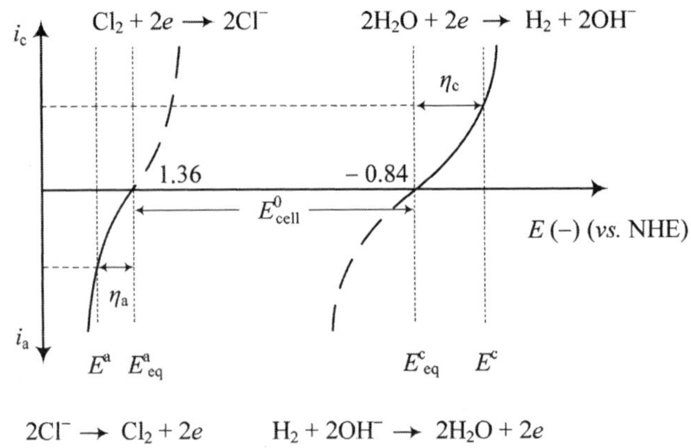

Fig. 1-18 The applied voltage (E_{appl}) between two electrodes for the brine electrolysis; the current-potential profiles, overpotentials, and thermodynamic decomposition voltage (E^0_{cell}). The standard state is assumed.

Since the overall reaction is not spontaneous ($\triangle G^0 > 0$), the reaction should be induced by supplying electric energy. Fig. 1-18 shows the current-potential profiles at the two electrodes. Under the assumption that the electrochemical reactions at both electrodes are charge transfer limited, the currents are plotted as an exponential function of the applied potential for each electrode. In fact, such an exponential current-potential behavior appears only in the Tafel region where the overpotential is relatively large (See Section 4-1). Here, however, they are plotted so as to change by the exponential function in the entire potential range to simplify the explanation. The equilibrium potential for the chlorine half-cell is 1.36 V (*vs.* NHE) as shown in Fig. 1-18. At this equilibrium potential (E^a_{eq}), the oxidation current for Cl_2 generation and the reduction current for Cl^- generation are the same, so apparently no current flows at all ($i_{net} = 0$). However, if the electrode potential is pushed to a more positive value than 1.36 V, the oxidation (Cl_2 evolution) current increases by an exponential function of overpotential ($\eta_a = E^a - E^a_{eq}$), where E^a and E^a_{eq} are the applied potential and the equilibrium potential for the chlorine half-cell, respectively; here, the standard state is assumed. In Fig. 1-18, the current curve indicated by the solid line at $E^a > 1.36$ V represents the current for Cl_2 evolution. On the contrary, if the electrode potential (E^a) is pushed to a more negative value than 1.36 V, the reduction current for Cl^- generation increases by an exponential function (the dotted line at $E^a < 1.36$ V). Meanwhile, the equilibrium potential for the hydrogen half-cell is -0.84

V *vs.* NHE. At the equilibrium potential (E^{c}_{eq}), the reduction current is the same with the oxidation current to yield the apparently observed current to be $i_{net} = 0$. However, both the reduction and oxidation current increase exponentially with the applied potential (E^{c}); when E^{c} moves to the negative direction from E^{c}_{eq}, the OH^{-} generation current increases by the exponential function of overpotential ($\eta_c = E^{c} - E^{c}_{eq}$) (the solid line at $E^{c} < -0.84$ V).

In the practical brine electrolysis, the Cl_2 generation (oxidation) reaction and the OH^{-} generation (reduction) reaction are paired; the chlorine half-cell is an anode, while the hydrogen half-cell is a cathode. In Fig. 1-18, the current curves indicated by the solid lines in both half-cells are the practical ones. The applied voltage (E_{appl}) between the two electrodes for brine electrolysis can be envisaged from the current-potential profiles shown in Fig. 1-18. The applied voltage (E_{appl}) changes according the current size in a constant-current electrolysis. If the electrolysis is carried out under the condition of zero current ($|i_a| = i_c = 0$), the voltage difference between the two electrodes is $E_{cell} = E^{a}_{eq} - E^{c}_{eq} = 1.36 - (-0.84)\ V = 2.2\ V$. E_{cell} is called the ***thermodynamic decomposition voltage***. Note that this is a thermodynamically determined value because it is derived from the thermodynamic values (two equilibrium potentials, E^{a}_{eq} and E^{c}_{eq}). The condition of $|i_a| = i_c = 0$ is not realistic for practical brine electrolysis. Current must flow for electrolysis; $|i_a| = i_c \neq 0$. For the oxidation current higher than zero ($|i_a| > 0$), as is seen in Fig. 1-18, the anode potential (E^{a}) should be more positive than the anodic equilibrium potential (E^{a}_{eq}). The difference between E^{a} and E^{a}_{eq} is called anodic overpotential; $\eta_a = E^{a} - E^{a}_{eq}$. Similarly, a cathodic overpotential ($\eta_c = E^{c} - E^{c}_{eq}$) is needed for $i_c > 0$. Two features should be noted on overpotential; firstly, the overpotential is the additionally required potential from the equilibrium potential to obtain any currents. Secondly, the overpotential increases with an increase in the current value.

The applied voltage (E_{appl}) at a specific current ($|i_a| = i_c \neq 0$) is determined by <1-6>.

$$E^{a} = E^{a}_{eq} + \eta_a,\ E^{c} = E^{c}_{eq} - \eta_c\ (|i_a| = i_c \neq 0)$$
$$E_{appl} = (E^{a} - E^{c}) + iR_{total}$$
$$E_{cell} = E^{a}_{eq} - E^{c}_{eq}\ (|i_a| = i_c = 0)$$
$$E_{appl} = E_{cell} + \eta_a + \eta_c + iR_{total} \qquad \text{<1-6>}$$

<1-6> illustrates that the applied voltage is determined by two parameters; one is the thermodynamic decomposition voltage (E_{cell}) and the other is the ***cell polarization*** ($= \eta_a + \eta_c + iR_{total}$). Here, η_a is the sum of the overpotentials (activation overpotential + concentration overpotential + crystallization overpotential + others if any) required for anodic reactions, whereas η_c is the sum of the overpotentials involved in cathodic reactions. If charge transfer determines the overall rate, the additionally required potential from an equilibrium potential for electrolysis is called an ***activation overpotential***. In the case of mass transfer limiting, the overpotential is also needed for current flow; it is called a ***concentration overpotential***. Let us assume that an electrochemical reaction proceeds by serially occurring mass transfer and charge transfer. If the charge transfer is slower, the activation overpotential is dominant over the concentration overpotential. The opposite is true if the mass transfer is slower. However, it is also possible that two steps show a comparable rate or the rate difference is not significant. If this is the case, the two steps participate in deciding the overall rate (current). Then, for example, the anodic overpotential (η_a) in <1-6> should be the sum of activation overpotential and concentration overpotential for the anodic reaction. In short, the overpotentials (η_a and η_c) in <1-6> should be represented by the sum of all possible overpotentials involved in electrode reactions.

iR_{total} is the voltage drop that is provoked by all the resistances that impede the current flow along a closed loop. $R_{total} = R_{solution} + R_{separator} + R_{circuit} + R_{others}$, where $R_{solution}$ is related with the ion conduction in electrolyte solution, $R_{separator}$ is related with ion conduction in the separator, and $R_{circuit}$ is the resistances for electrodes and external circuit. Frequently, $R_{separator}$ has the largest value.

In <1-6>, E_{cell} (= 2.2 V) is the thermodynamic value, whereas the cell polarization is the kinetic value because it depends on the current size (reaction rate, thus a kinetic parameter). The three components in cell polarization (η_a, η_c, and iR_{total}) increase with an increase in current, such that a larger E_{appl} is needed if one wants to perform any electrolysis at a higher rate (current).

Example 1-2

Why is the superscript [^{0}] added to the thermodynamic decomposition voltage E^0_{cell} in Fig. 1-18?

Solution

When all the reactants and products participating in electrode reactions in a half-cell are in the standard state (the activity of ions and the fugacity of gases are unity, and metals or solids are in a pure state), the standard electrode potential (E^0) of the half-cell is equal to the equilibrium potential (E_{eq}). For example, if the activity and the fugacity are unity in $Cl_2(g) + 2e = 2Cl^-$ electrode reaction, the equilibrium potential of this half-cell is $E_{eq} = E^0 = 1.36$ V(*vs.* NHE). $E_{cell} = E^a_{eq} - E^c_{eq}$ should be rewritten to $E^0_{cell} = E^{0a}_{eq} - E^{0c}_{eq}$ to specify the standard state. However, it is more common that the activity or fugacity are not unity. Hence, it is expressed as E_{cell} instead of E^0_{cell} in <1-6>.

The brine electrolysis is an important process for obtaining chlorine and NaOH, and a very large-scale facility is running around the world with an enormous amount of electric energy consumption. The electric energy required for electrolysis and the production quantity are expressed as follows.

$$\text{Electric energy} = Q \cdot V = it \cdot E_{appl}$$

$$\text{Production quantity} = \sigma \cdot Q$$

Here, σ is called a ***Coulombic efficiency***, which represents the fraction of electric charges utilized for electrolysis from the total charge (Q) consumed. Ideally, $\sigma = 1$, which means that all the electric charges supplied from a power supply are utilized for electrolysis without any side reactions. As shown in the above relation, in order to minimize the electric energy required, it is necessary to minimize the applied voltage (E_{appl}) for electrolysis. Considering the enormous electrical energy required for brine electrolysis worldwide, 1% reduction in E_{appl} will reduce the energy consumption by 1%, which is still an enormous amount. E_{appl} is expressed by <1-6>, from which a way to reduce this can be envisaged; it is a decrement of cell polarization. In general, the activation overpotential for electrochemical reactions depends on electrode; for example, H_2 evolution is very slow on mercury electrode but very fast on platinum electrode. Hence, the selection of electrodes with small activation overpotential is one way to reduce cell polarization and electric energy consumption. Meanwhile, if electrolyte solution is stirred to accelerate the mass transfer of reactants, the concentration overpotential can be reduced. Furthermore, since the separator resistance ($R_{separator}$) is commonly the largest in R_{total}, an electrolysis cell without separator is preferred. However, the separator is sometimes necessary to prevent solution mixing; in an electrolysis cell fabricated without a separator (called an undivided cell), an electroactive species generated at cathode can move to anode to be oxidized and those generated at anode can move to cathode to be reduced. In undivided cells, it is common that $R_{solution}$ is larger than $R_{circuit}$. Therefore, a sufficient amount of salts is dissolved to minimize the solution resistance. To achieve the same goal, the electrode area is maximized and the distance between two electrodes (called an inter-electrode gap) is minimized in cell design.

1-3-3 Galvanic Cell

In galvanic cells, electric energy is obtained through a spontaneous electrochemical reactions. Fuel cells and batteries are the typical examples. Corrosion belongs to galvanic cells even if electrical energy is not obtained because it occurs spontaneously. The principle of galvanic cell is explained by taking the fuel cell as an example.

$$2H_2 + O_2 \rightarrow 2H_2O \ (\Delta G^0 = -475 \text{ kJ/mol}) \qquad \text{<1-7>}$$

Hydrogen and oxygen react explosively in some cases. This is because ΔG^0 for the reaction is negative and the reaction is spontaneous. However, since ΔG^0 is a thermodynamic parameter, it only explains that it is spontaneous. How fast it proceeds should be explained in terms of kinetics. A mixture of hydrogen and oxygen does not necessarily cause an explosion, but reacts when the conditions to lower the activation energy are met.

Thermodynamic parameters are determined by the initial state of a reaction (reactant) and the final state (product), but independent on the reaction pathway. That is, even if

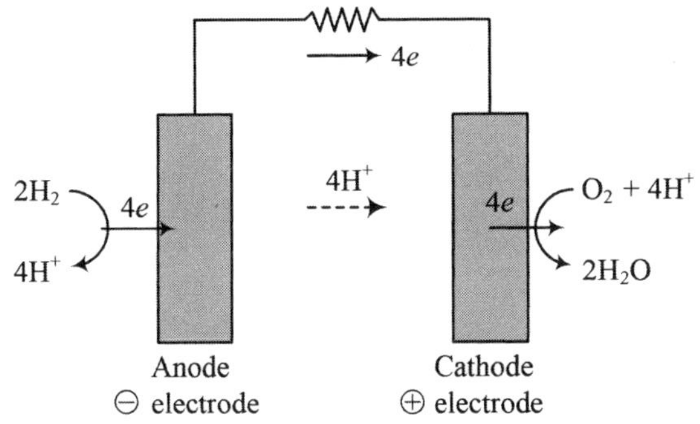

Fig. 1-19 Electrode reactions in H_2/O_2 fuel cell; reactions of hydrogen and oxygen through electrochemical pathway.

the reaction paths are different, the thermodynamic values are the same if the initial and final state are the same. There are diverse pathways for hydrogen and oxygen reaction, one of which is the electrochemical pathway (Fig. 1-19).

In the electrochemical pathway, H^+ and electrons are generated from H_2 at the anode, and the as-generated electrons move to the cathode through an external circuit, whereas H^+ moves to the cathode through the electrolyte. At the cathode, oxygen reacts with H^+ and electrons to form water.

Anode: $2H_2 \rightarrow 4H^+ + 4e$	$E^0 = 0.0$ V (*vs.* NHE)
Cathode: $O_2 + 4H^+ + 4e \rightarrow 2H_2O$	$E^0 = 1.23$ V (*vs.* NHE)
Overall: $2H_2 + O_2 \rightarrow 2H_2O$	($\Delta G^0 = -475$ kJ/mol)

The overall reaction is the production of water from hydrogen and oxygen. The initial and final state of the reaction are the same as in <1-7>, so the thermodynamic value should be the same ($\Delta G^0 = -475$ kJ/mol). Since ΔG^0 is negative, the overall reaction is spontaneous even through the electrochemical pathway. Through these paired electrochemical reactions, electric current flows across the external load as shown in Fig. 1-19, indicative of electric energy being obtained. The voltage difference between the two electrodes is called an ***electromotive force*** (E^0_{cell}). From the relation of $\Delta G^0 = -475$ kJ/mol $= -nFE^0_{cell}$, $E^0_{cell} = 1.23$ V is derived. Note that the same value is obtained from the standard electrode potential of two half-cells; $E^0_{cell} = E^{0c} - E^{0a} = 1.23$ V. This electromotive force is a value obtained from the thermodynamic parameters, but the working voltage

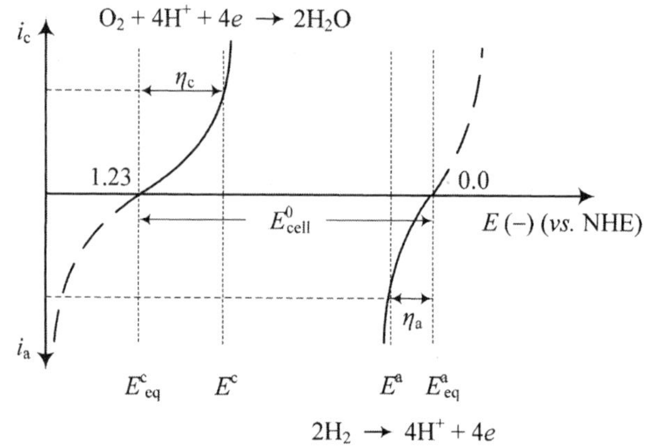

Fig. 1-20 Working voltage (E_{wk}) of H_2/O_2 fuel cell; the current-potential profiles at two electrodes, overpotentials, and open-circuit voltage (E^0_{cell}).

(E_{wk}) of the actual fuel cell is smaller than this. This is explained on Fig. 1-20.

In Fig. 1-20, if the potential of oxygen electrode (E^c) is more negative than the equilibrium potential of oxygen half-cell (E^c_{eq} = 1.23 V), the O_2 reduction current flows according to an exponential function of applied potential (E^c), whereas the reverse reaction takes place at $E^c > 1.23$ V. In actual fuel cells, O_2 reduction indicated by the solid line proceeds, so that the oxygen half-cell is a cathode. On the other hand, if the potential of hydrogen electrode (E^a) is controlled to be more positive than the equilibrium potential of hydrogen half-cell (E^c_{eq} = 0.0 V), the oxidation current for H^+ generation flows, which is the reaction occurring at the hydrogen electrode in fuel cells. The electrode reactions prevailing in fuel cells are indicated by the solid lines in Fig. 1-20. The voltage difference between the two electrodes at a current of zero is E^0_{cell} = 1.23 V, which is called an ***open-circuit voltage*** because the loop is open (under equilibrium condition). The working voltage (E_{wk}) observed during a constant-current fuel cell operation at $|i_a| = i_c \neq 0$ is given by <1-8>.

$$E^c = E^c_{eq} - \eta_c,\ E^a = E^a_{eq} + \eta_a \ (|i_a| = i_c \neq 0)$$
$$E_{wk} = (E^c - E^a) - iR_{total}$$
$$E_{cell} = E^c_{eq} - E^a_{eq} \ (|i_a| = i_c = 0)$$
$$E_{wk} = E_{cell} - (\eta_a + \eta_c + iR_{total})$$

<1-8>

<1-8> illustrates that the working voltage of fuel cell is given by subtracting the cell polarization ($\eta_a + \eta_c + iR_{total}$) from the thermodynamic value (E_{cell}). With an increase in current, the cell polarization increases to lower the working voltage. Accordingly, in order to bring the E_{wk} closer to the maximum value ($E^0_{cell} = 1.23$ V), it is necessary to minimize the cell polarization; the overpotentials at both electrodes and all the resistances along the closed loop should be minimized.

1-3-4 Cathode/anode or negative/positive electrode

The electrode where electrochemical oxidation reaction takes place is defined as anode, whereas the one for electrochemical reduction is defined as cathode. In spite of this definition, confusion about anode or cathode is not avoided, particularly in the case of secondary batteries. The electrode reactions involved in the discharge process of Li/MnO_2 secondary battery and the name of electrodes are illustrated in Fig. 1-21. When the cell is discharged, the lithium metal electrode is an anode because the oxidation reaction occurs, whereas the MnO_2 electrode is a cathode because the reduction reaction takes place. Meanwhile, in the discharge process, the amount of negatively charged electrons increases in the lithium electrode, whereas the electrons are consumed in the MnO_2 electrode. Since the negative charge in the lithium electrode is larger than that of the MnO_2 electrode,

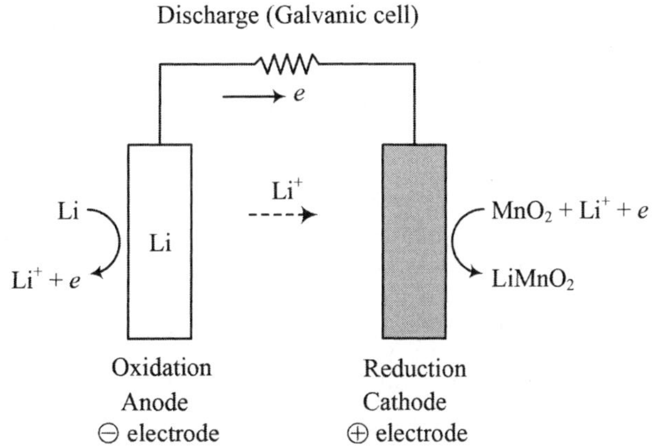

Fig. 1-21 Electrode reactions in the discharge process of Li/MnO_2 secondary battery and the name of electrodes.

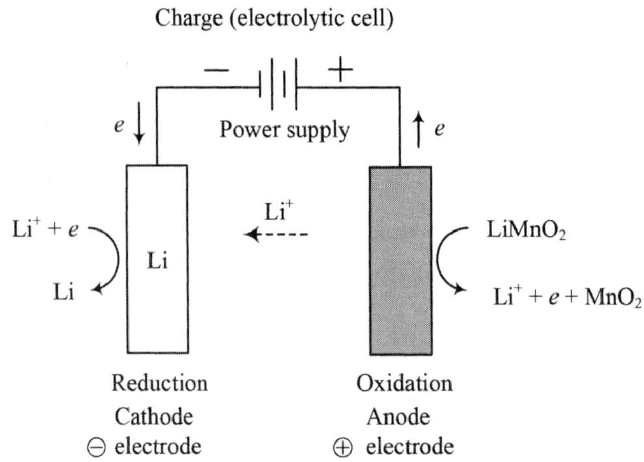

Fig. 1-22 Electrode reactions in the charging process of Li/MnO_2 secondary battery and the name of electrodes.

the potential of lithium electrode is more negative; lithium electrode is a negative electrode and the MnO_2 electrode is a positive electrode.

In the charging of Li/MnO_2 secondary battery (Fig. 1-22), the electrochemical reduction reaction occurs at the lithium electrode, so it is a cathode, whereas the MnO_2 electrode on which the electrochemical oxidation takes place is now an anode. Also, since the negative and positive potential is applied to the lithium metal and MnO_2 electrode, respectively, the lithium metal is a negative electrode and the MnO_2 is a positive electrode. In short, the lithium metal is an anode upon discharging but a cathode during charging, which leads to a confusion. However, the lithium metal is always a negative electrode regardless of charge and discharge, and the MnO_2 is always a positive electrode. Thus, the use of negative/positive electrode instead of anode/cathode is recommended to avoid confusion.

Example 1-3

The electrodes in an electrochemical cell are named on the basis of the electrochemical reactions occurring at the given electrode. Electrochemical reduction occurs on cathode, from which electrons are *emitted.* In opposite, electrons are *injected* into anode as a result of electrochemical oxidation. The terms of anode and cathode are also used in X-ray tube

(Coolidge tube) generating X-rays and hollow cathode lamps. What are the common features for the electrochemical cells, X-ray tube and hollow cathode lamps in terms of the movement (emission or injection) of electrons?

Solution

The working principle of X-ray tube is as follows. A tungsten filament is heated to generate electrons, which are then accelerated by applying electric field to be irradiated onto a copper target to generate X-rays. Note that electrons are emitted from the tungsten filament and electrons are injected into the copper target. The tungsten filament is called as a cathode and the copper as an anode. Note that it is the same as that in the electrochemical cells; electrons are emitted from a cathode and electrons are injected into an anode.

In hollow cathode lamps, argon is converted into Ar^+ and electrons within an electric field. The as-generated Ar^+ sputters a target metal to generate excited state metal atoms, from which the light of a specific wavelength is emitted. Here, Ar^+ are injected into the target metal and electrons are injected into another metal electrode. Since the movement direction of positive charges (Ar^+) and electrons is opposite, the injection of Ar^+ corresponds to the emission of negative charges, and thus it is called a cathode. In short, in all three cases, the electrode (or metal), from which electrons are emitted, is a cathode. The electrode (or metal), into which electrons are injected, is an anode.

1-4 Applications of electrochemistry

The application fields of electrochemistry can be classified as follows.

1) Electrochemical synthesis: Synthesis of chlorine (Cl_2) and NaOH by brine electrolysis, and extraction of aluminum in molten salt electrolyte are the examples. Chlorine is used for the production of poly (vinyl chloride) (PVC) and synthetic detergents. Fluorine (F_2), Na, Li, and Mg are produced by electrolysis in molten salt electrolytes. High-purity copper thin films are produced through electrorefining.

In addition, raw materials or intermediates for medicines, pesticides, and food additives are produced by electrolytic synthesis.

2) Electrochemical energy storage and conversion: Secondary batteries are widely used for power supplies for small electronic devices and electric vehicles, and energy storage systems. Fuel cells are being commercialized as power sources for automobiles and large-scale power generation facilities. Supercapacitors are expanding their markets in areas requiring high power.
3) Surface treatments: Electroplating of copper is used for wiring of semiconductor integrated circuits. Electrophoretic coating is used for surface treatment for corrosion prevention of automobile body.
4) Electrochemical analysis: The principle of electrochemistry is applied to pH meter, oxygen sensor of automobile engine, biosensor for the measurement of glucose and urea concentration. Electrochemical analysis is used to measure the concentration of trace elements or compounds.
5) Corrosion prevention: Corrosion of infrastructure, industrial facilities such as bridges, ports, etc., causes not only economic losses but also safety problems. Electrochemical principles are applied for corrosion prevention; e.g. cathodic protection and anodic protection.
6) Environmental field: Electrochemical principles are widely applied; the recovery of heavy metals by electrodialysis, purification of salts, decomposition of pollutants by electrolysis, and production of chlorine or sodium hypochlorite used for disinfection.
7) Bio-electrochemistry: Many enzymatic reactions for energy production are of an electrochemical nature. The movement of ions between cell membranes and subsequent formation of membrane potentials is an essential element for nerve action. Other processes associated with electrochemical reactions are photo-synthesis, vision, and smell. The principles and electrochemical methods are used to understand these phenomena. Electrochemical methods are also used to analyze biomaterials.

CHAPTER

02

Electrode potential

2-1 Standard electrode potential

The electrical potential is defined as the work required to bring a unit positive charge from an infinite distance to a certain phase. However, since the required work can not be defined as an absolute value, the electrical potential is always expressed as a relative value (as a difference); the electricity we use is 220 V, which is a relative value against the ground. Along the same line, if there is a difference in the type and quantity of electric charges in two interfacing phases, one can assume an electrical potential difference; if there is a difference in the amount of negative charges between two phases, the phase with larger negative charges has a lower electrical potential than the phase with smaller negative charges. The difference in the type and amount of electric charges is observed at the interface of two phases, so the potential difference also evolves at the interface; e.g. solid/solution or solution/solution. It should also be noted that the electrical potential difference is caused by only an *infinitesimally small* amount of charge separation. Further, since the electrical potential difference can be defined only when the electric charge distribution at the interface does not change with time, the electrical potential difference can be defined only in an equilibrium or steady state.

2-1-1 Electrical potential difference by an infinitesimally small charge separation

Assuming that one electric charge (i.e. one positive charge and one negative charge) is separated at an electrode/aqueous solution interface, the electric field can be calculated by the following equation.

$$\vec{E} = \frac{\vec{F}}{q} = \frac{1}{4\pi\varepsilon\varepsilon_0}\frac{q}{r^2}$$

By substituting the charge of unit charge ($q = 1.6\times10^{-19}$ C), the dielectric constant of water ($\varepsilon = 78.5$), the vacuum permittivity ($\varepsilon_0 = 8.85 \times 10^{-12}$ C^2 N^{-1} m^{-2}), and the distance between two charges (r = 10^{-9} m), the electric field strength is calculated to be:

$$\vec{E} = 2 \times 10^{7} \text{ V/m}$$

The electric field is much larger than what we imagine. From this, the electrical potential difference ($\Delta\phi$) is calculated to be 20 mV. This is the result obtained with an assumption that one charge is separated across an interface. However, considering that one equivalent (mole) of electrons is 6.02×10^{23}, one can recognize that electrical potential difference is caused even by a negligibly small charge separation.

2-1-2 Electrical potential difference under equilibrium state

Fig. 2-1 shows how the ***electrical potential difference*** develops at the interface between Ag electrode and aqueous solution containing Ag^{+}. The solution satisfies the electroneutrality condition of $[Ag^{+}] = [X^{-}]$ and the ion distribution is uniform, so that no electrical potential difference occurs anywhere in the solution. Assume that the Ag electrode maintains electroneutrality before being immersed into the solution. When the Ag electrode is put into the solution, Ag^{+} react with the electrons provided by the Ag electrode to form metallic Ag.

$$Ag^{+} + e \text{ (from Ag electrode)} \rightarrow Ag \qquad \langle 2\text{-}1 \rangle$$

As the above reaction proceeds, positive charges accumulate in the Ag electrode because electrons leave from the Ag electrode, while negative charges accumulate in the solution side because Ag^{+} are consumed for the reaction while X^{-} being idle, which eventually leads to a condition of $[Ag^{+}] < [X^{-}]$ in the solution. There was no charge separation

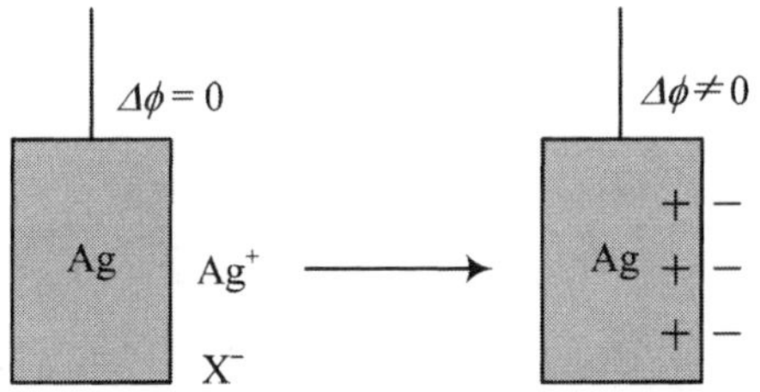

Fig. 2-1 Electrical potential difference ($\Delta\phi$) at the interface between Ag electrode and the solution containing Ag^{+}.

at the interface of Ag electrode/solution before the Ag electrode being immersed. However, after the immersion, an electrical potential difference occurs as a result of charge separation at the interface. As the reaction <2-1> proceeds to the right, the amount of positive charges in the Ag electrode increases, such that it tends not to lose additional electrons. In the solution, as the reaction progresses, the amount of negative charges (anions) increases, such that the tendency toward the right gradually decreases; the reaction tendency to the left gradually increases. Eventually, a dynamic equilibrium is reached, under which the two-way rate is equal to each other. In the non-equilibrium state where one side reaction is dominant over the other, the electrical potential difference can not be defined because the charge separation continuously changes. However, in the equilibrium state, the charge separation is kept constant; the potential difference can be defined.

2-1-3 Chemical potential and electrochemical potential

Since the electrical potential difference at an interface can be defined in the equilibrium state, the equilibrium state must first be defined. Equilibrium is defined as the state where the chemical potential of reactants is the same with the chemical potential of products. That is, in a chemical reaction of A + B $\rightleftharpoons$ C + D, the sum of chemical potential of A and B is the same as the sum of chemical potential of C and D.

Chemical potential is defined by the formula given in Fig. 2-2-a. It is seen that chemical potential has a unit of energy; e.g. J/mol. The chemical potential is defined as a change

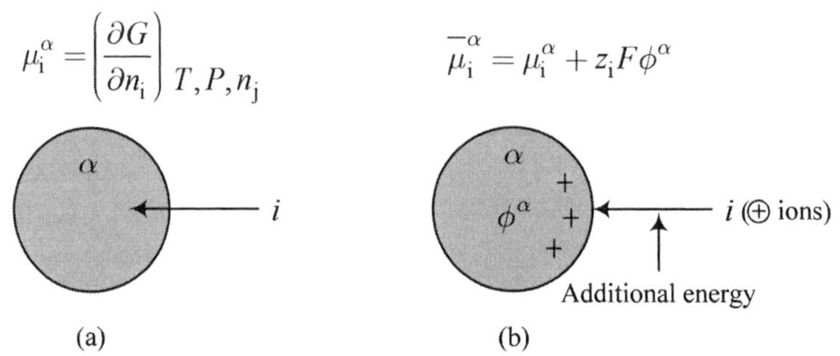

Fig. 2-2 Definition of chemical potential (a) and electrochemical potential (b).

in the free energy required to add a compound i into a phase (α) under a constant temperature and pressure. Also, the concentration of other chemical species (n_j) is assumed to be constant. Thus, a larger chemical potential means that more free energy is needed to add compound i, which in turn illustrates that the reaction tends to proceed to the opposite direction; removal of a compound i from the phase (α); the chemical potential is referred to as an escaping tendency. Namely, if the sum of chemical potential of reactants A and B is larger than the sum of chemical potential of C and D, the reaction tends to proceed to the right because the reactants have a higher chemical potential; higher escaping tendency. However, as the reaction proceeds, the escaping tendency of the reactants decreases while the escaping tendency of the products increases, which eventually approaches an equilibrium state, where the escaping tendency (chemical potential) is the same for the reactants and products. A dynamic equilibrium is reached when the chemical potentials of both sides are identical.

The charged species (electrons, cations, and anions) participate in electrochemical reactions. In this case, when a charged species i is added into a phase (α) that has electrical potential of ϕ^{α}, an electric energy ($z_i F\phi^{\alpha}$) is required in addition to the free energy to add compound i. For example, assuming that the phase (or electrode) has an excess of positive charges and a cation i is added thereto, not only the free energy due to an increase in the concentration of i in the phase increases but also the additional electrical energy is required (Fig. 2-2-b). Therefore, in electrochemical reactions, the escaping tendency and equilibrium should be described using ***electrochemical potential*** instead of chemical potential. In electrochemical reactions, the equilibrium is defined as the state where the electrochemical potential of reactants and products are identical to each other.

Electrochemical potential is defined in Fig. 2-2-b, where z_i is the charge of ionic species i and F is the Faraday constant. Chemical and electrochemical potentials have the following characteristics.

1) Chemical potential: $\mu_i = \mu_i^0 + RT \ln a_i$

   Here, a_i is the activity of i. If $a_i = 1.0$, $\mu_i = \mu_i^0$. μ_i^0 is referred to as the standard chemical potential. For an uncharged species ($z_i = 0$): $\overline{\mu}_i^{\alpha} = \mu_i^{\alpha}$.

2) For a pure phase at unit activity (z_i = 0 and a_i = 1.0); e.g. Ag, AgCl, and H_2 at unit fugacity. $\overline{\mu}_i^{\alpha} = \mu_i^{0\alpha} + RT \ln a_i + z_i F \phi^{\alpha} = \mu_i^{0\alpha}$

3) For electrons in a metal (z_i = −1): $\overline{\mu}_e^{\alpha} = \mu_e^{0\alpha} + RT \ln a_e^{\ \alpha} + z_e F \phi^{\alpha} = \mu_e^{\alpha'} - F\phi^{\alpha}$

Since the population of electrons in metals is very high, the change of electron concentration can be neglected (a_e^{a} is constant). Therefore, since the first and the second term in the right-hand side of the above expression are constants, they can be expressed $\mu_e^{a'}$ as a sum.

2-1-4 Electrical potential difference at interfaces according to potential-determining equilibrium

Using the definition of electrochemical potential in Fig. 2-2, the electrical potential difference at the Ag/Ag^+ solution interface (Fig. 2-1) can be quantified under the dynamic equilibrium state. Under equilibrium, the electrochemical potentials of both sides are the same for the reaction of Ag = Ag^+ + *e*.

$$\overline{\mu}_{Ag^0}^{M} = \overline{\mu}_{Ag^+}^{S} + \overline{\mu}_{e}^{M}$$

Here, *M* and *S* mean the metal and solution, respectively. This can be summarized as <2-2> in consideration of the characteristics for electrochemical potential.

$$\mu_{Ag^0}^{0\,M} + RT \ln a_{Ag^0}^{M} + z_{Ag^0} F \phi^{M}$$

$$= \mu_{Ag+}^{0\,S} + RT \ln a_{Ag^+}^{S} + z_{Ag^+} F \phi^{S} + \mu_{e}^{0\,M} + RT \ln a_{e}^{M} + z_e F \phi^{M}$$

$$\mu_{Ag^0}^{0\,M} = \mu_{Ag+}^{0\,S} + RT \ln a_{Ag^+}^{S} + F\phi^{S} + \mu_{e}^{0'M} - F\phi^{M} \qquad \text{<2-2>}$$

By rearranging <2-2>, the electrical potential difference is derived as <2-3>, in which $\Delta\phi^0(M,S)$ is the electrical potential difference under standard state (the activity of Ag^+ is unity), such that it is a constant.

$$(\phi^{M} - \phi^{S}) = \Delta\phi(M,S) = \Delta\phi^{0}(M,S) + \frac{RT}{F} \ln a_{Ag^+}^{S} \qquad \text{<2-3>}$$

It is seen in <2-3> that the electrical potential difference at the interface is determined only by the temperature and the activity of Ag^+. This is because the activity of Ag metal is constant in <2-1> and the activity of electrons is also constant because electron concentration is invariant. The electrochemical reaction that determines the electrical potential difference called the ***potential-determining equilibrium***; e.g. $Ag = Ag^+ + e$. Meanwhile, a system in which Ag metal, Ag^+, and electrons participate in electrochemical oxidation and reduction reactions such as $Ag = Ag^+ + e$ is called a half-cell. Hence, the potential-determining equilibrium is also called a ***half-cell reaction***.

2-1-5 Types of half-cells

Half-cells are classified as follows according to the cell configuration and potential-determining equilibrium.

1) Half-cells composed of metal/metal ions: e.g. Ag/Ag^+ as described above.
2) Half-cells composed of metal/insoluble metal salt/anion: e.g. $Ag/AgCl(s)/Cl^-$ (Fig. 2-3). Here, $AgCl(s)$ is an insoluble metal salt containing the same cation (Ag^+) as Ag metal, and Cl^- is the same anion as the anion contained in the insoluble metal salt (AgCl). The potential-determining equilibrium of this half-cell is

$$AgCl(s) + e = Ag(s) + Cl^-(aq)$$

The electrical potential difference is derived as follows.

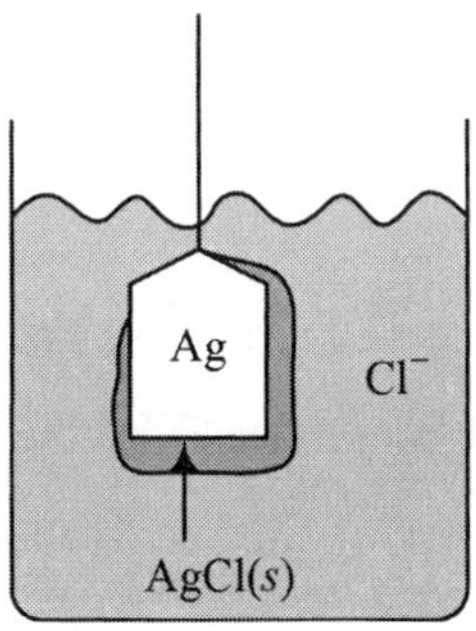

$$Ag^+ + e = Ag$$
$$AgCl(s) = Ag^+ + Cl^-$$
$$\overline{AgCl(s) + e = Ag + Cl^-}$$

Fig. 2-3 The potential-determining equilibrium (or half-cell reaction) for $Ag/AgCl(s)/Cl^-$ half-cell.

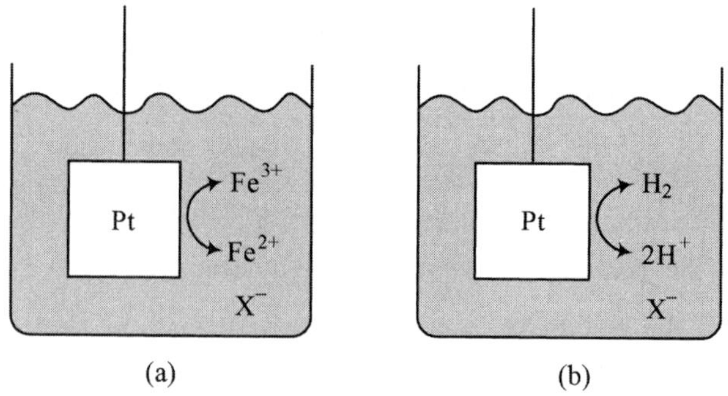

Fig. 2-4 (a); A half-cell composed of an inert electrode/redox pair and (b); half-cell consisting of an inert electrode/gas/ion.

$$\Delta\phi(M,S) = \Delta\phi^0(M,S) + \frac{RT}{F}\ln\frac{1}{a^{S}_{Cl^-}}$$

Here, the electrical potential difference is determined only by the activity of Cl^- and temperature. This is because the activity of AgCl(*s*), Ag(*s*) and electrons is constant but only the activity of Cl^- is variable.

3) Half-cells composed of an inert electrode/redox pair: The inert electrode is not electrochemically oxidized or reduced, while the redox pair is active for electrochemical reduction/oxidation (in short, redox) reaction. The Pt/Fe^{2+}, Fe^{3+} belongs to this type of half-cell (Fig. 2-4-a). The potential-determining equilibrium of this half-cell is

$$Fe^{3+}(aq) + e = Fe^{2+}(aq)$$

The electrical potential difference at the interface of Pt/solution containing Fe^{2+} and Fe^{3+} is derived as follows.

$$\Delta\phi(M,S) = \Delta\phi^0(M,S) + \frac{RT}{F}\ln\frac{a^{S}_{Fe^{3+}}}{a^{S}_{Fe^{2+}}}$$

4) Half-cells composed of an inert electrode/gas/ion: The typical example is Pt/H_2/H^+ (Fig. 2-4-b). Here, the ion is H^+ that can be produced by the oxidation of the gas (H_2). The potential-determining equilibrium of this half-cell is

$$2H^+(aq) + 2e = H_2(g)$$

The electrical potential difference at the interface is derived as follows.

$$\Delta\phi(M,S) = \Delta\phi^0(M,S) + \frac{RT}{2F}\ln\frac{(a^{S}_{H^+})^2}{f_{H_2}}$$

When the potential-determining equilibrium is expressed as <2-4> as a general form, the electrical potential difference is expressed as <2-5>. Among the chemical species participating in the half-cell reaction of <2-4>, except for the metals and solids having unit activity, the activity of ions and the fugacity of gases are the variables in <2-5>.

$$\nu_A A + \nu_B B + \cdots + ne = \nu_C C + \nu_D D + \cdots \qquad \text{<2-4>}$$

$$\Delta\phi(M,S) = \Delta\phi^0(M,S) + \frac{RT}{nF}\ln\frac{a_A^{\nu_A} a_B^{\nu_B}\cdots}{a_C^{\nu_C} a_D^{\nu_D}\cdots} \qquad \text{<2-5>}$$

Self learning 2-1

There are other types of half-cells in addition to those described above. For example, the half-cell with a potential-determining equilibrium of $PbO_2(s) + 4H^+(aq) + 2e = Pb^{2+}(aq) + 2H_2O$ does not belong to any of the above four types. Derive the electrical potential difference at the interface ($PbO_2(s)$ / aqueous solution) for this half-cell. Explain how the electrochemical reaction differs from the above four.

2-1-6 Standard electrode potential

The electrical potential difference derived so far is the value that evolves at the interface of two phases. The electrical potential of an electrode (in short, electrode potential) in a half-cell such as Pt/Fe^{2+}, Fe^{3+} can be measured by using a voltmeter. One of two terminals of the voltmeter is connected to the Pt electrode of Pt/Fe^{2+}, Fe^{3+} half-cell and the other terminal to another half-cell that plays as a reference of electrical potential, and the voltage difference (V) is read between two terminals. This is explained in Fig. 2-5, where the electrode potential of Pt/Fe^{2+}, Fe^{3+} half-cell is measured by using the $Pt/H_2/H^+$ half-cell as a reference. In the $Pt/H_2/H^+$ half-cell, the activity and fugacity are assumed

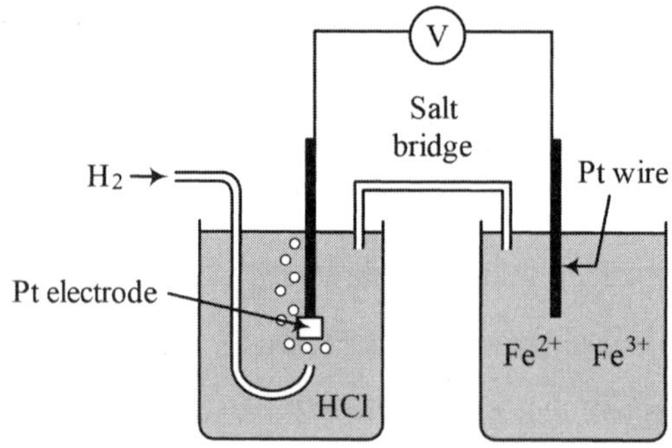

Fig. 2-5 Measurement of electrode potential (voltage) of Pt/Fe^{2+}, Fe^{3+} half-cell against Pt/H_2/H^+ reference electrode.

to be unity, such that it is called ***standard hydrogen electrode*** (SHE) or ***normal hydrogen electrode*** (NHE).

When a full-cell is constructed by using Pt/H_2/H^+ half-cell and Pt/Fe^{2+}, Fe^{3+} half-cell, the electrical potential difference evolves at the interfaces formed between two phases. In this case, four interfaces are found as shown in Fig. 2-5; the Pt/solution containing H_2 and H^+, the solution containing H_2 and H^+/the solution of salt bridge, the solution of salt bridge/the solution containing Fe^{2+} and Fe^{3+}, and the solution containing Fe^{2+} and Fe^{3+}/Pt electrode. All the potential differences developed are depicted in Fig. 2-6, wherein the potential difference occurred at both ends of the salt bridge is summed and indicated

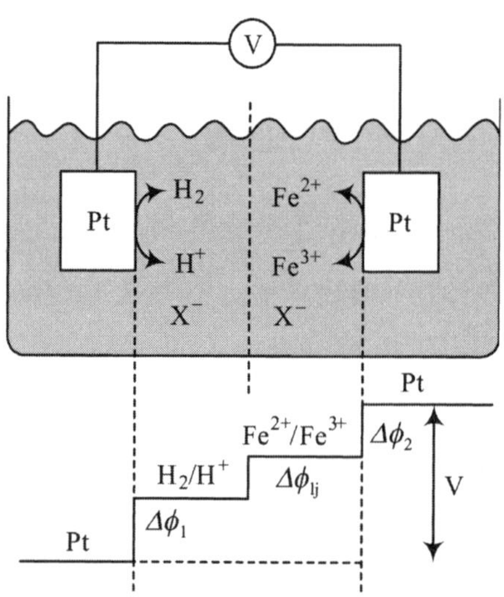

Fig. 2-6 The electrical potential difference evolved at the interfaces shown in Fig. 2-5.

as a liquid junction potential ($\Delta\phi_{1j}$). The potential difference (V) measured by the voltmeter is the sum of three values; $\Delta\phi_1$, $\Delta\phi_{1j}$, and $\Delta\phi_2$ as shown in Fig. 2-6. In the experimental setup in Fig. 2-5, as current can not flow between two half-cells due to infinite resistance (R $\rightarrow \infty$) of voltmeter, two half-cells are electrically separated to each other. Then, each half-cell approaches the equilibrium by itself (without any influences from the other half-cell); an electrical potential difference develops in the equilibrium state at the interface of Pt/solution containing H_2 and H^+, which is represented as $\Delta\phi_1$. Another electrical potential difference develops at the interface of Pt/solution containing Fe^{2+} and Fe^{3+}, which is represented as $\Delta\phi_2$. Here, $\Delta\phi_1$ is constant because the fugacity of H_2 and the activity of H^+ are assumed to be unity (standard state); $\Delta\phi_1 = \Delta\phi^0$ (Pt/H_2/H^+). The liquid junction potential can be minimized by using the salt bridge; $\Delta\phi_{1j} \approx 0$. Here, the sum of two constant values can be assumed to be zero ($\Delta\phi_1 + \Delta\phi_{1j}$ $\equiv 0$), which is valid because the electrical potential is a relative value. Then, the remnant variable is $\Delta\phi_2$. Ultimately, $V = \Delta\phi_2 = \Delta\phi(\mathrm{Pt/Fe^{2+}, Fe^{3+}})$ $(vs.\mathrm{NHE})$.

$$V = \Delta\phi_1 + \Delta\phi_{lj} + \Delta\phi_2$$

$$V = \Delta\phi\,(\mathrm{Pt/H_2, H^+}) + \Delta\phi_{lj} + \Delta\phi\,(\mathrm{Pt/Fe^{2+}, Fe^{3+}})$$

$$\Delta\phi\,(\mathrm{Pt/H_2/H^+}) + \Delta\phi_{lj} \equiv 0$$

$$V = \Delta\phi\,(\mathrm{Pt/Fe^{2+}, Fe^{3+}})\ (vs.\ \mathrm{NHE})$$

The electrode potential of the Pt/Fe^{2+}, Fe^{3+} half-cell against the NHE reference electrode is given as <2-6>.

$$V = E_{eq} = \Delta\phi\,(\mathrm{Pt/Fe^{2+}, Fe^{3+}}) = \Delta\phi^0\,(\mathrm{Pt/Fe^{2+}, Fe^{3+}}) + \frac{RT}{F}\ln\frac{a_{\mathrm{Fe^{3+}}}}{a_{\mathrm{Fe^{2+}}}}$$

$$E_{eq} = E^0\,(\mathrm{Pt/Fe^{2+}, Fe^{3+}}) + \frac{RT}{F}\ln\frac{a_{\mathrm{Fe^{3+}}}}{a_{\mathrm{Fe^{2+}}}} \qquad \text{<2-6>}$$

<2-6> is called the ***Nernst equation***, where E_{eq} is the equilibrium potential, which is determined by charge separation at the equilibrium state of Pt/Fe^{2+}, Fe^{3+} half-cell. E^0 (Pt/Fe^{2+}, Fe^{3+}) is referred to as the ***standard electrode potential***, which is the equilibrium potential when the activity (or fugacity) of all the species participating in the potential-de-

termining equilibrium is unity; $E_{eq} = E^0$ in <2-6> if the activity of Fe^{2+} and Fe^{3+} is unity. When the potential-determining equilibrium is expressed as <2-4> as a general form, the Nernst equation is expressed as <2-7>. Here, since the activity of metals and solids is unity, <2-7> can be simplified by using the activity of ions and fugacity of gases.

$$E_{eq} = E^0 + \frac{RT}{nF} \ln \frac{a_A^{\nu_A} a_B^{\nu_B} \cdots}{a_C^{\nu_C} a_D^{\nu_D} \cdots} \qquad \text{<2-7>}$$

Table 2-1 lists the E^0 values of half-cells, which are calculated from the thermodynamic values at 298 K and are the relative values against the NHE reference electrode; the electrode potential of NHE is assumed to be zero. Meanwhile, <2-8> is derived when the equilibrium potential is expressed as a function of concentration instead of activity in the Nernst equation. $E^{0'}$ is called ***formal potential***. In general, the activity coefficients of O and R do not differ greatly ($\gamma_O \approx \gamma_R$), so E^0 and $E^{0'}$ are comparable to each other. Therefore, when calculating the equilibrium potential using <2-8>, there is no big error even if the E^0 values listed in Table 2-1 are used instead of $E^{0'}$. In <2-8>, R is the gas constant (8.314 J mol^{-1} K^{-1}), T is temperature, n is the number of electrons involved in the potential-determining equilibrium, F is the Faraday constant (9.6485×10^4 C mol^{-1}), and C_O/C_R are the concentration of O and R (in M, moles/L).

$$E_{eq} = E^0 + \frac{RT}{nF} \ln \frac{a_O}{a_R} = E^0 + \frac{RT}{nF} \ln \frac{\gamma_O}{\gamma_R} + \frac{RT}{nF} \ln \frac{C_O}{C_R}$$

$$E_{eq} = E^{0'} + \frac{RT}{nF} \ln \frac{C_O}{C_R} \qquad \text{<2-8>}$$

The E^0 values of Li/Li^+ and Cu/Cu^{2+} half-cell are as follows.

$Li^+ + e = Li$ $\qquad E^0 = -3.045$ V (*vs.* NHE at 25°C)

$Cu^{2+} + 2e = Cu$ $\qquad E^0 = 0.34$ V (*vs.* NHE at 25°C)

The E^0 of Li/Li^+ half-cell is more negative than that of Cu/Cu^{2+} one. This indicates that even if the activity of Li^+ and Cu^{2+} is the same at unity, a larger amount of electrons is present in Li metal in the equilibrium state as compared to that in Cu; this illustrates that a larger amount of Li compared to Cu dissolves to produce electrons; this means that the Li half-cell has a higher oxidation tendency (to the left in the potential-determin-

Table 2–1 Standard electrode potential (V *vs.* NHE) of half–cells in aqueous solution at 298 K

E^0/V			
$Ag^+ + e = Ag$	+0.80	$I_2 + 2e = 2I^-$	+0.54
$Ag^{2+} + e = Ag^+$	+1.98	$I_3^- + 2e = 3I^-$	+0.53
$AgBr + e = Ag + Br^-$	+0.07	$In^+ + e = In$	–0.13
$AgCl + e = Ag + Cl^-$	+0.22	$In^{3+} + 2e = In^+$	–0.44
$AgI + e = Ag + I^-$	–0.15	$In^{3+} + 3e = In$	–0.34
$Al^{3+} + 3e = Al$	–1.68	$K^+ + e = K$	–2.93
$Au^+ + e = Au$	+1.83	$Li^+ + e = Li$	–3.04
$Au^{3+} + 3e = Au$	+1.52	$Mg^{2+} + 2e = Mg$	–2.36
$Ba^{2+} + 2e = Ba$	–2.92	$Mn^{2+} + 2e = Mn$	–1.18
$Be^{2+} + 2e = Be$	–1.97	$Mn^{3+} + e = Mn^{2+}$	+1.51
$Br_2 + 2e = 2Br^-$	+1.09	$MnO_2 + 4H^+ + 2e = Mn^{2+} + 2H_2O$	+1.23
$BrO^- + H_2O + 2e = Br^- + 2OH^-$	+0.76	$MnO_4^- + e = MnO_4^{2-}$	+0.56
$2HOBr + 2H^+ + 2e = Br_2 + 2H_2O$	+1.60	$NO_3^- + 2H^+ + e = NO_2 + H_2O$	+0.80
$Ca^{2+} + 2e = Ca$	–2.84	$NO_3^- + 4H^+ + 3e = NO + 2H_2O$	+0.96
$Cd(OH)_2 + 2e = Cd + 2OH^-$	–0.82	$NO_3^- + H_2O + 2e = NO_2^- + 2OH^-$	+0.01
$Cd^{2+} + 2e = Cd$	–0.40	$Na^+ + e = Na$	–2.71
$Ce^{3+} + 3e = Ce$	–2.34	$Ni^{2+} + 2e = Ni$	–0.26
$Ce^{4+} + e = Ce^{3+}$	+1.72	$Ni(OH)_2 + 2e = Ni + 2OH^-$	–0.72
$Cl_2 + 2e = 2Cl^-$	+1.36	$NiO_2 + 4H^+ + 2e = Ni^{2+} + 2H_2O$	+1.59
$ClO^- + H_2O + 2e = Cl^- + 2OH^-$	+0.89	$O_2 + 2H_2O + 4e = 4OH^-$	+0.40
$2HOCl + 2H^+ + 2e = Cl_2 + 2H_2O$	+1.63	$O_2 + 4H^+ + 4e = 2H_2O$	+1.23
$ClO_3^- + 2H^+ + e = ClO_2 + H_2O$	+1.17	$O_2 + e = O_2^-$	–0.33
$ClO_4^- + 2H^+ + 2e = ClO_3^- + H_2O$	+1.20	$O_2 + H_2O + 2e = HO_2^- + OH^-$	–0.08
$Co^{2+} + 2e = Co$	–0.28	$O_2 + H^+ + e = HO_2$	–0.13
$Co^{3+} + e = Co^{2+}$	+1.92	$O_2 + 2H^+ + 2e = H_2O_2$	+0.70
$Co(NH_3)_6^{3+} + e = Co(NH_3)_6^{2+}$	+0.06	$Pb^{2+} + 2e = Pb$	–0.13
$Cr^{2+} + 2e = Cr$	–0.90	$PbO_2 + 4H^+ + 2e = Pb^{2+} + 2H_2O$	+1.70
$Cr^{3+} + 3e = Cr$	–0.74	$PbSO_4 + 2e = Pb + SO_4^{2-}$	–0.36
$Cs^+ + e = Cs$	–2.92	$Pt^{2+} + 2e = Pt$	+1.19
$Cu^+ + e = Cu$	+0.52	$Rb^+ + e = Rb$	–2.93
$Cu^{2+} + 2e = Cu$	+0.34	$S + 2e = S^{2-}$	–0.48
$Cu^{2+} + e = Cu^+$	+0.16	$2SO_2 + 2H^+ + 4e = S_2O_3^{2-} + H_2O$	–0.40
$CuCl + e = Cu + Cl^-$	+0.12	$SO_2 + 4H^+ + 4e = S + 2H_2O$	+0.50
$Cu(NH_3)_4^{2+} + 2e = Cu + 4NH_3$	–0.00	$S_4O_6^{2-} + 2e = 2S_2O_3^{2-}$	+0.08

E^0/V			
$F_2 + 2e = 2F^-$	+2.87	$2SO_4^{2-} + 4H^+ + 2e = S_2O_6^{2-} + 2H_2O$	−0.25
$Fe^{2+} + 2e = Fe$	−0.44	$S_2O_8^{2-} + 2e = 2SO_4^{2-}$	+1.96
$Fe^{3+} + 3e = Fe$	−0.04	$Sn^{2+} + 2e = Sn$	−0.14
$Fe^{3+} + e = Fe^{2+}$	+0.77	$Sn^{4+} + 2e = Sn^{2+}$	+0.15
$Fe(CN)_6^{3-} + e = Fe(CN)_6^{4-}$	+0.36	$Sr^{2+} + 2e = Sr$	−2.89
$Fe(CN)_6^{4-} + 2e = Fe + 6CN^-$	−1.16	$Ti^{2+} + 2e = Ti$	−1.63
$2H^+ + 2e = H_2$	0.00	$Ti^{3+} + e = Ti^{2+}$	−1.37
$2H_2O + 2e = H_2 + 2OH^-$	−0.83	$TiO^{2+} + 2H^+ + e = Ti^{3+} + H_2O$	+0.10
$H_2O_2 + H^+ + e = OH^- + H_2O$	+0.71	$V^{2+} + 2e = V$	−1.13
$H_2O_2 + 2H^+ + 2e = 2H_2O$	+1.76	$V^{3+} + e = V^{2+}$	−0.26
$Hg^{2+} + 2e = Hg$	+0.80	$VO_2 + 2H^+ + 2e = V^{2+} + H_2O$	+0.34
$Hg_2Cl_2 + 2e = 2Hg + 2Cl^-$	+0.27	$VO_2^+ + 2H^+ + e = VO^{2+} + H_2O$	+1.00
$2Hg^{2+} + 2e = Hg_2^{2+}$	+0.91	$Zn^{2+} + 2e = Zn$	−0.76

ing equilibrium reaction of $Li^+ + e = Li$).

In short, the half-cells having a more negative E^0 in Table 2-1 show a higher oxidation tendency. In opposite, the half-cells of a more positive E^0 have a larger reduction tendency. Note here, however, that E^0 corresponds to E_{eq} at the standard state, such that the above criterion is only valid in the standard state. In the more general case, the ***oxidation/reduction tendency*** of half-cells should be compared on the basis of E_{eq} value (See below).

Self learning 2-2

Assuming that $[Fe^{2+}] = [Fe^{3+}] = 0.1\ M$ and $[Ag^+] = 10^{-5}\ M$, calculate the E_{eq} of the Fe^{2+}/Fe^{3+} half-cell and Ag/Ag^+ half-cell, and compare the oxidation/reduction tendency of the two half-cells on the basis of the calculated E_{eq} values. Here, the half-cell of more negative E_{eq} shows a larger oxidation tendency. Note that the tendency is opposite to that judged from the E^0 value. Then, of E_{eq} and E^0, which one is the right criterion for determining the oxidation/reduction tendency of half-cells?

2-1-7 Types of reference electrode

So far, the standard electrode potential of half-cells was defined by setting the NHE as the ***reference electrode***. However, NHE is not practical because it uses hydrogen gas. Thus, in practical experiments, a saturated calomel electrode (SCE, $Hg/Hg_2Cl_2(s)$, KCl (4.2 *M*, saturated)) or Ag/AgCl(*s*), KCl (4.2 *M*, saturated) electrode are used in aqueous electrolytes. For non-aqueous electrolytes, Pt/$FeCp_2$, $FeCp_2^+$ or Ag/Ag^+ electrode are used. The electrode potential of these reference electrodes against NHE is shown in Fig. 2-7. The electrode potential (voltage) of SCE is 0.2412 V higher than that of NHE.

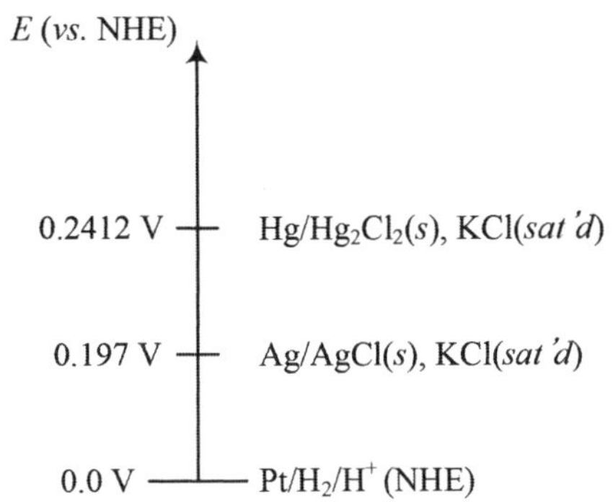

Fig. 2-7 Relative electrode potential (voltage) of other reference electrodes against NHE (normal hydrogen electrode).

Self learning 2-3

Let us calculate the equilibrium potential of a half-cell composed of a Pt electrode and Fe^{2+}/Fe^{3+} redox pair using <2-8>. If $[Fe^{3+}]$ = 0.1 M and $[Fe^{2+}]$ = 0.0 M, what is the equilibrium potential? The actually measured value differs from the calculated one. Give an explanation on this.

Self learning 2-4

E^0 = 0.34 V (*vs.* NHE) for Cu/Cu^{2+} half-cell. What is E^0 when SCE is used as the reference electrode?

Self learning 2-5

Ensure that the SCE belongs to the half-cell consisting of metal/insoluble metal salt/anion. Explain why KCl is used as a saturated solution (4.2 M).

Self learning 2-6

Reference electrodes shall be an ideally non-polarizable electrode. Explain why.

Self learning 2-7

$[Cu^{2+}]$ can be calculated using <2-8> after obtaining the potential of Cu/Cu^{2+} half-cell against NHE. Assume that $[Cu^{2+}]$ = 0.001 M. In Cu/Cu^{2+} half-cell, the equilibrium state is reached by $Cu^{2+} + 2e = Cu$ reaction, such that $[Cu^{2+}]$ in the equilibrium state may not be 0.001 M, which is the initial concentration. If this is the case, a measurement error may occur. However, the measured $[Cu^{2+}]$ is very close to 0.001 M. Explain why.

Self learning 2-8

Assume that 0.7 V is applied to the Ag/Ag^+ half-cell against NHE at 25 °C. Silver may be dissolved or plated depending on $[Ag^+]$. What is the minimum $[Ag^+]$ for plating?

Self learning 2-9

For the fabrication of Ag/AgCl(*s*) electrode, AgCl is electrodeposited on the surface of Ag wire and then contacted with aqueous KNO_3 solution instead of aqueous KCl solution. Can this reference electrode work?

2-2 Standard electrode potential of metal complexes

An inspection of Table 2-1 reveals that E^0 are more negative when metal ions and ligands form complexes as shown in the following examples. The metal ions for which the ligands are not represented are assumed to be hydrated; water coordination in aqueous solution; e.g. $Co(H_2O)_n^{3+}$. The more negative standard electrode potential for the complexes ($E^0_{complex}$) illustrates a higher oxidation tendency as compared to the water-coordinated redox pair.

$$Co^{3+} + e = Co^{2+} \qquad E^0 = 1.92\ V\ (vs.\ NHE)$$
$$Co(NH_3)_6^{3+} + e = Co(NH_3)_6^{2+} \qquad E^0 = 0.06\ V\ (vs.\ NHE)\ (E^0_{complex})$$
$$Au^{3+} + 3e = Au \qquad E^0 = 1.50\ V\ (vs.\ NHE)$$
$$AuCl_4^- + 3e = Au + 4Cl^- \qquad E^0 = 1.00\ V\ (vs.\ NHE)\ (E^0_{complex})$$

When Au^{3+} form a complex with Cl^-, an equilibrium constant (K_f, formation constant) is given as follows.

$$Au^{3+} + 4Cl^- = AuCl_4^-$$

$$K_f = \frac{a_{AuCl_4^-}}{a_{Au^{3+}} a_{Cl^-}^4}$$

The Nernst equation derived from the potential-determining equilibrium of $Au^{3+} + 3e = Au$ is as follows. Substituting the activity of Au^{3+} and K_f in this equation yields <2-9>.

$$E_{eq} = E^0 + \frac{RT}{3F}\ln a_{Au^{3+}} = E^0 + \frac{RT}{3F}\ln\left(\frac{a_{AuCl_4^-}}{K_f a_{Cl^-}^4}\right)$$

$$= E^0 + \frac{RT}{3F}\ln\left(\frac{1}{K_f}\right) + \frac{RT}{3F}\ln\left(\frac{a_{AuCl_4^-}}{a_{Cl^-}^4}\right) \qquad \text{<2-9>}$$

If we sum the first two terms on the right side of <2-9>, <2-10> is obtained, and <2-11> is derived from <2-10>.

$$E^0_{\text{complex}} = E^0 + \frac{RT}{3F}\ln\left(\frac{1}{K_f}\right)$$ <2-10>

$$E_{\text{eq}} = E^0_{\text{complex}} + \frac{RT}{3F}\ln\left(\frac{a_{\text{AuCl}_4^-}}{a^4_{\text{Cl}^-}}\right)$$ <2-11>

<2-11> is identical with the Nernst equation derived from the potential-determining equilibrium of $AuCl_4^-$ + 3*e* = Au + $4Cl^-$. It can be seen from <2-10> that the larger the K_f is, i.e., the stronger the complex formation between metal ion and ligand, the more negative the E^0_{complex} value becomes.

In polarography, mercury (Hg) electrodes are used to analyze trace amounts of electro-active compounds. The usable potential range for Hg electrodes is 0.0 V ~ −1.2 V (*vs.* SCE) because mercury itself is oxidized at the positive voltage higher than 0.0 V (*vs.* SCE), and hydrogen evolution becomes substantial at the voltage lower than −1.2 V (Fig. 2-8).

Let us assume that Cu^{2+} concentration is measured by measuring the reduction current of Cu^{2+} using polarography. The half-wave potential ($E_{1/2}$, which is similar to E^0) for Cu^{2+} reduction is 0.02 V (*vs.* SCE), which is outside of the usable potential range of mercury electrode. However, if Cu^{2+} forms strong complexes with Cl^- or NH_3, the $E_{1/2}$ value shifts to be more negative than 0.0 V (*vs.* SCE). Then, the reduction current of Cu^{2+} can be measured without interference of the mercury oxidation current.

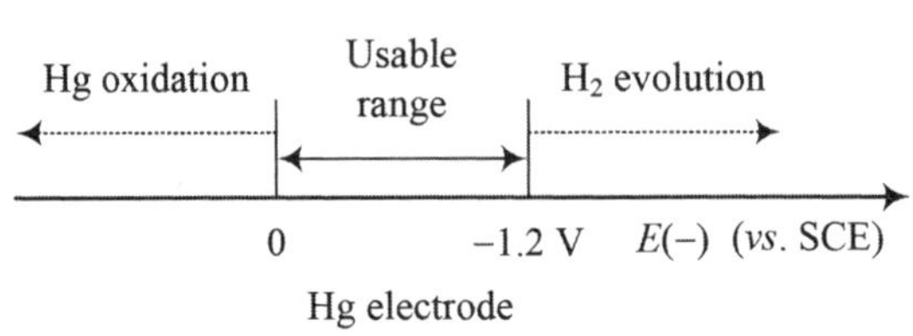

$Cu^{2+} + 2e = Cu$ $E_{1/2}$ = 0.02 V (SCE)

Cu^{2+} in KCl solution $E_{1/2}$ = − 0.22 V (SCE)

Cu^{2+} in 1 *M* NH_3/NH_4Cl $E_{1/2}$ = − 0.15 V (SCE)

Fig. 2-8 Changes of ligands for Cu^{2+} to electrochemically reduce in the usable range of mercury electrode.

2-3 Thermodynamic parameters from standard electrode potentials

Suppose that <2-12> is in equilibrium in aqueous solution. Since the equilibrium constant (K_{eq}) is a thermodynamic value, it has the identical value regardless of the reaction path if the initial state (reactants) and final state (products) are the same. Electrochemical pathway is one of the reaction paths. Here, if the initial state and the final state are the same with those given in <2-12> when the reaction proceeds through the electrochemical pathway, the equilibrium constant can be obtained from the electrochemical parameters. This procedure is shown in Fig. 2-9.

$$Fe^{2+}_{(aq)} + Ce^{4+}_{(aq)} \underset{K_{eq}}{=} Fe^{3+}_{(aq)} + Ce^{3+}_{(aq)} \qquad \text{<2-12>}$$

In general, the configuration of electrochemical cells is expressed as follows. The interface of two phases is represented by a single bar, the separator is represented by two bars, and the parentheses indicate the activity of ions. The full-cell in Fig. 2-9 can be represented as

$$\mathrm{Pt}\,|\,Fe^{2+}(a=1),\ Fe^{3+}(a=1)\,||\,Ce^{3+}(a=1),\ Ce^{4+}(a=1)\,|\,\mathrm{Pt}$$

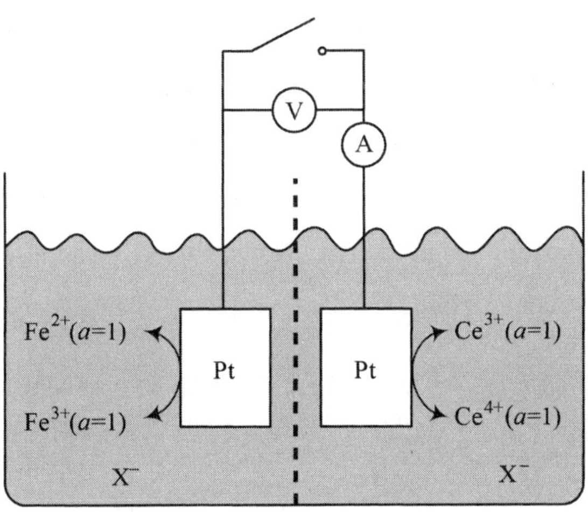

Fig. 2-9 Electrochemical cell (full-cell) consisting of two half-cells.

The standard electromotive force (E^0_{cell}) of a full-cell is defined as follows. Here, E^0_{RHS} and E^0_{LHS} refer to the equilibrium potential for the half-cell placed at the right-hand side and left-hand side, respectively. Since they assume a standard state, they each correspond to the standard electrode potential.

$$E^0_{cell} = E^0_{RHS} - E^0_{LHS}, \quad \Delta G^0 = -nFE^0_{cell}$$

When the half-cell with a higher oxidation tendency (more negative E^0) places on the left (LHS, left-hand side) and the other with a higher reduction tendency (more positive E^0) places on the right (RHS, right-hand side), the condition of $E^0_{cell} > 0$ is always satisfied and $\Delta G^0 < 0$. A ***galvanic cell*** forms, and the paired reaction of the oxidation on LHS and reduction on RHS takes place spontaneously. Also a current flows through an external circuit.

In Fig. 2-9, when the switch connecting the two half-cells is open (open-circuit), current cannot flow. Further, the current flow through the voltmeter (V) is negligible because of $R \to \infty$ for the voltmeter. Two half-cells remain electrically separated from each other and each reaches its equilibrium. Here, since the activity of all ions is 1.0, E_{eq} of the left half-cell = E^0 (Pt/Fe^{2+}, Fe^{3+}) = 0.77 V (*vs*. NHE), and E_{eq} of the right half-cell = E^0 (Pt/Ce^{3+}, Ce^{4+}) = 1.72 V. (*vs*. NHE). When the switch is closed (closed-circuit), current (electrons) can flow. The left half-cell has a greater oxidation tendency than the right one (E_{eq} is more negative), such that the oxidation reaction proceeds and the generated electrons move to the right half-cell through the closed circuit. Since the right half-cell has a greater reduction tendency than the left one (E_{eq} is more positive), the reduction reaction proceeds by taking electrons transferred from the left. Overall, the oxidation reaction in the left half-cell and the reduction reaction in the right half-cell are paired, and the paired reactions proceed spontaneously. The spontaneity can also be confirmed from $\Delta G^0 < 0$ for the overall reaction.

LHS; $Fe^{2+} \rightarrow Fe^{3+} + e$ $\quad E^0 = 0.77$ V (*vs*. NHE)
RHS; $Ce^{4+} + e \rightarrow Ce^{3+}$ $\quad E^0 = 1.72$ V (*vs*. NHE)
Overall; $Fe^{2+} + Ce^{4+} \rightarrow Fe^{3+} + Ce^{3+}$ $\quad (\Delta G^0 < 0)$

In Fig. 2-9, since the activity of all ions was initially assumed to be 1.0, the equilibrium potential of the two half-cells has their respective standard electrode potential values when the switch is open. Hence, the difference in equilibrium potential between two half-cells corresponds to the standard electromotive force ($E^0_{cell} = E^0_{RHS} - E^0_{LHS} = 1.72 - 0.77 = 0.95$ V). When the above spontaneous reaction proceeds once the circuit is closed, however, the activity of ions should change, such that the voltage difference between two half-cells also changes. As shown in <2-13>, the equilibrium potential of the left half-cell (E_{LHS}) increases due to an increase in the $a_{Fe^{3+}}/a_{Fe^{2+}}$ value as a result of oxidation reaction. On the other hand, the equilibrium potential of the right half-cell (E_{RHS}) decreases due to a decrease in $a_{Ce^{4+}}/a_{Ce^{3+}}$ as the reduction reaction proceeds on RHS.

$$E_{LHS} = E^0(Fe^{2+}, Fe^{3+}) + \frac{RT}{F}\ln\frac{a_{Fe^{3+}}}{a_{Fe^{2+}}}, \quad E_{RHS} = E^0(Ce^{3+}, Ce^{4+}) + \frac{RT}{F}\ln\frac{a_{Ce^{4+}}}{a_{Ce^{3+}}} \quad \text{<2-13>}$$

Here, a question arises as to how long the oxidation and reduction reactions proceed on the two half-cells. In the initial open-circuit condition, E_{LHS} [$= E^0_{LHS} = E^0$ (Fe^{2+}, Fe^{3+})] $<$ E_{RHS} [$= E^0_{RHS} = E^0(Ce^{3+}, Ce^{4+})$]. However, since E_{LHS} increases while E_{RHS} decreases with the continued reactions, the condition for $E_{LHS} = E_{RHS}$ approaches. Under this condition, the equilibrium potential at both electrodes is identical; the oxidation/reduction tendency is also equal; eventually, the rate of oxidation and reduction is the same on LHS, and also on RHS. That is, an equilibrium is reached in both half-cells and also the equilibrium is reached in the whole electrochemical cell (full-cell).

The overall reaction through the electrochemical pathway is identical with that shown in <2-12>, such that the equilibrium constant (K_{eq}) can be obtained from the electrochemical parameters. Under equilibrium, $E_{LHS} = E_{RHS}$, so the electromotive force of the full-cell $E_{cell} = E_{RHS} - E_{LHS} = 0$. If $E_{RHS} = E_{LHS}$ in <2-13>, <2-14> is derived. <2-15> is derived by substituting the equilibrium constant (K_{eq}).

$$E_{cell} = (E^0_{RHS} - E^0_{LHS}) + \frac{RT}{F}\ln\left(\frac{a_{Ce^{4+}}a_{Fe^{2+}}}{a_{Ce^{3+}}a_{Fe^{3+}}}\right) = 0 \quad \text{<2-14>}$$

$$K_{eq} = \frac{a_{Fe^{3+}}a_{Ce^{3+}}}{a_{Ce^{4+}}a_{Fe^{2+}}}$$

$$E^0_{RHS} - E^0_{LHS} = \frac{RT}{nF}\ln K_{eq} \quad \text{<2-15>}$$

Meanwhile, from the relationship between the standard electromotive force (E^0_{cell}) and the free energy difference (ΔG^0), <2-16> is derived.

$$E^0_{cell} = E^0_{RHS} - E^0_{LHS} = -\frac{\Delta G^0}{nF}$$

$$\Delta G^0 = -RT \ln K_{eq} = -nF\,(E^0_{RHS} - E^0_{LHS}) \qquad \text{<2-16>}$$

<2-16> illustrates that the equilibrium constant (K_{eq}), which is a thermodynamic parameter, can be obtained from the standard electrode potential values (E^0_{RHS} and E^0_{LHS}) of two half-cells.

Example 2-1

What are the conceptual differences between the electromotive force and the thermodynamic decomposition potential?

Solution

Both refer to the difference in equilibrium potential for two half-cells and are determined from the thermodynamic values. In an electrolytic cell, it is the thermodynamic decomposition voltage, which is the minimum voltage required for electrolysis as in <1-6>. In a galvanic cell, it is the electromotive force, which corresponds to the maximum working voltage as shown in <1-8>.

Example 2-2

Show that the solubility product (K_{sp}) is 1.74×10^{-10} (25°C) for $AgCl(s) = Ag^+ + Cl^-$.

Solution

The ***solubility product*** (K_{sp}) is an equilibrium constant, and can be obtained from electrochemical parameters. To do this, two half-cells must first be selected. The overall reaction from two half-cell reactions should be $AgCl(s) = Ag^+ + Cl^-$.

LHS: $AgCl(s) + e = Ag + Cl^-$ $\qquad E^0 = 0.22$ V (*vs.* NHE)

RHS: $Ag^+ + e = Ag$ $\qquad E^0 = 0.799$ V (*vs.* NHE)

Two half-cells as above are selected. The half-cell with more negative E_{eq} (E^0 with a unit activity for all ions) places on the left side and the other with more positive E_{eq} on the right side. In this full-cell, the oxidation on LHS and the reduction on RHS occur spontaneously. The overall reaction is opposite to the AgCl(s) dissociation reaction.

LHS: $Ag + Cl^- \rightarrow AgCl(s) + e$ (Oxidation)

RHS: $Ag^+ + e \rightarrow Ag$ (Reduction)

Overall: $Ag^+ + Cl^- \rightarrow AgCl(s)$

When the switch of the full-cell is closed, the oxidation and reduction take place spontaneously, and the equilibrium potential of the two half-cells changes as shown in <2-17>.

$$E_{RHS} = E^0_{RHS} + \frac{RT}{F}\ln a_{Ag^+},\quad E_{LHS} = E^0_{LHS} + \frac{RT}{F}\ln\frac{1}{a_{Cl^-}} \qquad \text{<2-17>}$$

<2-18> is valid when the equilibrium is reached in the full-cell. K_{sp} is obtained by substituting $E^0{}_{RHS}$ and $E^0{}_{LHS}$ (Table 2-1) into <2-18>.

$$E_{cell} = E_{RHS} - E_{LHS} = (E^0{}_{RHS} - E^0{}_{LHS}) + \frac{RT}{F}\ln(a_{Ag^+}a_{Cl^-}) = 0 \qquad \text{<2-18>}$$

$$K_{sp} = a_{Ag^+}a_{Cl^-} = 1.74\times10^{-10}\,(25°C)$$

2-4 Mixed potential

The electrode potential of Zn/Zn^{2+} half-cell against NHE should be -0.76 V at 25 °C if $a_{Zn^{2+}} = 1.0$, which is the standard electrode potential. However, the practically observed value is more positive. This is explained in Fig. 2-10.

Two half-cells can be combined to form a full-cell. Among the two half-cells, the half-cell with a more negative E_{eq} has a greater oxidation tendency. When an external circuit is connected between two half-cells, the electrons generated as a result of oxidation are transferred to the other half-cell (E_{eq} has a more positive value, so the reduction tendency is higher) to be consumed for an electrochemical reduction.

The standard electrode potentials (E^0) are given for the following half-cells.

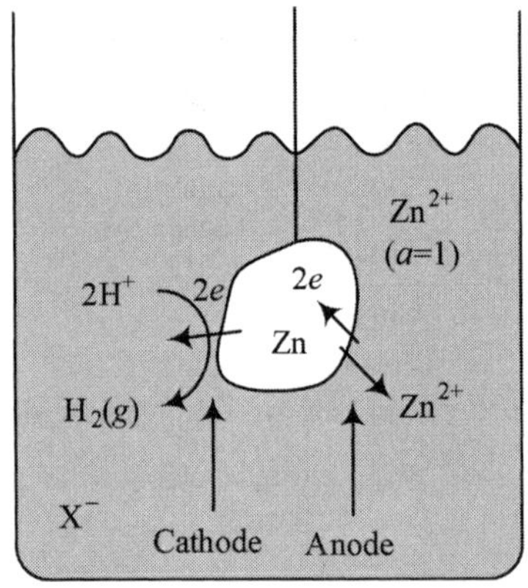

Fig. 2-10 Formation of a full-cell (galvanic cell) by Zn dissolution and hydrogen evolution.

$$Zn^{2+} + 2e = Zn \qquad E^0 = -0.76 \text{ V } (vs. \text{ NHE})$$
$$2H^+ + 2e = H_2 \qquad E^0 = 0.0 \text{ V } (vs. \text{ NHE})$$

Under the standard conditions, E_{eq} of the zinc half-cell = E^0 = -0.76 V ($vs.$ NHE), and E_{eq} of the hydrogen half-cell = 0.0 V ($vs.$ NHE). Clearly, the zinc half-cell has a higher oxidation tendency, while the hydrogen half-cell a higher reduction tendency. Therefore, if an external circuit through which electrons can be transferred is connected, electrochemical oxidation occurs in the zinc half-cell, and the generated electrons are transferred to the hydrogen half-cell to be used for an electrochemical reduction (hydrogen evolution).

$$\text{Anode: } Zn \rightarrow Zn^{2+} + 2e$$
$$\text{Cathode: } 2H^+ + 2e \rightarrow H_2$$
$$\text{Overall: } Zn + 2H^+ \rightarrow Zn^{2+} + H_2 \ (\Delta G^0 < 0)$$

In Fig. 2-10, a part of Zn metal can serve as an anode where Zn dissolves, whereas the other part of Zn metal can be a cathode where H_2 evolves. The electrons generated at the anode part can be transferred to the cathode part through the Zn metal. Note that Zn is an excellent electron conductor. In addition, there is an electrolyte solution capable of ion conduction. A full-cell (galvanic cell) forms and the whole reactions proceed spontaneously. It can be seen from <2-19> that the overall reaction is thermodynamically spontaneous.

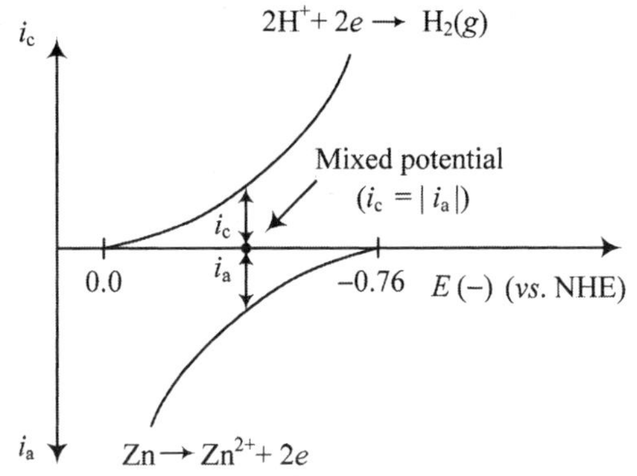

Fig. 2-11 Mixed potential by Zn dissolution and hydrogen evolution.

$$E^0_{cell} = E^0_{RHS} - E^0_{LHS} = 0.0 - (-0.76) = 0.76 \text{ V}, \ \Delta G^0 < 0 \qquad \text{<2-19>}$$

Fig. 2-11 shows the current-potential profiles of hydrogen evolution and Zn dissolution, wherein the current varies by the exponential function of applied potential, assuming that charge transfer determines the overall rate. Also note that E_{eq} are 0.0 V and −0.76 V (*vs.* NHE) for the hydrogen and Zn half-cell, respectively. In a galvanic cell, current flows through a closed loop with $i_c = |i_a|$, such that the cathodic current (i_c) for hydrogen evolution must be equal to the anodic current (i_a) for Zn dissolution. The electrode potential is determined by satisfying this condition (Fig. 2-11). As-determined electrode potential is called a ***mixed potential*** or ***corrosion potential*** because it is related with metal corrosion. As shown in Fig. 2-11, the mixed potential is more positive than the standard electrode potential (-0.76 V), which is the practically observed potential for the Zn half-cell.

Self learning 2-10

Let us consider that a 3-electrode cell is formed by using the Zn electrode of Fig. 2-10 as the working electrode. Sketch the net current ($i_{net} = i_c - |i_a|$) that is detected when the potential of Zn electrode is changed to either the positive or negative direction from the mixed potential shown in Fig. 2-11. What electrochemical reactions are responsible for these net currents (i_{net})?

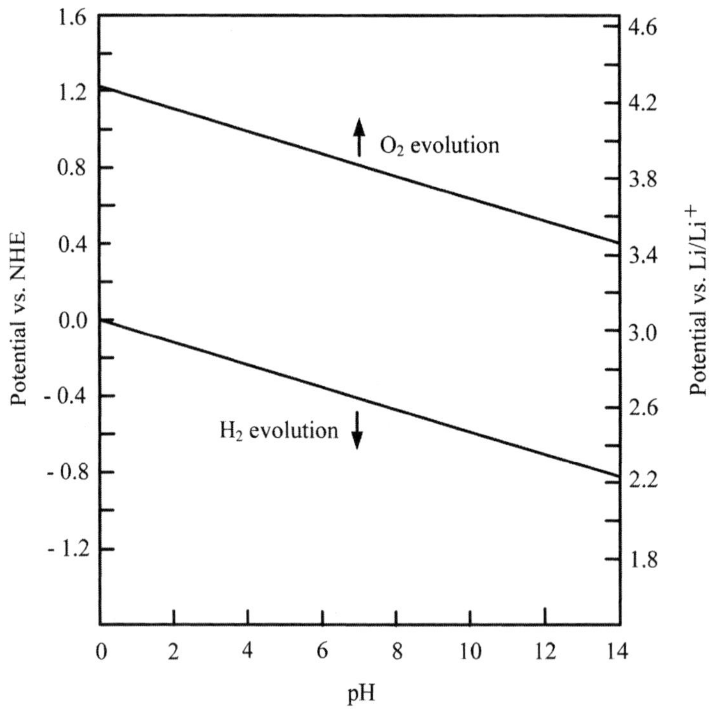

Fig. 2-12 Change of the equilibrium potential (E_{eq}) for hydrogen and oxygen half-cells according to pH in aqueous solution.

The alkaline batteries use Zn powder as an anode, MnO_2 cathode, and aqueous electrolyte. If the aqueous electrolyte is acidic, the Zn dissolution and H_2 evolution take place on the Zn powder surface as described above. This means that the Zn anode is self-discharged. The internal pressure of the cell increases to cause a leakage of the aqueous electrolyte. The Zn dissolution/H_2 evolution can be minimized by using alkaline KOH solution as described below. As the solution pH increases, E_{eq} of hydrogen half-cell becomes more negative as shown in <2-20> and Fig. 2-12. In <2-20>, the fugacity of H_2 was assumed to be unity and R = 8.314 J/mol K, F = 96,485 C/mol, and T = 298K.

$$2H^+ + 2e = H_2$$

$$E_{eq} = E^0 + \frac{RT}{2F}\ln\frac{a_{H^+}^2}{f_{H_2}} = \text{Constant} + \frac{2.3RT}{F}\log a_{H^+}$$
$$= \text{Constant} + 0.059\ \text{V} \times \log a_{H^+} = (0-0.059\times\text{pH})\ \text{V}\ (vs.\ \text{NHE}) \qquad \text{<2-20>}$$

Self learning 2-11

Confirm that E_{eq} of oxygen half-cell changes as shown in Fig. 2-12 from the following potential-determining equilibrium of oxygen half-cell. Assume unit activity and fugacity.

$$O_2 + 4H^+ + 4e = 2H_2O$$

Self learning 2-12

Assuming the pH of aqueous solution is 7.0 in Fig. 2-12, indicate the potential range where hydrogen evolution is thermodynamically allowed and the potential range where oxygen evolution (due to oxidation of water) is thermodynamically possible.

As shown in Fig. 2-13, E_{eq} of hydrogen half-cell moves to the negative direction on the x-axis with an increase in the solution pH. The curves representing H_2 evolution current also move to the negative direction, such that the mixed potential satisfying $i_c = |i_a|$ also shifts to the negative direction as indicated by the dot on the x-axis in Fig. 2-13. Note that E_{eq} of Zn half-cell does not change with pH. In addition, as the pH increases, the rate of hydrogen evolution and Zn dissolution decreases as indicated by the arrows in Fig. 2-13. This illustrates that the self-discharge of Zn anode and H_2 evolution can be suppressed by using the aqueous solutions of high pH. In real, the alkaline batteries

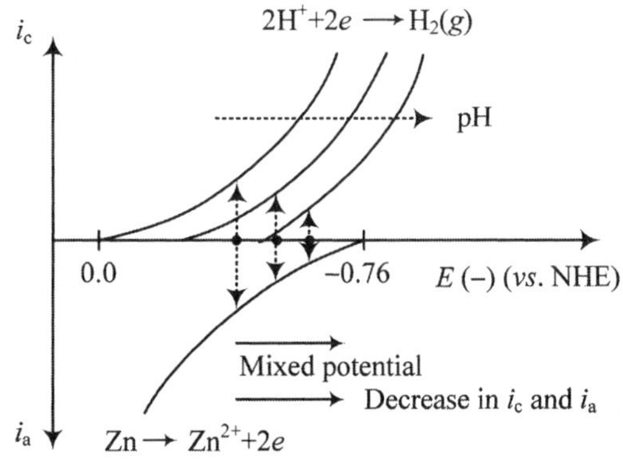

Fig. 2–13 Zn dissolution and H_2 evolution according to the pH of aqueous solution and change of mixed potential.

use aqueous KOH solution (pH = 14), hence, referred to an ***alkaline battery***. However, even with the use of an alkaline electrolyte, the evolution of a trace amount of hydrogen and Zn self-discharge can not be completely prevented. In order to further suppress the Zn dissolution/H_2 evolution, mercury (Hg) is added to the Zn electrode. The H_2 evolution rate is extremely slow on Hg electrode as illustrated in Figs. 1-15 and 1-16. This means that H_2 evolution on the Hg-coated Zn powders is very slow in Fig. 2-10. The suppression of H_2 evolution leads to a suppression of Zn dissolution because the condition of $i_c = |i_a|$ should be met in the galvanic cell. In the past, mercury was added into alkaline cells. Due to the mercury pollution, however, the mercury-free alkaline batteries have been developed, in which zinc alloys are employed as the anode instead of pure zinc powders. Here, the composition of zinc alloys is optimized to suppress either H_2 evolution or Zn dissolution.

Self learning 2-13

Metal corrosion is an electrochemical oxidation reaction, so that an electrochemical reduction reaction must be paired to form a galvanic cell. Refer to Table 2-1 and consider what kind of electrochemical reduction reactions can be paired with Fe corrosion ($Fe \rightarrow Fe^{2+} + 2e$).

*2-5 Pourbaix diagram and corrosion

2-5-1 Thermodynamic estimation of corrosion by using Pourbaix diagram

The standard electrode potential of Pt/Fe^{2+}, Fe^{3+} half-cell is 0.77 V (*vs*. NHE), which is the equilibrium potential (E_{eq}) at the standard state. Here, when the potential of Pt electrode moves to the negative direction from E_{eq}, the reduction current is observed, while the oxidation current is detected when $E_{appl} > E_{eq}$. The applied electrode potential is a decisive factor to decide the type of electrochemical reaction (oxidation or reduction). Meanwhile, the equilibrium potential changes according to solution pH for Pt/H_2, H^+ half-cell (Fig. 2-12). This is because H^+ is involved in the potential-determining equilibrium (half-cell reaction); $2H^+ + 2e = H_2$. Since the solution pH affects the equilibrium potential, the solution pH indirectly but ultimately determines the type of electrochemical reactions (oxidation or reduction reaction). The ***Pourbaix diagrams*** illustrate how the electrode potential and the solution pH determine the type of electrochemical reactions. Note in advance that the Pourbaix diagrams are drawn from the thermodynamics viewpoint, such that they can not provide the kinetic values such as current, but only the possibility of the reactions.

Fig. 2-14 shows the Pourbaix diagram for iron (Fe) in aqueous solution. In aqueous solution, there are numerous Fe-containing chemical species, and there is a variety of chemical and electrochemical equilibriums as shown below. Note that only (3) and (7) are chemical equilibrium and the rest are electrochemical equilibrium.

(1) $Fe^{2+} + 2e = Fe$

(2) $Fe^{3+} + e = Fe^{2+}$

(3) $Fe(OH)_2 + 2H^+ = Fe^{2+} + 2H_2O$

(4) $Fe(OH)_3 + 3H^+ + e = Fe^{2+} + 3H_2O$

(5) $Fe(OH)_3 + H^+ + e = Fe(OH)_2 + H_2O$

(6) $Fe(OH)_2 + 2H^+ + 2e = Fe + 2H_2O$

(7) $FeO_2H^- + H_2O = Fe(OH)_2 + OH^-$

(8) $FeO_2H^- + H_2O + 2e = Fe + 3OH^-$

(9) $Fe(OH)_3 + e = FeO_2H^- + H_2O$

(10) $FeO_4^{2-} + 5H^+ + 3e = Fe(OH)_3 + H_2O$

(11) $FeO_4^{2-} + 8H^+ + 3e = Fe^{3+} + 4H_2O$

In Fig. 2-14, the oxygen and hydrogen line correspond to the pH-dependent equilibrium potential for oxygen and hydrogen half-cell, which were shown in Fig. 2-12. Several features should be noted in Fig. 2-14. Firstly, E_{eq} appears as a horizontal line regardless of the pH if neither H^+ nor OH^- participate in electrochemical reactions; e.g. $Fe^{2+} + 2e$

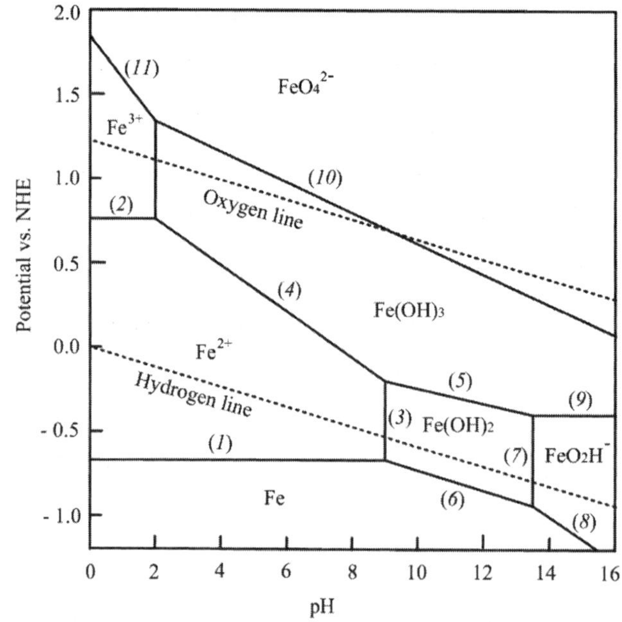

Fig. 2-14 A simplified Pourbaix diagram for iron (Fe) in pure water at 25°C. The activity of all the ions is assumed to be 10^{-6}.

= Fe (1). Secondly, in the case of potential-independent reactions such as $Fe(OH)_2 + 2H^+ = Fe^{2+} + 2H_2O$ (3), a vertical line appears in the Pourbaix diagram. This is because H^+ is involved in the chemical equilibrium. Finally, the sloped lines appear for the reactions that are affected by both electrode potential and pH. For example, E_{eq} varies not only with the activity of Fe^{2+} but also with pH for $Fe(OH)_3 + 3H^+ + e = Fe^{2+} + 3H_2O$ (4). Thus, the solution pH is a factor determining whether the reaction proceeds to the left or to the right.

Fig. 2-14 illustrates how the Fe corrosion characteristics vary according to the solution pH and electrode potential. The corrosion reaction of iron is $Fe \rightarrow Fe^{2+} + 2e$ or $Fe \rightarrow Fe^{3+} + 3e$, and the electrochemical reduction reactions must be accompanied to form a galvanic cell. Possible reduction reactions are:

Under acidic conditions

$2H^+ + 2e \rightarrow H_2$ (Hydrogen generation)

or $O_2 + 4H^+ + 4e \rightarrow 2H_2O$ (Oxygen reduction)

Under alkaline conditions,

$2H_2O + 4e \rightarrow H_2 + 2OH^-$ (Hydrogen generation)

or $O_2 + 2H_2O + 4e \rightarrow 4OH^-$ (Oxygen reduction)

The equilibrium potentials for the four half-cells listed above change according to pH, so that the potential range for the reduction reaction also changes with pH.

The Pourbaix diagrams can be utilized to predict the following features. Firstly, iron compounds are thermodynamically stable only in a given potential and pH. For example, when the potential of Fe electrode is −0.5 V, Fe^{2+} is the most stable one at pH < 9, but $Fe(OH)_2$ is most stable at pH = 9.0~13.5. Secondly, from the Pourbaix diagram, one can tell whether iron corrosion is thermodynamically feasible in a given potential and pH. For example, at pH < 9.0, the horizontal line *(1)* places in a more negative position than the hydrogen and oxygen line. From the fact that the half-cells of more negative E_{eq} have a higher oxidizing tendency, it can be said that the Fe corrosion ($Fe \rightarrow Fe^{2+} + 2e$) is thermodynamically feasible at pH < 9.0 when electrochemical reduction (hydrogen evolution or oxygen reduction listed above) is paired. Thirdly, it can be predicted which reactions can occur in iron (Fe) in a given potential and pH. As shown in Fig. 2-15, for example, iron does not corrode under conditions where the potential is −1.0 V and pH = 7.0 (***immunity***). At a potential of −0.5 V and pH = 11.0, iron is corroded to generate Fe^{2+}, which is immediately converted into an insoluble $Fe(OH)_2$ by reaction with OH^-. The as-generated insoluble species precipitates on Fe surface to form a passivation layer, which inhibits further corrosion (***passivation***).

Self learning 2-14

What are the requirements for $Fe(OH)_2$ and $Fe(OH)_3$ to be a completely passivating film?

The Pourbaix diagram has the following limitations, such that it is frequent to misinterpret the diagram. Firstly, since the activity of ions is assumed to be constant (it is assumed to be 10^{-6} in Fig. 2-14), it may be different from actual conditions. Secondly, it is only

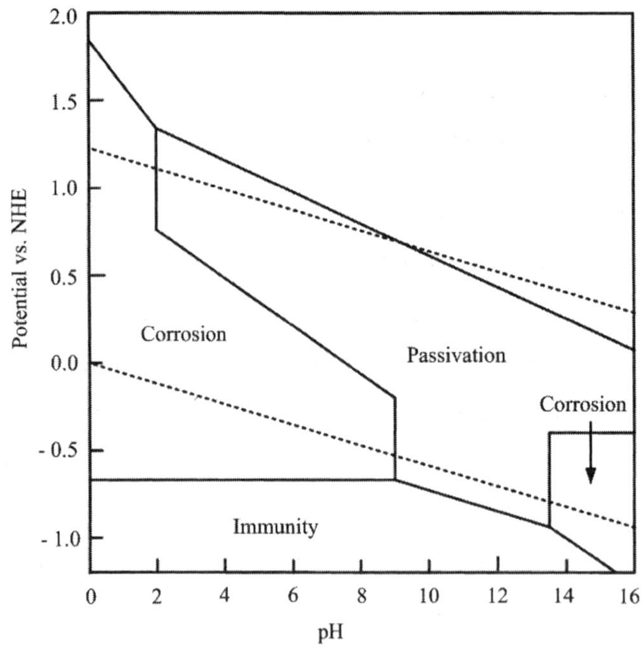

Fig. 2-15 The diagram in terms of the type and reaction that can occur, which is inferred from the Pourbaix diagram of iron (Fe).

a thermodynamic prediction. Kinetic factors may give different results. Therefore, only a rough estimate for corrosion is available.

2-5-2 Corrosion prevention

Various chemical and electrochemical methods are being used to inhibit corrosion. One example is illustrated from the data shown in Fig. 2-15. Let us assume that a Fe piece is dropped into an oxygen-free aqueous solution of pH = 8.0. Here, the electrochemical reduction reaction that should be paired with Fe corrosion ($Fe \rightarrow Fe^{2+} + 2e$) is the hydrogen evolution because O_2 is absent. Under these circumstances, the corrosion potential (E_{corr}) of iron locates between the E_{eq} of hydrogen half-cell (−0.47 V from the hydrogen line at pH = 8.0 on Fig. 2-14) and the E_{eq} of Fe/Fe^{2+} half-cell (−0.62 V). The actual E_{corr} varies within this range according to the kinetic parameters of two half-cell reactions. E_{corr} is a kind of equilibrium potentials, such that it can be a criterion for determining the net oxidation or reduction reaction. If the potential of the iron specimen is controlled to be more negative than E_{corr} ($-0.47 \sim -0.62$ V), the Fe corrosion ($Fe \rightarrow Fe^{2+} + 2e$) can be prevented. This is called a ***cathodic protection***, in which two methods are utilized. Firstly, a metal having a higher oxidizing tendency than iron (for example, Zn, Mg, and Al alloy, which is referred to as a ***sacrificial anode***) is brought into electrical contact with the to-be-protected steel (iron) structure. The working principle of sacrificial anode can be explained by using the Zn corrosion phenomenon shown in Fig. 2-11. If the oxidation tendency of a given sacrificial anode is higher than that of Zn, its E^0 is more negative and thus its E_{eq} can be assumed to be more negative than that of Zn half-cell. When the oxidation current of this sacrificial anode is overlapped with that of Zn while the hydrogen evolution current being left in Fig. 2-11, one can recognize that the given sacrificial anode has a more negative E_{corr} value. The same principle is applied when a metal having a higher oxidation tendency than iron is used as a sacrificial anode in a steel structure. Here, the dissolution of sacrificial anode (oxidation), which is thermodynamically favored over the dissolution of iron, and hydrogen evolution (reduction) are paired to form a galvanic cell. Since the steel structure is in electrical contact with the sacrificial anode, the open-circuit potential of the sacrificial anode/steel structure is the E_{corr} that is determined by the sacrificial anode and H_2 evolution. Since this E_{corr} is more negative than that determined by the iron dissolution (oxidation) and H_2 evolution, the iron dissolution (corrosion) can be suppressed. In fact, Zn is used as a sacrificial anode for steel structures such as pipelines and ships. Another cathodic protection method is to construct an electrolytic cell to induce electrochemical oxidation at an inert anode (for example, O_2 evolution from water) and reduction at the to-be-protected steel structure (for example, H_2 evolution). By impressing current in such an electrolytic cell, the potential of steel structure can be adjusted to be more negative than E_{corr}. That is, it is possible to hold the potential at a value where the to-be-protected metal is thermodynamically immune from corrosion.

Meanwhile, in the case of a metal such as Fe, which can be passivated, if the voltage of the metal specimen can be controlled to have a positive value and eventually move to the passivation region, the corrosion can be protected due to the presence of passivating films. This principle is called ***anodic protection***, and various methods have been developed. For example, an electrolytic cell may be fabricated with an inert cathode and a steel structure anode to control the electrode potential of the steel structure to have a positive value. Also, there is a method of contacting with a strong oxidizing agent (nitrite, chromate). Here, since iron is oxidized to lose electrons, the electrode potential of the steel structure changes to a positive value and can move to the passivation region.

In addition to the examples described above, there are many ways to prevent corrosion; plating with a corrosion-resistant metal, coating an oxide film by anodizing, and so on. Corrosion inhibitors are used, some of which are adsorbed on metal surfaces to inhibit metal dissolution (amines, carbonyls, and sulfur compounds) or to inhibit the accompanying reduction reactions (P, As, and Sb compounds).

Self learning 2-15

If Fe^{2+} can form a complex with ligands, how does the position of line (1) in Fig. 2-14 change? How does the corrosion behavior of iron change?

Self learning 2-16

Aluminum half-cell ($Al^{3+} + 3e = Al$) has a very negative value with $E^0 = -1.68$ V (*vs.* NHE), but Al is free from corrosion. Give an explanation on this.

Problems

01 Give the potential-determining equilibrium and derive the Nernst equation that represents the equilibrium potential of the following half-cells.

(a) $Ag/AgCl(s)/Cl^-$

(b) $Pt/Cr_2O_7^{2-}$, Cr^{3+}

(c) Fe_2O_3/FeO

(d) IO_3^-/I_2

(e) MnO_2/Li_xMnO_2

(f) SiO_2/Si in acidic medium

02 Calculate the standard electrode potential of the following half-cell from the two standard electrode potential values given below.

$CuI + e = Cu + I^-$ $\quad E^0 = ?$

$Cu^{2+} + 2e = Cu$ $\quad E^0 = 0.34$ V (*vs.* NHE)

$Cu^{2+} + I^- + e = CuI$ $\quad E^0 = 0.86$ V (*vs.* NHE)

03 The potential-determining equilibrium for a half-cell comprising an inert electrode and redox pair (Pt/Co^{2+}, Co^{3+}) is

$$Co^{3+}(aq) + e = Co^{2+}(aq)$$

The standard electrode potential of this half-cell is $E^0 = 1.92$ V (*vs.* NHE). A neutral ligand L forms a strong complex with Co^{3+} and Co^{2+}, and the equilibrium constant (K_f) is as follows.

$Co^{3+} + 6L = CoL_6^{3+}$ $\quad K_f' = 2.0 \text{ x } 10^{19}$

$Co^{2+} + 6L = CoL_6^{2+}$ $\quad K_f'' = 2.0 \text{ x } 10^{5}$

(a) Calculate the standard electrode potential ($E^0_{complex}$) of the half-cell (Pt/CoL_6^{2+},

CoL_6^{3+}).

(b) Compare the oxidation tendency between Co^{2+} and CoL_6^{2+}.

04 The formal potential ($E^{0'}$) of Fe^{3+}/Fe^{2+} redox pair was measured in three different acids of pH = 0 and the following values were obtained.

HCl ; 0.52 V (*vs.* SCE)

H_2SO_4; 0.04 V (*vs.* SCE)

H_3PO_4; 0.44 V (*vs.* NHE)

In which acid does Fe^{3+} have the strongest oxidizing power? In which acid does Fe^{3+} form the strongest complex?

05 The voltage between the two Cd electrodes in the following electrochemical cell was −0.41 V at 25°C. If Cd^{2+} and CN^- form $Cd(CN)_4^{2-}$, what is the equilibrium constant (K_f) for the complex formation? Neglect the liquid junction potential.

$Cd \mid Cd^{2+}(0.01\ M) \parallel Cd^{2+}\ (0.01\ M),\ CN^-(0.1\ M) \mid Cd$

06 The standard electrode potentials of Cu^{2+}/Cu^+ and Cu^+/Cu half-cells are 0.17 V and 0.52 V (*vs.* NHE) in aqueous solution, whereas 1.19 V and −0.38 V (*vs.* NHE) in acetonitrile solution.

(a) Calculate K_{eq} for $2Cu^+ = Cu^{2+} + Cu$ in two solutions.

(b) Explain how the stability of Cu^+ differs in two solutions from the above calculation.

07 Assume that Cu^{2+} (0.01 *M*) and Sn^{2+} (0.01 *M*) are dissolved in an aqueous solution of pH = 0.0.

(a) Calculate the potential at which the plating of Cu and Sn can start when the potential of an inert electrode (Pt) is gradually changed from 1.6 V (*vs.* NHE) to the negative values.

$$Cu^{2+} + 2e = Cu \qquad E^{0'} = 0.34\ V(vs.\ NHE)$$

$$Sn^{2+} + 2e = Sn \qquad E^{0'} = -0.14\ V(vs.\ NHE)$$

(b) When the potential of Pt electrode is swept from 1.6 V to the negative value at pH = 14, what happens in the order of Cu plating, Sn plating and H_2 evolution? Describe the reason why the pH of aqueous solution should be kept high enough in the industrial plating process of Sn.

CHAPTER

03

Electrolytes and ion conduction

Solutions in which salts are dissolved are called electrolytes or supporting electrolytes. In the electrolytes, the activity coefficient and molar conductivity change due to ion-ion interactions and ion-solvent interactions.

The activity (a_j) of ion j in an electrolyte solution is

$$a_j = \gamma_j c_j$$

Here, γ_j is the activity coefficient and c_j is the free-ion concentration. $\gamma_j = 1.0$ in an infinitely diluted solution, but decreases with increasing ion concentration; however, it shows an increase after a minimum. Meanwhile, molar conductivity (Λ) shows a maximum value when infinitely diluted, but gradually decreases with an increase in ion concentration. Here, the activity coefficient is the activity (a_j) per ion concentration

$$\gamma_j = a_j/c_j$$

and the molar conductivity is also the ionic conductivity (κ) per ion concentration.

$$\Lambda = \kappa/c$$

Hence, both (γ_j and Λ) are the *characteristics (or ability) per mole of ions*. This concept can further be extended to the characteristics per a single ion. Hence, the decrease of γ_j and Λ with an increase in ion concentration means that the ionic properties (or abilities) per a single ion decrease with increasing ion concentration. That is, in an infinitely diluted solution, the ion exerts all its ability ($\gamma_j = 1.0$ and $\Lambda = \Lambda^0$, where Λ^0 is the molar conductivity at infinite dilution), but the ability decreases ($\gamma_j < 1.0$ and $\Lambda < \Lambda^0$) with increasing ion concentration. Here, a question arises as to why the ability per a single ion decreases in more concentrated solutions. This is due to ion-ion interactions and ion-solvent interactions.

3-1 Activity coefficient of ions

Fig. 3-1 illustrates a thought experiment that explains why the activity coefficient of an ion decreases by ***ion-ion interactions***. There are ion-ion interactions in actual electrolyte solutions because positive and negative ions coexist to maintain electrical neutrality. If all ions lose their charges, there is no ion-ion interaction. The change of free energy in this process is as follows. When the ions have charges in actual electrolyte solutions, the chemical potential is expressed by <3-1>.

$$\mu_j = \mu_j^0 + RT\ln a_j = \mu_j^0 + RT\ln c_j + RT\ln \gamma_j \tag{3-1}$$

In ideal electrolyte solutions wherein the charges are removed, the chemical potential is given by <3-2> because $\gamma_j = 1.0$.

$$\mu_j = \mu_j^0 + RT\ln c_j \tag{3-2}$$

Hence, the free energy change upon a switch from an actual electrolyte solution with ion-ion interaction to an ideal solution without ion-ion interaction is calculated by <3-3>.

$$\begin{aligned}\Delta G &= [\text{free energy of ideal electrolyte}] - [\text{free energy of actual electrolyte}] \\ &= [\mu_j^0 + RT\ln c_j] - [\mu_j^0 + RT\ln c_j + RT\ln\gamma_j] \\ &= -RT\ln\gamma_j\end{aligned} \tag{3-3}$$

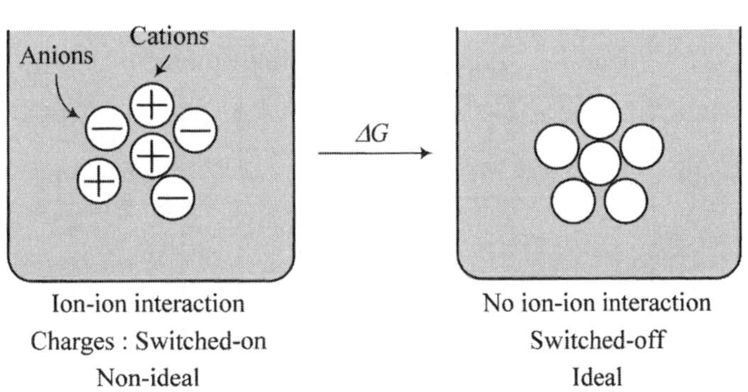

Fig. 3-1 Free energy change (ΔG) when switching from the actual non-ideal situation with charges present to an imaginary situation with ionic charges absent.

Fig. 3-2 shows how ΔG changes with an increase in ion concentration (with a decrease in γ_j) in the above thought experiment. When the switch is made from the actual solutions to the ideal solution, ΔG from the more concentrated actual solution is larger (4.6 RT) than that from the less concentrated solution (2.3 RT). This manifests itself that the free energy of actual solutions is smaller when concentrated. The smaller free energy for the more concentrated electrolyte solutions indicates that the ions are *stabilized* to a higher degree when concentrated. The stabilization can be accounted for by ion-ion interaction. Namely, ions are stabilized by electrostatic attractive interactions between ions of different charges; the attractive interactions become stronger with increasing ion concentration. A de-stabilization by repulsive interactions between the same ions can be considered. However, the stabilization by cation-anion interaction is stronger than de-stabilization since the population of counter ions is larger than the same ions around a given ion as illustrated in Fig. 3-3-a.

3-1-1 Debye-Hückel theory

The ***Debye-Hückel theory*** explains why the activity coefficient (γ_j) decreases with an increase in ion concentration. Consider an 1:1 electrolyte, in which the charge for cation and anion is 1 and −1, respectively (for example, Na^+ and Cl^-). Let us specify a cation. Anions will distribute around this cation to satisfy electrical neutrality (+1 = −1). However, the anions interacting with the designated cation also interact with other neighboring cati-

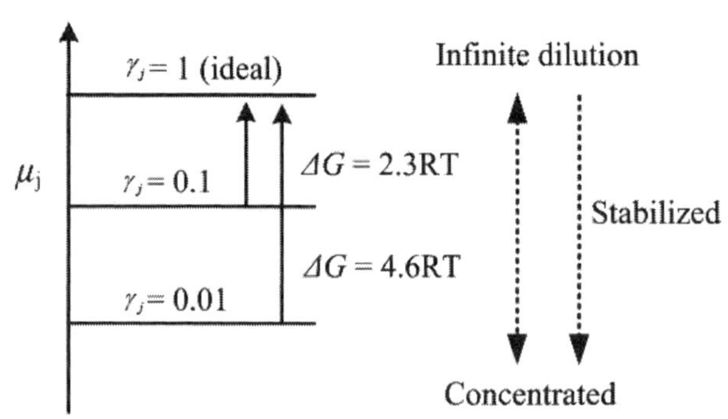

Fig. 3-2 Free energy change (ΔG) according to ion concentration in the thought experiment in Fig. 3-1.

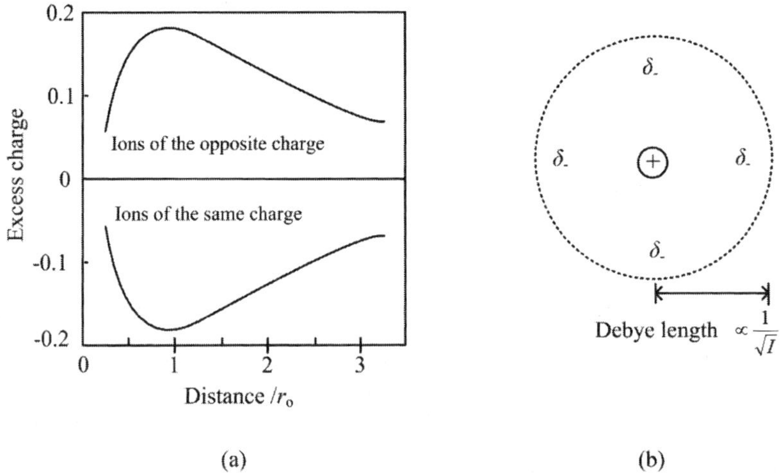

Fig. 3-3 (a): The distribution of positive and negative ions around a given positive ion, and (b): ionic atmosphere and Debye length.

ons, such that the charge of anions interacting with the designated cation can be considered to be smaller than -1. To meet the electrical neutrality ($\delta^+ = \delta^-$), the positive charge ($\delta^+ = 1$) for the designated cation should be balanced by the negative charge ($\delta^- = -1$), which is the sum of negative charges of the interacting anions that are distributed in a three-dimensional space around the designated cation. The three-dimensional space, where anions are distributed with their total charge of -1, is called an ***ionic atmosphere***. ***Debye length*** is referred to as the radius of the ionic atmosphere (Fig. 3-3-b).

Debye length is inversely proportional to $I^{1/2}$, where I is the ionic strength defined in <3-4>, in which z_j is the ionic charge.

$$I = \frac{1}{2}\sum_j c_j z_j^2 \quad \text{<3-4>}$$

In an infinitely diluted electrolyte solution, the distance between cation and anion is infinitely long, which means an infinite Debye length. However, as ion concentration increases, anions that are populated at a closer distance to the designated cation can satisfy $\delta^+ = \delta^-$ ($+1 = -1$) condition, so that Debye length becomes shorter (Fig. 3-3-b). The key point here is that the distance between the cation and anion decreases with increasing ion concentration, so that the electrostatic attraction increases to stabilize ions, which

Table 3-1 Crystallographic radius, hydrated radius, and hydration number for Li^+, Na^+ and K^+.

	Li^+	Na^+	K^+
Crystal radius[a]/Å	0.60	0.95	1.33
Hydrated radius[b]/Å	2.4	1.8	1.3
Hydration number	2-22	2-13	1-6

[a] From x-ray diffraction analysis, [b] Using Stokes' law

eventually results in a decrease of activity coefficient (γ_j). This relationship is explained by the ***Debye-Hückel limiting law*** (DHL) shown in <3-5>, where $\log_{10}\gamma_j$ linearly decreases to $I^{1/2}$. <3-5> is referred to as a limiting law because this relation is matched with the real one only in a limited concentration range (less than 10^{-2} *M*) (Fig. 3-4).

$$\log_{10} \gamma_j = -Az_j^2\sqrt{I} \qquad \text{<3-5>}$$

The reason why DHL coincides with experimental data only in the diluted condition (less than 10^{-2} *M*) is that it makes various assumptions in the derivation. For instance, ions are assumed to be a point charge with a radius of zero. In real, however, ions are solvated in solutions, such that the size of ions can not be ignored (Table 3-1).

The ***extended Debye-Hückel law*** (EDHL), which takes account of ionic size, adds a parameter related to the solvated (hydrated for aqueous solutions) ionic size (a) as in <3-6>, wherein A and B are constant. Since ionic size is assumed to be closer to the real one, EDHL coincides with experimental data in the higher ion concentration range (Fig. 3-4); the ion concentration range for coincidence is extended.

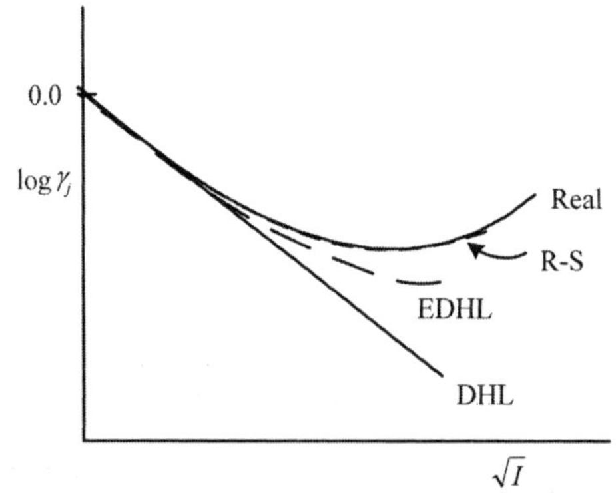

Fig. 3-4 Activity coefficient with an increase in ionic strength; Comparison between real and various theoretical derivations.

$$\log_{10} \gamma_j = \frac{-Az_j^2 \sqrt{I}}{1+Ba\sqrt{I}}$$ <3-6>

Fig. 3-4 shows that the activity coefficient in actual electrolytes increases after showing a minimum. ***Robinson-Stokes (R-S) equation*** explains this feature by taking ***ion-solvent interactions*** into account. In aqueous solutions, ions are solvated by water (hydrated) as shown in Fig. 3-5. Here, the water molecules located near the cation (M^+) are strongly bound to the cation by ion-dipole interactions. This is called an inner solvation shell or primary solvation shell. It is assumed that when the ion moves, these solvated water molecules move together. The water molecules on the far side of the ion are partially aligned because the ion-dipole interactions are relatively weak due to longer distance. This is called the outer solvation shell. As shown in Table 3-1, among Li^+, Na^+, and K^+, the crystal radius is the smallest for Li^+, but the hydrated radius is the largest because Li^+ is hydrated by the largest number of water molecules.

Just as the ions are stabilized by attractive ion-ion interactions, ions are stabilized by ion-solvent interactions even if the degree of stabilization is relatively weak. Water molecules carry a permanent dipole, such that ion-dipole interactions between ions and water molecule can prevail; e.g. the negatively polarized (δ_-) oxygen atom in H_2O can interact with cations. Due to such an attractive interaction, the cations are stabilized. In aqueous

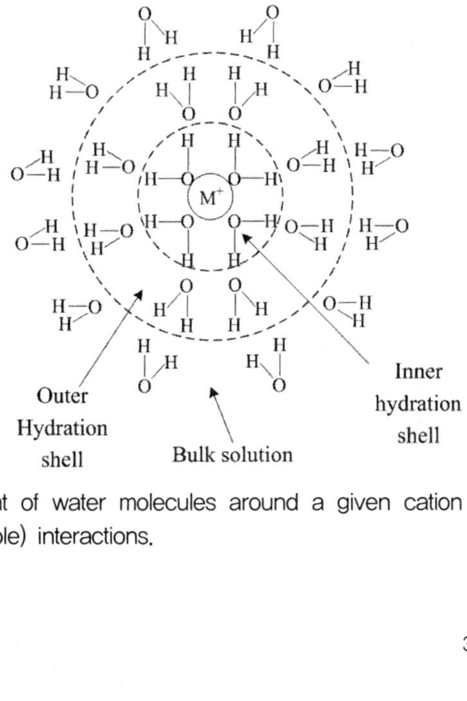

Fig. 3-5 Alignment of water molecules around a given cation showing ion-solvent (ion-dipole) interactions.

electrolytes, all ions can be hydrated to the maximum hydration number in infinitely diluted solutions. However, above a certain ion concentration, there must appear a shortage of water molecules for the maximum hydration. In other words, the average hydration number gradually decreases as the ion concentration increases; the average hydration number is maximum in infinitely diluted solution, but becomes smaller in a more concentrated solution. The decrease in the average hydration number means a decrease in ion-dipole interactions, and thus a decrease in ion stabilization. As a result, the activity coefficient, which is an indicator for ion stabilization per a single ion, rather increases in the region of high ion concentration. In Robinson-Stokes (R-S) equation <3-7>, cI is added to EDHL, wherein c is a constant. As shown in Fig. 3-4, the activity coefficient derived from R-S equation approaches the real value even in the region of high ionic strength.

$$\log_{10} \gamma_j = \frac{-Az_j^2\sqrt{I}}{1+Ba\sqrt{I}} + cI \tag{3-7}$$

Self learning 3-1

Assuming the maximum hydration number of cation and anion in an 1:1 aqueous electrolyte is 10, estimate the ion concentration where the activity coefficient begins to increase after showing a minimum.

3-1-2 Mean activity coefficient

In actual electrolytes, the ions of different charge coexist to maintain electrical neutrality. Therefore, in actual electrolytes, it is necessary to define the ***mean activity coefficient*** ($\gamma_\pm$). The mean activity coefficient of NaCl solution is derived as follows. The chemical potentials of Na^+ and Cl^- are as shown in <3-8>, and the sum of them is <3-9>.

$$\begin{aligned} \mu_{Na^+} &= \mu^0_{Na^+} + RT\ln c_{Na^+} + RT\ln \gamma_{Na^+} \\ \mu_{Cl^-} &= \mu^0_{Cl^-} + RT\ln c_{Cl^-} + RT\ln \gamma_{Cl^-} \end{aligned} \tag{3-8}$$

$$\mu_{Na^+} + \mu_{Cl^-} = (\mu^0_{Na^+} + \mu^0_{Cl^-}) + RT\ln c_{Na^+}c_{Cl^-} + RT\ln \gamma_{Na^+}\gamma_{Cl^-} \quad \text{<3-9>}$$

From this, $\mu_\pm$ is derived as <3-10>.

$$\mu_\pm \equiv \frac{1}{2}(\mu_{Na^+} + \mu_{Cl^-}) = \frac{1}{2}(\mu^0_{Na^+} + \mu^0_{Cl^-}) + RT\ln\sqrt{c_{Na^+}c_{Cl^-}} + RT\ln\sqrt{\gamma_{Na^+}\gamma_{Cl^-}}$$

$$= \mu^0_\pm + RT\ln c_\pm + RT\ln \gamma_\pm$$

$$\gamma_\pm = \sqrt{\gamma_{Na^+}\gamma_{Cl^-}} \quad \text{<3-10>}$$

Applying this to general electrolytes results in <3-11.>

$$M_aX_b \rightarrow aM^{z+} + bX^{z-}$$

$$az^+ = bz^-$$

$$\gamma_\pm = (\gamma_+^a\gamma_-^b)^{\frac{1}{a+b}} \quad \text{<3-11>}$$

Self learning 3-2

Derive $\gamma_\pm$ for aqueous $Al_2(SO_4)_3$ solution by using the activity coefficient of cation and anion (γ_+ and γ_-).

3-2 Use of Debye-Hückel limiting law

When a salt is dissolved in a solvent, the solubility product (K_{sp}) is defined as <3-12>, from which <3-13> is derived. The substitution of <3-14> to <3-13> yields <3-15>, which shows that the concentration of free ion increases with increasing ionic strength (Fig. 3-6). It is generally accepted that cation-anion pairs can form in concentrated electrolyte. Unlike ion pairs, the free ions are completely dissociated and exist in a free state.

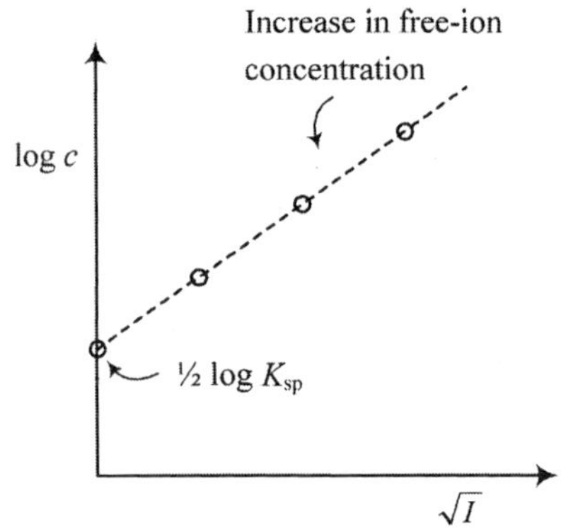

Fig. 3-6 Change of free ion concentration (c) according to ionic strength (I).

$$MX(s) \leftrightarrow M^{z+} + X^{z-}$$

$$K_{sp} = a_{M^{z+}} a_{X^{z-}} = \gamma_{M^{z+}}[M^{z+}]\ \gamma_{X^{z-}}[X^{z-}] = \gamma_{\pm}^{2}[M^{z+}][X^{z-}] \qquad \text{<3-12>}$$

$$[M^{z+}] = [X^{z-}] = c \text{ (free ion)}$$

$$\log K_{sp} = 2\log\gamma_{\pm} + 2\log c \qquad \text{<3-13>}$$

$$\log\gamma_{\pm} = -A\left|z^{+}z^{-}\right|\sqrt{I} \qquad \text{<3-14>}$$

$$\log K_{sp} = -2Az^{2}\sqrt{I} + 2\log c$$

$$\log c = \frac{1}{2}\log K_{sp} + Az^{2}\sqrt{I} \qquad \text{<3-15>}$$

The solubility of AgCl is poor in water. However, the solubility of AgCl can be increased by adding other salts; e.g. KNO_3. This is called ***salting-in*** phenomenon. The underlying role of other salts is to increase the ionic strength in the solution. As shown in Fig. 3-6, the free ion concentration (c) increases with increasing ionic strength (I). This is due to the fact that the Debye length becomes shorter with an increase in ionic strength. Consequently, the ions (Ag^+ and Cl^-) are more stabilized to push the reaction to the rightward side, which is the dissociation of AgCl; the solubility of AgCl increases by adding other salts.

$$AgCl(s) = Ag^{+}(aq) + Cl^{-}(aq)$$

3-3 Ion migration and ionic conductivity

As copper wire shows a good electrical conductivity, the brine (NaCl dissolved in water) also shows electrical conductivity. The electrical conductivity of metals is due to the movement of electrons, whereas ions are responsible for the electrical conduction in electrolytes. Ion movement within an electric field is called a ***migration***.

As shown in <3-16>, ***ionic conductivity*** (κ) is defined by I/E, in which I is the current density (i/A) and E is the electric field (V/l); A is the area of two electrodes, V and l are the applied voltage and the distance between two electrodes, respectively (Fig. 3-7). The free ions in electrolytes participate in the conduction by migration regardless of their size and charges, so the ion conductivity is defined as the sum of the conductivity contributions from each ion. The z_j, c_j, and u_j are the ionic charge, free ion concentration and mobility of ion j, respectively. The unit of ionic conductivity is $\Omega^{-1}m^{-1}$ or Sm^{-1}, hence 1 S (siemens) = 1 Ω^{-1}.

$$\kappa \equiv \frac{I}{E} = \frac{i/A}{V/l} = \frac{1}{R} \cdot \frac{l}{A} = \sum_j |z_j| F c_j u_j \qquad \text{<3-16>}$$

<3-17>, which is derived from <3-16>, represents the resistance of electrolyte solution.

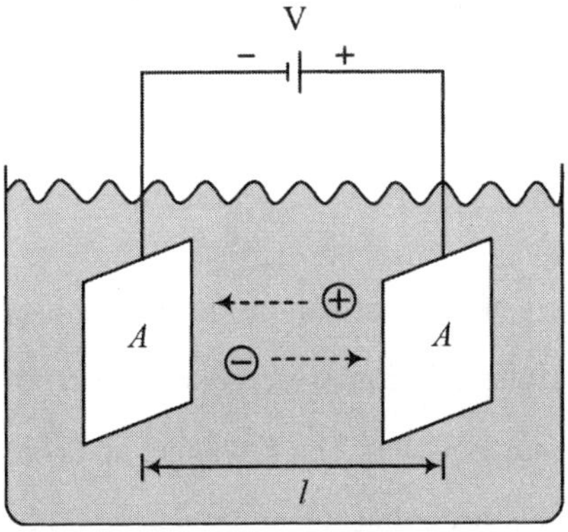

Fig. 3–7 Ion migration within an electric field and measurement of ionic conductivity.

As described in Introduction, the electrolyte solution resistance ($R_{solution}$) greatly influences the cell performances in both electrolytic and galvanic cells. As expected from <3-17>, in order to minimize the solution resistance, the distance (l) between two electrodes should be shortened, the electrode area (A) should be enlarged. Also, a higher concentration of free ions (c_j) and a larger ionic mobility (u_j) lead to a higher ionic conductivity in solution (κ), which eventually gives a lower solution resistance ($R_{solution}$).

$$R_{solution} = \frac{l}{\kappa A} \quad \text{<3-17>}$$

Self learning 3-3

As shown in <1-8>, even if a battery is discharged at the same current, the discharge voltage (E_{wk}) is smaller when the solution resistance ($R_{solution}$) is larger. Generally, ionic conductivity of non-aqueous electrolytes is poorer than that of an aqueous ones. If so, how do you design a cell to use a non-aqueous electrolyte to compensate for its poorer ionic conductivity? (refer to self learning 12-11).

Free ions in electrolytes all participate in ion conduction, and the contribution of a given ion is represented by ***transport number*** or ***transference number***, which is defined as <3-18>. The transport number of cation in an 1:1 electrolyte can be calculated from the mobility values of cation and anion (u_+ and u_-) as shown in <3-19>.

$$t_i = \frac{|z_i| F c_i u_i}{\sum_j |z_j| F c_j u_j} \quad \text{<3-18>}$$

$$t_+ = \frac{u_+}{u_+ + u_-} \quad \text{<3-19>}$$

Table 3-2 lists the transport number of cations according to ion concentration in aqueous solutions. The values at infinite dilution were calculated from the mobility values at infinite dilution (Table 3-3) using <3-19>. The t_+ values in Table 3-2 illustrate that transport numbers do not greatly change according to ion concentration; for HCl solution, the transport number of Cl^- is 0.17~0.18 and the transport number of H^+ is 0.82 ~ 0.83. From

Table 3-2 Transference (transport) number of cations (t_+) according to ion concentration in aqueous solutions (25℃)

	Concentration, C_{eq}[a]				
Electrolyte	0.00	0.01	0.05	0.1	0.2
HCl	0.8208	0.8251	0.8292	0.8314	0.8337
NaCl	0.3962	0.3918	0.3876	0.3854	0.3821
KCl	0.4906	0.4902	0.4899	0.4898	0.4894
NH_4Cl	0.4903	0.4907	0.4905	0.4907	0.4911
KNO_3	0.5072	0.5084	0.5093	0.5103	0.5120
Na_2SO_4	0.3857	0.3848	0.3829	0.3828	0.3828
K_2SO_4	0.4795	0.4829	0.4870	0.4890	0.4910

[a] Moles of positive (or negative) charge per liter.

Table 3-3 Characteristics of ions in infinitely diluted aqueous solution (25℃)

	$\lambda_0{}^a$, $cm^2\ \Omega^{-1}\ equiv^{-1}$	u^b, $cm^2\ sec^{-1}\ V^{-1}$
H^+	349.82	3.625×10^{-3}
K^+	73.52	7.619×10^{-4}
Na^+	50.11	5.193×10^{-4}
Li^+	38.69	4.010×10^{-4}
$NH_4{}^+$	73.4	7.61×10^{-4}
$1/2Ca^{2+}$	59.50	6.166×10^{-4}
OH^-	198	2.05×10^{-3}
Cl^-	76.34	7.912×10^{-4}
Br^-	78.4	8.13×10^{-4}
I^-	76.85	7.96×10^{-4}
$NO_3{}^-$	71.44	7.404×10^{-4}
OAc^-	40.9	4.24×10^{-4}
$ClO_4{}^-$	68.0	7.05×10^{-4}
$1/2SO_4{}^{2-}$	79.8	8.27×10^{-4}
$HCO_3{}^-$	44.48	4.610×10^{-4}
$1/3Fe(CN)_6{}^{3-}$	101.0	1.047×10^{-3}
$1/4Fe(CN)_6{}^{4-}$	110.5	1.145×10^{-3}

[a] Equivalent ionic conductivity from D.A. MacInnes, "The Principles of Electrochemistry", Dover, New York, 1961, p. 342.

[b] Calculated from λ_0.

<3-19>, one can know that the mobility of hydrated H^+ is about four times larger than that of Cl^-, which is confirmed from the mobility data listed in Table 3-3.

Self learning 3-4

Calculate t_+ for an infinitely diluted $CaCl_2$ aqueous solution.

Ionic conductivity can be measured using the apparatus shown in Fig. 3-7. Since the l/A value (***cell constant***) in <3-16> is a unique value for the conductivity meter, the ionic conductivity in solution (κ) can be obtained by measuring the resistance (R). The order of measurement is first to determine the cell constant by measuring the resistance in a KCl solution, of which the ionic conductivity is already known, and then measure the solution resistance by using the same apparatus. Since the cell constant (l/A) is already obtained, the ion conductivity (κ) can be calculated by substituting the as-measured solution resistance ($R_{solution}$) in <3-17>.

Fig. 3-8 shows the change of ion conductivity (κ) according to salt concentration. In general, the concentration of free ions (c_j) increases with increasing salt concentration in the low concentration region, so that ionic conductivity increases by the relation of <3-16>. Thereafter, it shows a maximum value at a certain concentration and then decreases. The decrease of ionic conductivity at high concentration region is due to the decrease of ion mobility (u_j) in <3-16>. This feature can be accounted for by the ***relaxation effect*** and the ***electrophoretic effect***. As shown in Fig. 3-9, when an external electric field ($E_{external}$) is applied for ionic conductivity measurement, the positive ions move toward the negative electrode. Here, the anions forming ionic atmosphere around the cations tend to move toward the opposite electrode, so that the movement of positive ions toward the negative electrode is retarded. Similarly, the anions movement toward the positive

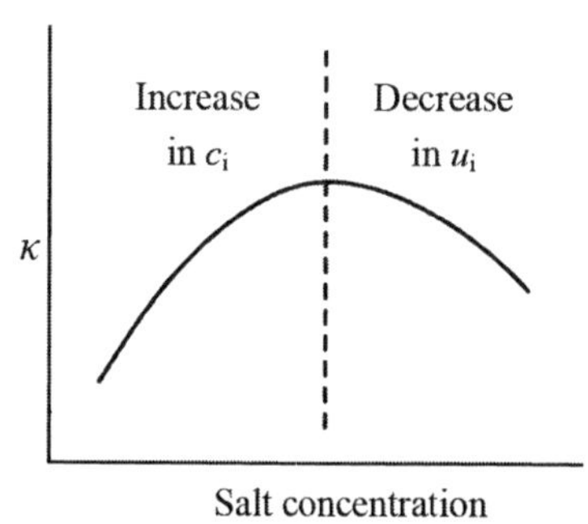

Fig. 3-8 Ionic conductivity according to salt concentration in electrolytes.

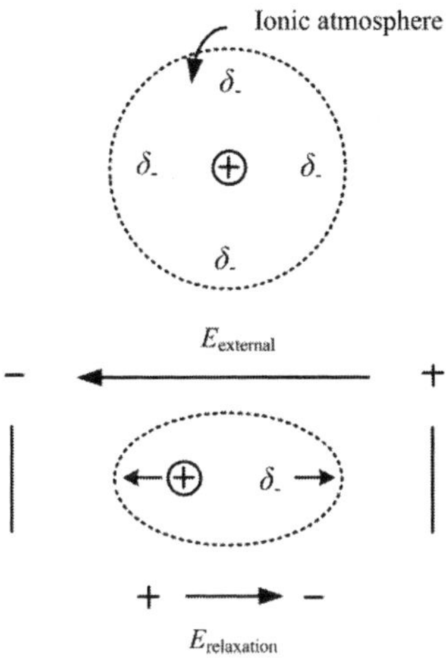

Fig. 3-9 Decrease of ionic mobility due to relaxation effect.

electrode within $E_{external}$ is retarded by the cations forming ionic atmosphere. The retardation of cations by the anions within ionic atmosphere appears that there develops an electric field, of which the direction is opposite to $E_{external}$; this is called a relaxation field ($E_{relaxation}$). The key feature here is that such a retardation becomes more severe with an increase in ionic concentration, which must be due to higher attractive interactions between the ions of different charge with a decrease in Debye length; a higher ion concentration leads to a shorter Debye length and a stronger $E_{relaxation}$. Resultantly, ion mobility decreases due to such a retardation effect, which is more serious as the salt concentration increases. This is called the relaxation effect. Meanwhile, ionic mobility is defined by the ratio of electric field (E) to drift velocity (v) of ions as shown in <3-20>. In general, the viscosity (η) of electrolyte solutions increases with increasing salt concentration, as a brine has higher viscosity than ordinary water. As shown in <3-20>, the ionic mobility (u_j) is inversely proportional to viscosity, such that ion mobility decreases with an increase in salt concentration. This is called electrophoretic effect. Meanwhile, ionic mobility (u_j) and diffusion coefficient (D_j) are proportional to each other as shown in <3-21>, which is called the ***Einstein relation***.

$$u_j \equiv \frac{v}{E} = \frac{|z_j|e}{6\pi\eta r} \qquad \text{<3-20>}$$

Table 3-4 Physicochemical properties of organic solvents used in lithium-ion batteries

Solvent	Dielectric constant (ε)[a]	Viscosity (η/cP)[a]	Freezing point (°C)	Boiling point (°C)
Ethylene carbonate (EC)	95.3	1.9	36.4	238
Propylene carbonate (PC)	64.9	2.5	−54.5	242
Dimethyl carbonate (DMC)	3.12	0.59	3.0	90
Diethyl carbonate (DEC)	2.82	0.75	−43.0	127

[a] 25°C

$$u_j = \frac{|z_j|F}{RT} D_j \qquad \text{<3-21>}$$

In general, the electrolyte solutions used in lithium-ion batteries are prepared by dissolving a lithium salt ($LiPF_6$) in a mixture of two solvents; one has a high dielectric constant (ε) and the other has a low viscosity (η). Table 3-4 lists the physicochemical properties of some organic solvents that are used in lithium-ion batteries. Since the dielectric constant of cyclic carbonates (PC and EC) is larger than that of linear carbonates (DMC and DEC), the dissociation of lithium salt ($LiPF_6$) is advantageous and the free ion concentration (c_j) can be increased. However, since the viscosity of cyclic carbonates is high, the ionic mobility (u_j) is disadvantageous. To overcome this problem, a mixture of two solvents, one with a high dielectric constant (e.g. EC) and the other with a low viscosity (e.g. DEC), is commonly used.

3-4 Molar conductivity

Molar conductivity (Λ) is defined as <3-22>. Kohlrausch proposed in his empirical formula <3-23> that molar conductivity decreases with an increase in electrolyte concentration. In <3-23>, Λ^0 means the molar conductivity at infinite dilution, s is a constant, and c is the electrolyte concentration. The Kohlrausch formula can be explained by using the concept that molar conductivity is the ability to transfer charge per mole of ions (or per a single ion). Namely, in an infinitely diluted solution, the ions fully exert their ion conducting ability. But as ion concentration increases, the mobility decreases due to the relaxation effect and the electrophoretic effect, so that the ions can not exhibit their latent conducting ability; the ion conducting ability per a single ion (molar conductivity, Λ) decreases.

$$\Lambda = \frac{\kappa}{c} \qquad \text{<3-22>}$$

$$\Lambda = \Lambda^0 - sc^{\frac{1}{2}} \qquad \text{<3-23>}$$

Electrolytes can be divided into two groups: ***strong electrolytes*** and ***weak electrolytes***. The salts are fully dissociated (ionized) in the former, but only partially in the latter. Fig. 3-10 shows the molar conductivity according to the electrolyte concentration. The ideal electrolyte shows a constant molar conductivity irrespective of electrolyte concentration and can not be said to exist. Both the strong electrolytes and weak electrolytes show a decrease in molar conductivity as suggested in the ***Kohlrausch empirical formula*** <3-23>. This feature can be explained by the ion mobility decrease caused by the relaxation effect and electrophoretic effect. For the weak electrolytes, the decrease of molar conductivity is more significant (Fig. 3-10), which is due to ion pair formation that causes a decrease in free ion concentration.

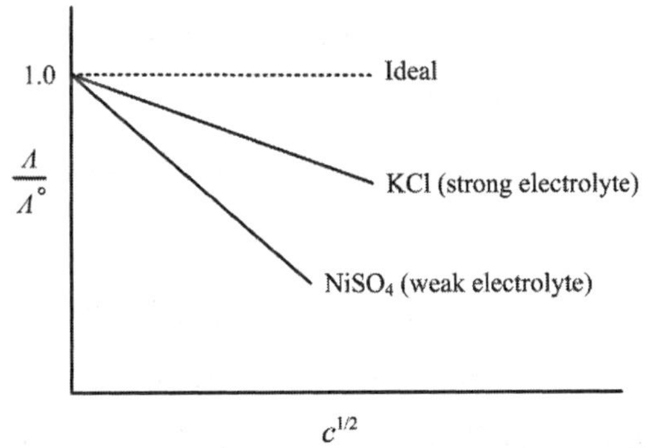

Fig. 3-10 Molar conductivity according to electrolyte concentration in strong electrolyte and weak electrolyte.

Example 3-1

How can the degree of dissociation of acetic acid, a weak electrolyte, be calculated by measuring its molar conductivity?

Solution

The equilibrium constant for the dissociation of acetic acid in water is shown in <3-24>, in which c is the initial concentration of acetic acid (HOAc), and α is the degree of dissociation. 100 % dissociation can be assumed at infinite dilution, such that α can be expressed as <3-25>, wherein Λ is the molar conductivity of acetic acid to be measured. <3-26> is valid because the contribution to ion conduction is the same for all ions in infinitely diluted solutions; ions fully exert their conduction ability. Λ^0(HOAc) can be calculated from the Λ^0 values of HCl, NaOAc, and NaCl. α can be obtained by substituting the measured molar conductivity (Λ) of acetic acid and Λ^0(HOAc) into <3-25>.

$$HOAc + H_2O = H_3O^+ + OAc^-$$

$$K_c = \frac{(c\alpha)^2}{c(1-\alpha)} = \frac{\alpha^2 c}{1-\alpha} \quad \text{<3-24>}$$

$$\alpha = \frac{\Lambda}{\Lambda^0} \quad \text{<3-25>}$$

$$\Lambda^0(\text{HOAc}) = \Lambda^0(\text{HCl}) + \Lambda^0(\text{NaOAc}) - \Lambda^0(\text{NaCl}) \quad \text{<3-26>}$$

3-5 Diffusion potential

In Chap. 2, it is mentioned that the electrical potential difference develops by a very small amount of charge separation at interfaces, and the electrical potential difference can be defined only in the equilibrium state where the degree of charge separation is unchanged. However, if the degree of charge separation does not change with time (called a steady state), the electrical potential difference can be defined/measured even if it is not an equilibrium state.

Consider that two $Hg/Hg_2Cl_2(s)$ (calomel) electrodes are positioned with a separator between them. Here, the two half-cells reach their own equilibrium potential because they are electrically isolated with $R \rightarrow \infty$ for the voltmeter. As seen in Fig. 3-11, 0.1 *M* HCl solution is contained in one side while 0.1 *M* KCl solution in the other. Since the Cl^- concentration is the same, the voltage difference between the two electrodes should be V = 0.0 V. Note that the equilibrium potential of these half-cells is determined only by the activity (concentration) of Cl^-. The measured V = 27 mV. The discrepancy is explained in Fig. 3-12. The interface between the two liquids is present because the different solutions are placed. Here, if any charge separation occurs at the interface, an electrical potential difference must evolve. Initially, there is 0.1 *M* H^+ in α phase but H^+ is not present in β phase, so H^+ will diffuse from α to β phase. Similarly, K^+ will diffuse from β to α phase due to the concentration difference. Here, the diffusion rate of H^+ is higher with a higher mobility as compared to K^+ (see Einstein relation <3-21>). From the ionic mobility data in Table 3-3, the mobility of H^+ is 4.8 times larger than that of K^+. The

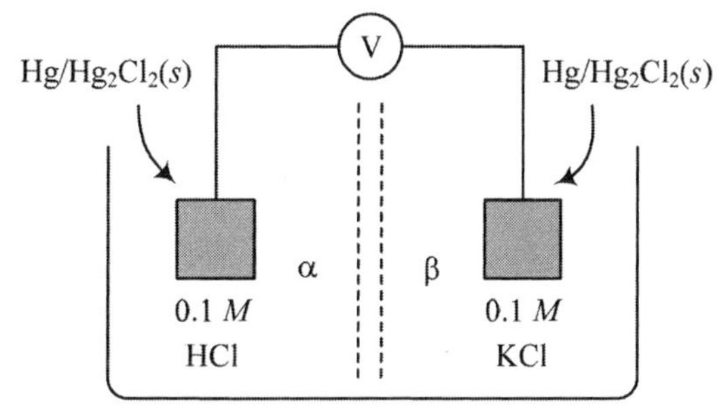

Fig. 3-11 Voltage difference (V) between two calomel electrodes containing different electrolyte.

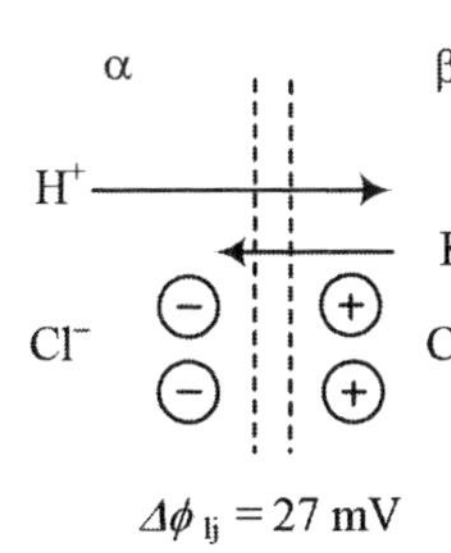

Fig. 3-12 Evolution of diffusion potential at the interface of two electrolyte solutions containing different ions.

difference in diffusion rate leads to a charge separation at the interface; α phase is enriched by negative ions while β phase by positive ions; this charge separation means a development of an electric field across the interface. The as-generated electric field now drives the migration of H^+ in the direction opposite to its diffusion direction and migration of K^+ in the same direction with its diffusion direction; the electric field retards the movement of H^+ from α to β but speeds up the passage of K^+ from β to α phase until two ions cross the interface at an equal rate. At this point, the charge separation at the interface does not change with time (steady state), and thus electrical potential difference can be defined. Note that this is not an equilibrium state since the concentration of ions (for example, H^+) is not the same for two phases. This electrical potential difference is referred to as the ***liquid junction potential*** (E_{lj}). Also, from its origin, it is called a ***diffusion potential***.

The liquid junction potential can be calculated using ***Henderson equation*** given in <3-27>, wherein $c_j(\alpha)$ and $c_j(\beta)$ are the initial concentration of ion j in α and β phase, respectively.

$$E_{lj} = (\phi^\beta - \phi^\alpha) = \frac{\sum_j \frac{|z_j| u_j}{z_j}[c_j(\beta) - c_j(\alpha)]}{\sum_j |z_j| u_j [c_j(\beta) - c_j(\alpha)]} \frac{RT}{F} \ln \frac{\sum_j |z_j| u_j c_j(\alpha)}{\sum_j |z_j| u_j c_j(\beta)} \qquad \text{<3-27>}$$

Example 3-2

By substituting the ion mobility data at infinite dilution in Table 3-3 into <3-27>, confirm that $\Delta\phi_{lj}$ = 27 mV in Fig. 3-12.

Solution

In order to calculate E_{lj} using Henderson equation, not only the ion concentration in two phases but also the ionic mobility should be provided. However, since the ionic mobility varies depending on ion concentration, it is difficult to determine the ion mobility for a given concentration. Hence, E_{lj} can be roughly calculated by using the ion mobility data at infinite dilution (Table 3-3).

A ***salt bridge*** is used to minimize the liquid junction potential. A typical salt bridge

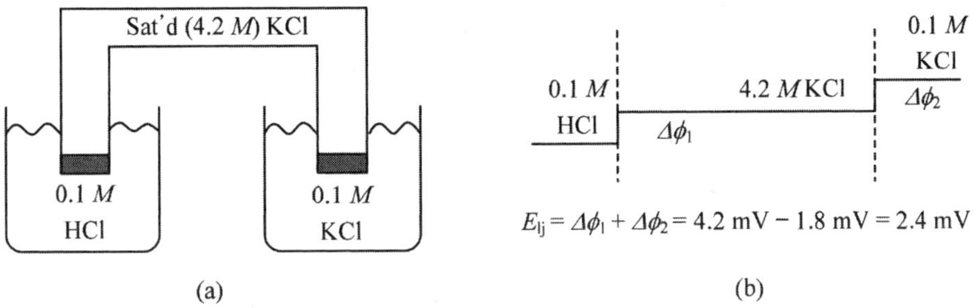

Fig. 3-13 Minimization of liquid junction potential by using a salt bridge.

is shown in Fig. 3-13-a, which connects two solutions by a saturated KCl solution. Note that the concentration of KCl is 4.2 *M* when saturated. The use of salt bridge leads to a formation of two new liquid-liquid junctions (Fig. 3-13-b). Using the Henderson equation, the liquid junction potentials are calculated as $\Delta\phi_1 = 4.2$ mV and $\Delta\phi_2 = -1.8$ mV. The total liquid junction potential is $E_{lj} = \Delta\phi_1 + \Delta\phi_2 = 2.4$ mV. It is seen that E_{lj} decreases from 27 mV to 2.4 mV by using the salt bridge.

Self learning 3-5

Commonly, the saturated KNO_3 or KCl solution are used for the salt bridge. Explain this by considering the Henderson equation.

3-6 Donnan potential

Until now, it is shown that the electrical potential difference is defined if the charge separation at interfaces does not change; in either equilibrium state or steady state. The ***Donnan potential*** is developed under equilibrium state but its origin is somewhat different to those mentioned previously. Let us assume that aqueous KCl and NaCl solution are placed with a boundary as shown in Fig. 3-14. When the boundary is removed, the ions

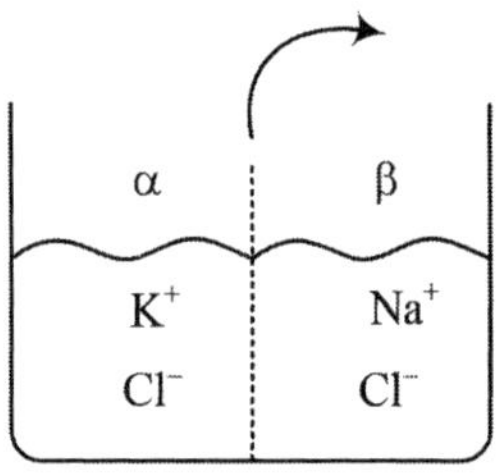

Fig. 3-14 Ion distribution when the boundary between KCl and NaCl solution is removed.

move easily because they are mobile, and eventually an equilibrium state is reached, where the concentration of every ion is the same anywhere in the solution. Obviously, electrical potential difference does not develop in any place in the solution.

The scenario is totally different when ions are immobile. Let us assume that there is a polymer membrane in β phase with cations in its backbone; the cations are fixed at the polymer chains, such that they can not move to a long distance (immobilized). The equivalent amount of Cl^- must coexist for charge neutrality in β phase. In α phase, 0.1 M KCl solution is placed. Note that both K^+ and Cl^- are mobile (not fixed). When the boundary (the dotted line in Fig. 3-15) between two phases is removed, the mobile ions can freely move, but the fixed ions can not.

Let us do the following thought experiment. Since K^+ exists only in α phase initially, it will diffuse to β phase upon removal of the boundary until the equilibrium state, wherein K^+ concentration is the same (0.05 M) for two phases. Then, the concentration of Cl^- in α phase must be 0.05 M to meet the electroneutrality condition in the equilibrium state ($[K^+] = [Cl^-] = 0.05\ M$). To meet this condition, 0.05 M Cl^- should move from α to β phase, and then Cl^- concentration in β phase is 0.15 M. This scenario is impossible because Cl^- concentration is different; 0.05 M in α phase and 0.15 M in β phase in the equilibrium state. This thought experiment can not actually happen. In other words, even when equilibrium is reached, K^+ can not have the same concentration (0.05 M) in two phases.

When the concentration of a given ion is different across an interface in the equilibrium state, an electrical potential difference develops across the interface. This is called the Donnan potential. This may develop when, as described above, there is a fixed charge

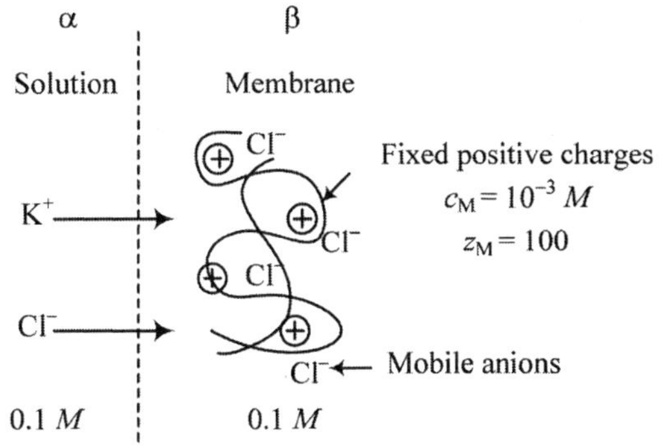

Fig. 3–15 Evolution of Donnan potential due to fixed (immobilized) charges in equilibrium state.

in any phases. In the polymer membrane shown in Fig. 3-15, the cation is fixed.

Since the Donnan potential occurs under equilibrium state, it can be derived considering the equilibrium condition; the electrochemical potential of a particular ion is identical for two phases. <3-28> explains this equilibrium condition, from which <3-29> is derived. Similarly, <3-30> is derived from the same electrochemical potential of Cl^- for two phases. <3-29> and <3-30> illustrate that an electrical potential difference develops if there is a concentration difference for any ions between two phases in equilibrium state.

$$\bar{\mu}^{\alpha}_{K^+} = \bar{\mu}^{\beta}_{K^+}$$

$$\mu^0_{K^+} + RT\ln a^{\alpha}_{K^+} + F\phi^{\alpha} = \mu^0_{K^+} + RT\ln a^{\beta}_{K^+} + F\phi^{\beta}$$ <3-28>

$$\Delta\phi = (\phi^{\beta} - \phi^{\alpha}) = \frac{RT}{F}\ln\frac{a^{\alpha}_{K^+}}{a^{\beta}_{K^+}}$$ <3-29>

$$\bar{\mu}^{\alpha}_{Cl^-} = \bar{\mu}^{\beta}_{Cl^-}$$

$$\Delta\phi = (\phi^{\beta} - \phi^{\alpha}) = \frac{RT}{F}\ln\frac{a^{\beta}_{Cl^-}}{a^{\alpha}_{Cl^-}}$$ <3-30>

$\Delta\phi$ in <3-29> and in <3-30> should be the same, and thus <3-33> is derived if <3-32> is assumed to be valid.

$$\frac{a^{\alpha}_{K^+}}{a^{\beta}_{K^+}} = \frac{a^{\beta}_{Cl^-}}{a^{\alpha}_{Cl^-}}$$

$$a^{\alpha}_{K^+}a^{\alpha}_{Cl^-} = a^{\beta}_{K^+}a^{\beta}_{Cl^-} \quad \text{<3-31>}$$

$$\gamma_{K^+} = \gamma_{Cl^-} = 1 \quad \text{<3-32>}$$

$$c^{\alpha}_{K^+}c^{\alpha}_{Cl^-} = c^{\beta}_{K^+}c^{\beta}_{Cl^-} \quad \text{<3-33>}$$

Meanwhile, for electroneutrality, <3-34> in α phase and <3-35> in β phase should be satisfied. Here, z_M and c_M are the number of cations fixed per mole of polymer and moles of polymer, respectively.

$$c^{\alpha}_{K^+} = c^{\alpha}_{Cl^-} \quad \text{<3-34>}$$

$$c^{\beta}_{K^+} + z_M c_M = c^{\beta}_{Cl^-} \quad \text{<3-35>}$$

Example 3-3

Calculate the Donnan potential with z_M = 100 and c_M = 10^{-3} M in Fig. 3-15.

Solution

By using the equilibrium condition <3-33>, electroneutrality condition <3-34> and <3-35>, and the following initial condition,

$$c^{\alpha}_{K^+} + c^{\beta}_{K^+} = 0.1\ M$$

one can obtain in the equilibrium state that $c^{\beta}_{K^+}$ = 0.033 M, $c^{\alpha}_{K^+} = c^{\alpha}_{Cl^-}$ = 0.067 M, and $c^{\beta}_{Cl^-}$ = 0.133 M. The Donnan potential is calculated to be 17.8 mV by using <3-29> or <3-30>. Note that the concentration of K^+ and Cl^- is not the same in α and β phase even under equilibrium state. However, the electroneutrality condition is met in α and β phase.

3-7 Ion-selective electrodes

Fig. 3-16 shows the configuration of a ***pH meter*** that measures H^+ concentration (activity). The SCE is used as the external reference electrode, and a fritted glass between the SCE and the test solution (a_{H^+}) prevents physical mixing of two solutions. This separator (fritted glass) is a porous material capable for ion movement. Meanwhile, an Ag/AgCl(s) electrode is used as an indicator electrode, and 0.1 M HCl is filled as an internal solution. A glass membrane is placed between the test solution and the internal solution. This glass membrane has a composition of $Na_2O-CaO-SiO_2$, in which only Na^+ is mobile, but Ca^{2+} and Si^{4+} are strongly bound to O^{2-}. Note that O^{2-} is fixed (immobile) in the glass matrix. A unique feature for this glass membrane is that the mobile Na^+ can come out of the membrane easily and other cations can penetrate into the membrane; cation exchange is possible.

There are five interfaces in Fig. 3-16, where electrical potential differences can develop. The five interfaces are depicted in Fig. 3-17. Since the Cl^- concentration in the SCE is constant as a saturated state, $\Delta\phi_1$ has a constant value. $\Delta\phi_{1j}$ is constant and has a very small value due to an involvement of KCl solution. $\Delta\phi_4$ is also constant because the Cl^- concentration is fixed at 0.1 M in the internal solution. Therefore, when the volt-

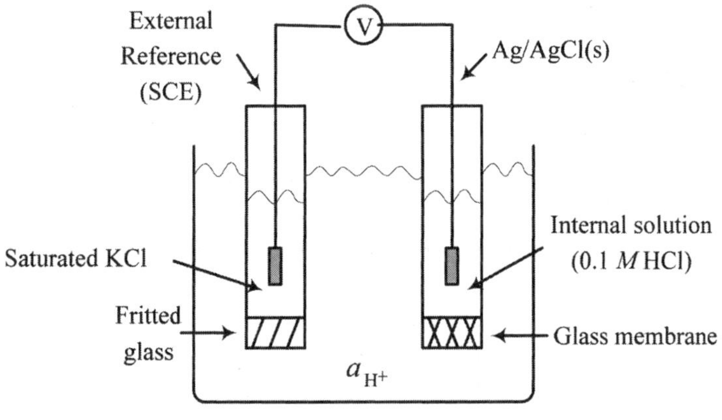

Fig. 3-16 Configuration of a pH meter.

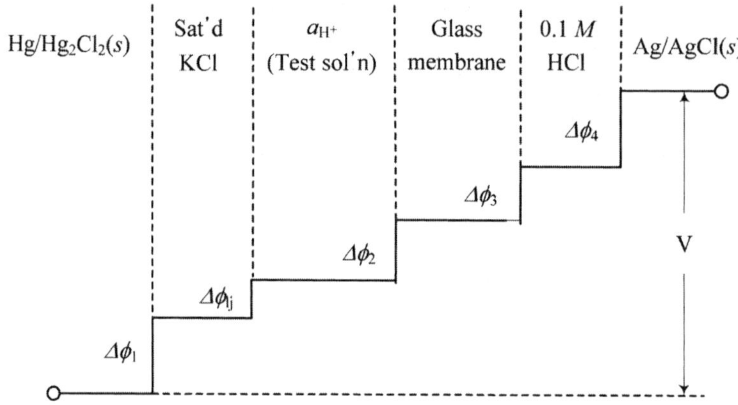

Fig. 3-17 Evolution of electrical potential difference at five interfaces in the pH meter shown in Fig. 3-16.

age (V) between two electrodes is measured, only $\Delta\phi_2$ and $\Delta\phi_3$ are variable.

$$V = \text{constant} + \Delta\phi_2 + \Delta\phi_3$$

Fig. 3-18 shows an enlarged view for the interface between the glass membrane and the sample solution (α), and the glass membrane and the internal solution (β). Another feature for the glass membrane is that it is hydrated when it comes into contact with aqueous solutions. Therefore, the glass membrane can be divided into the hydrated phases (m′ and m″) and non-hydrated phase (m). There are four interfaces as shown in Fig. 3-18.

As described above, the measured voltage V = constant + $\Delta\phi_2$ + $\Delta\phi_3$ and $\Delta\phi_2$ +

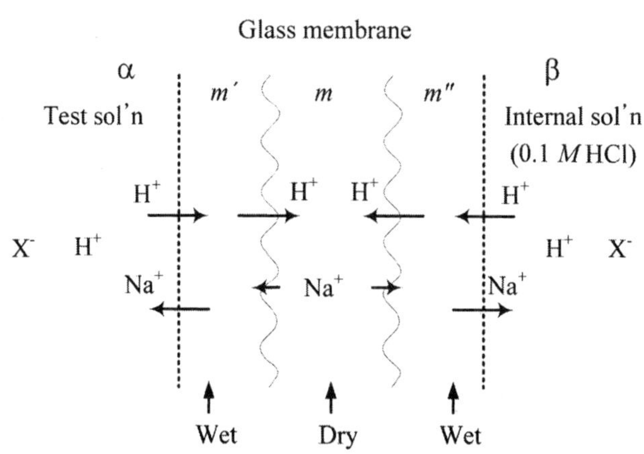

Fig. 3-18 Evolution of membrane potential (E_m) across the glass membrane.

$\Delta\phi_3 = E_m$, which is called the ***membrane potential***. E_m is the sum of the electrical potential difference at the four interfaces shown in Fig. 3-18.

$$V = \text{Constant} + E_m$$

$$E_m = (\phi^{\beta} - \phi^{m''}) + (\phi^{m''} - \phi^{m}) + (\phi^{m} - \phi^{m'}) + (\phi^{m'} - \phi^{\alpha})$$ <3-36>

(Donnan) (Diffusion) (Diffusion) (Donnan)

The four interfaces can be categorized into two groups. At the interface between β and m" phase, a Donnan potential develops because of the involvement of mobile H^+ and Na^+, and O^{2-} being fixed on the glass membrane (see Fig. 3-15). Using the equilibrium condition that the electrochemical potential of H^+ on β and m" phase are the same, Donnan potential is derived as <3-37>. Similarly, Donnan potential is derived as <3-38> at the interface between α and m' phase.

$$\bar{\mu}^{\beta}_{H^+} = \bar{\mu}^{m''}_{H^+}$$

$$\mu^{0}_{H^+} + RT\ln a^{\beta}_{H^+} + z_{H^+}F\phi^{\beta} = \mu^{0}_{H^+} + RT\ln a^{m''}_{H^+} + z_{H^+}F\phi^{m''}$$

$$\phi^{\beta} - \phi^{m''} = \frac{RT}{F}\ln\frac{a^{m''}_{H^+}}{a^{\beta}_{H^+}}$$ <3-37>

$$\phi^{m'} - \phi^{\alpha} = \frac{RT}{F}\ln\frac{a^{\alpha}_{H^+}}{a^{m'}_{H^+}}$$ <3-38>

In the interior of m, m' and m" phase, O^{2-} is fixed with the same concentration. Therefore, there is no O^{2-} diffusion across the interfaces. However, the concentration of H^+ and Na^+ differs at the m/m' and m/m" interfaces, hence diffusion potentials develop as shown in <3-39> and <3-40> due to the mobility difference for the two ions (H^+ and Na^+). The membrane potential (E_m), which is the sum of the potential differences at the four interfaces, is summarized as <3-41>.

$$\phi^{m} - \phi^{m'} = \frac{RT}{F}\ln\frac{u_{H^+}a^{m'}_{H^+} + u_{Na^+}a^{m'}_{Na^+}}{u_{Na^+}a^{m}_{Na^+}}$$ <3-39>

$$\phi^{m''} - \phi^{m} = \frac{RT}{F}\ln\frac{u_{Na^+}a^{m}_{Na^+}}{u_{Na^+}a^{m''}_{Na^+} + u_{H^+}a^{m''}_{H^+}}$$ <3-40>

$$E_m = \frac{RT}{F}\ln\frac{a^{\alpha}_{H^+}a^{m''}_{H^+}}{a^{\beta}_{H^+}a^{m'}_{H^+}} + \frac{RT}{F}\ln\frac{\left(\frac{u_{Na^+}}{u_{H^+}}\right)a^{m'}_{Na^+} + a^{m'}_{H^+}}{\left(\frac{u_{Na^+}}{u_{H^+}}\right)a^{m''}_{Na^+} + a^{m''}_{H^+}} \qquad \text{<3-41>}$$

Meanwhile, an equilibrium state is reached through ion exchange at the α/m′ and β/m″ interfaces. The equilibrium constant is expressed as <3-42> for the ion exchange at the α/m' interface. Here, the ***potentiometric selectivity constant*** is defined as <3-43>. Note that $k^{pot}_{H^+,Na^+}$ is proportional to the equilibrium constant for the ion exchange.

$$Na^+(\alpha) + H^+(m') \rightleftharpoons H^+(\alpha) + Na^+(m')$$
$$Na^+(\beta) + H^+(m'') \rightleftharpoons H^+(\beta) + Na^+(m'')$$

$$K_{H^+,\,Na^+} = \frac{a^{m'}_{Na^+}a^{\alpha}_{H^+}}{a^{m'}_{H^+}a^{\alpha}_{Na^+}} \qquad \text{<3-42>}$$

$$k^{pot}_{H^+,\,Na^+} = K_{H^+,\,Na^+}\left(\frac{u_{Na^+}}{u_{H^+}}\right) \qquad \text{<3-43>}$$

The substitution of <3-42> and (3-43> into <3-41> yields <3-44>.

$$E_m = \frac{RT}{F}\ln\frac{a^{\alpha}_{H^+} + k^{pot}_{H^+,Na^+}a^{\alpha}_{Na^+}}{a^{\beta}_{H^+} + k^{pot}_{H^+,Na^+}a^{\beta}_{Na^+}} = \text{constant} + \frac{RT}{F}\ln\,(a^{\alpha}_{H^+} + k^{pot}_{H^+,Na^+}a^{\alpha}_{Na^+}) \qquad \text{<3-44>}$$

<3-44> illustrates that the membrane (E_m) is determined by the activity of H^+ and Na^+ in the test solution (α phase), and the potentiometric selectivity constant. If $k^{pot}_{H^+,Na^+}$ is large enough, there evolves an error in the pH measurement even if a small amount of Na^+ is present in the test solution. This is called an ***alkali error***. Therefore, the pH meter with a good performance should use the glass membrane that shows a very small potentiometric selectivity constant. In this case, even an appreciable amount of Na^+ in the test solution does not cause an error in the pH measurement.

The potentiometric selectivity constant, defined by <3-43>, is proportional to the equilibrium constant of the ion exchange reaction. The smaller the value, the better the performance of the glass membrane. For this, the equilibrium constant should be small; $H^+(\alpha)$ and Na^+(m′) in the equilibrium state should be small while $Na^+(\alpha)$ and H^+(m′) should be large as is seen in <3-42>. In order to meet this condition, the glass membrane should be able to easily release Na^+ but accept H^+; the glass membrane should favor H^+ over Na^+ for ion exchange. $k^{pot}_{H^+,Na^+}$ of the glass membrane used in pH meters is as small as 10^{-11}. Therefore, the alkali error is not serious as is seen from <3-44>. Here, the value of $k^{pot}_{H^+,Na^+}$ means the preference of the latter ion (Na^+) over the former (H^+). That is, the ion exchange preference of H^+ is about 10^{11} times larger than that of Na^+. The commonly used glass membranes for pH meters have a much higher preference for H^+ over the other cations, so that even if other ions coexist in the test solution as an impurity, the errors caused by the impurity ions is not serious.

The membrane having a high preference for a specific ion is referred to as an ***ion-selective membrane***, and the electrode using the ion-selective membrane is referred to as an ***ion-selective electrode***. A variety of ion-selective membranes is used for ion-selective electrodes. The solid-state LaF_3 doped with EuF_2 has an excellent selectivity for F^- and thus used as an electrode for measuring the concentration of F^-. Many types of ion-selective membranes have internal fixed charges, but some are neutral. A representative example of the neutral membrane is ***valinomycin***. This has a ring structure, and the oxygen atom acts as a chelating agent capable of binding to K^+, so that K^+ is selectively bound. Valinomycin has an ion exchange preference for K^+, such that the concentration of K^+ may be different between the membrane phase and solution phase, resulting in a Donnan potential. Therefore, this is a K^+ selective membrane and is used as an ion-selective electrode for K^+ analysis (Fig. 3-19).

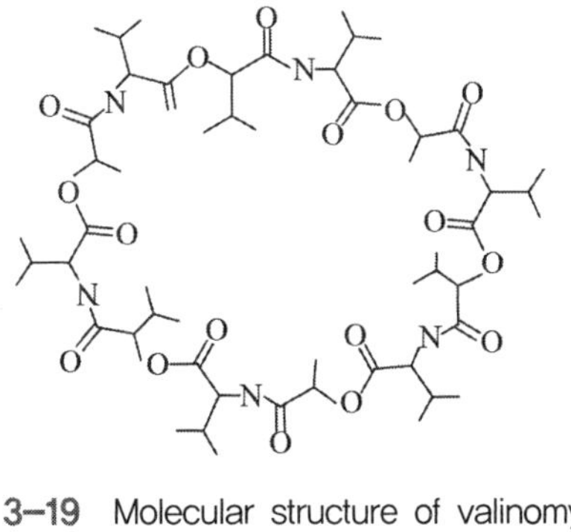

Fig. 3-19 Molecular structure of valinomycin.

3-8 Electrochemical sensors

Electrochemical sensors are classified into potentiometric and amperometric ones. The former measures the voltage of an indicator electrode against a reference electrode. The typical example is the pH meter mentioned above. The amperometric sensors measure the current that is proportional to the concentration of to-be-sensed materials.

Fig. 3-20 shows the configuration of a potentiometric urea sensor that uses a NH_4^+ selective electrode. The urease being immobilized in the membrane decomposes urea to generate NH_4^+, which is then analyzed by the NH_4^+ selective electrode.

$$\text{Urea} \xrightarrow{\text{Urease} + 2H_2O} 2NH_4^+ + CO_3^{2-}$$

Fig. 3-21 illustrates the working principle of an amperometric glucose sensor. The Pt electrode has a coating layer of a polymeric membrane, in which GOD (glucose oxidase) is immobilized. Note that GOD is a redox enzyme that can oxidize glucose into gluconolactone. When glucose is oxidized, the redox center inside GOD is reduced, which is then oxidized by oxygen to produce hydrogen peroxide. A potential is applied on the Pt electrode to detect the oxidation current of hydrogen peroxide. The concentration of glucose is measured by measuring the oxidation current of hydrogen peroxide. Here, it

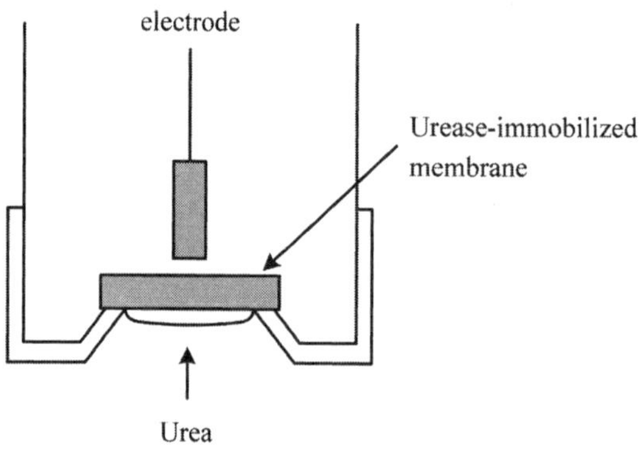

Fig. 3-20 Configuration of a potentiometric urea sensor.

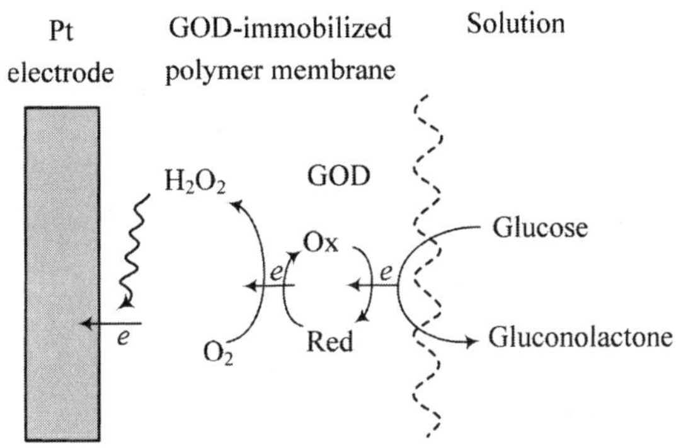

Fig. 3-21 Working principle of an amperometric glucose sensor.

should be guaranteed that the concentration of hydrogen peroxide is proportional to the glucose concentration.

3-9 Structure of electric double-layer

The electron energy inside an electrode can be controlled by changing the potential of that electrode, and excessive charge is accumulated on the electrode surface. If the potential of a metal electrode is adjusted to accumulate a negative charge on the surface, an equal amount of positive charge should be arranged on the solution side for electrical neutrality.

$$\sigma^{M} = -\sigma^{S}$$

Here, *M* and *S* mean the metal electrode and solution, respectively, and σ is the electrical charge density (charge per area). When the electrode potential is changed in the negative direction, σ^{M} increases with more negative value. In opposite, σ^{M} increases with more positive value when the electrode potential is adjusted in the positive direction. Here, one can expect an electrode potential, at which the electrode has neither negative nor positive charges; $\sigma^{M} = 0$. The electrode is not polarized. This is called the ***potential of***

zero charge (E_z).

When the electrode potential is deviated from the potential of zero charge (when the electrode is polarized, see Section 1-2), charge separation occurs at the electrode/solution interface, and the negative and positive charges are strongly bound by electrostatic attraction force. Such an attractive interaction is, however, disrupted by thermal energy, such that the charges in the electrode and solution are distributed within a certain distance on both sides of the interface. The region where the excess charges are populated is called a ***space charge region***, and its thickness is determined by temperature, the density of charge carriers (electrons or ions), and others. The higher the temperature, the greater the relaxation in the electrostatic attractive interaction due to thermal energy, thereby widening the space charge region. As the density of charge carriers decreases, the space charge region becomes wider; this is the same phenomenon that the Debye length increases as ion concentration decreases. In the case of metal, since the density of free electrons as a charge carrier is very high, the space charge region is very thin, about several Å on the metal surface. However, since the concentration of charge carrier in electrolyte solution, which is ions, is relatively small compared to the free electrons in metals, the thickness of space charge region in the solution side, where excess ions are populated, is about 10 ~ 1000 Å; much wider than that on metal surface. Therefore, we ignore the space charge region in the metal as in Fig. 3-22 and Fig. 3-23, and consider only the space charge region on the solution side.

Self learning 3-6

In the case of an *n*-type semiconductor electrode doped with 1 ppm of phosphorus (P) in silicon (Si), the concentration of electrons, which are the major charge carrier, is $5 \times 10^{16}/cm^3$. Calculate the ion concentration (unit: number/cm^3) for 0.1 *M* NaCl solution. Assuming that the *n*-type semiconductor electrode and the 0.1 *M* NaCl solution form an interface, compare the thickness of the space charge region for two sides in consideration of the charge carrier density.

For electrochemical reactions, the heterogeneous charge transfer reaction takes place only near electrode surface; only the reactants approaching the electrode surface can give or take electrons. Hence, it can be assumed that the charge transfer rate is influenced by the structure of electric double-layer near electrode surface. For example, any compounds that are adsorbed on electrode surface may accelerate or retard the charge transfer reaction. Even if the structure of electric double-layer has a great influence on the charge transfer kinetics, it is difficult to obtain detailed information on the electric double-layer. The structure and reaction characteristics of solid/gas interfaces can be seen in greater detail using surface analysis equipments. However, the experimental tools to investigate the solid/liquid interfaces have recently been developed and used, but is still limited. Therefore, the theoretical approach has been the mainstream for the structure of electric double-layer, and the ***Gouy-Chapman-Stern model*** is the combined one for several theoretical models.

The structure of electric double-layer is depicted in Fig. 3-22, which is the case for the electrode potential being more negative than E_z. According to the theoretical models, electrode surface is polarized by a strong electric field (10^5~10^7 V/cm) and water molecules are adsorbed. In the case of mercury (Hg) electrode, anions can be adsorbed on Hg surface even if excessive negative charges are accumulated on the Hg electrode

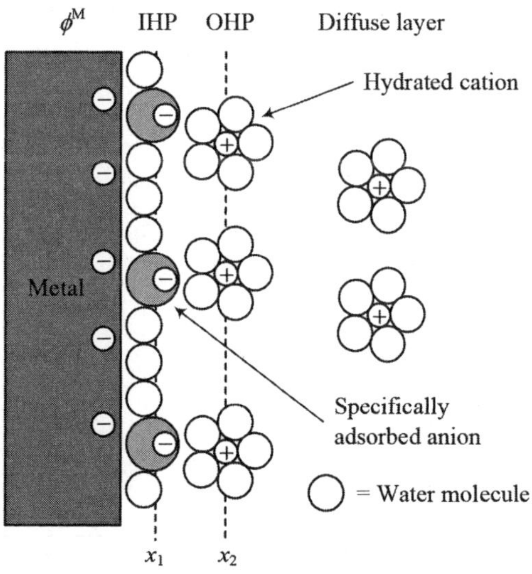

Fig. 3-22 A model of electric double layer with specifically adsorbed anion.

surface. This is called specific adsorption because the anions are adsorbed by overcoming the electrostatic repulsion. The plane formed along the center of the specifically adsorbed anions is called the inner Helmholtz plane (IHP). Meanwhile, since excess negative charges are accumulated on the electrode surface, there are more positive ions than the negative ones on the solution side near the electrode surface. The excessive positive ions are distributed within the space charge region. Since the cations are solvated (hydrated here), the distance between the electrode surface and the center of hydrated cations corresponds to the diameter of one to two water molecules even if the hydrated cations are closest to the electrode surface. This is called the outer Helmholtz plane (OHP). The combination of IHP and OHP is called a compact layer or Helmholtz layer or Stern layer.

In the outer region of OHP, excessive cations are arranged due to the electrostatic attraction with the negative charges on the electrode surface. The population of cations becomes smaller with an increase in the distance from the electrode surface. The region where the excess cations are distributed is referred to as a ***diffuse layer***. Note that the formation of diffuse layer is the same phenomenon as the ionic atmosphere formation described in the Debye-Hückel theory. As the ionic strength (ion concentration) increases, the thickness of diffuse layer becomes thinner as the Debye length decreases with an increase in ion concentration.

Fig. 3-23 shows the electrical potential profile on the solution side of electrical double layer. Here, the potential profile in the electrode side is ignored because the space charge

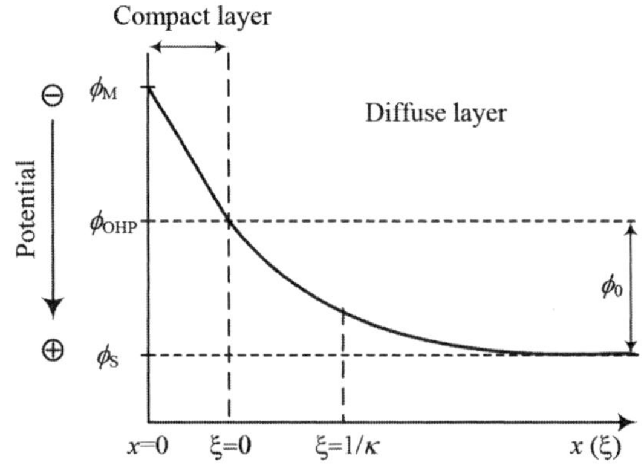

Fig. 3-23 Electrical potential profile in the space charge region of solution side.

region in the metal electrode is negligibly thin. The Helmholtz-Perrin model proposes that the electrical potential changes linearly in the compact layer. The Gouy-Chapman model, however, illustrates that the electrical potential changes by an exponential function of the distance from the electrode surface to the solution side. The Gouy-Chapman-Stern model combines the two models. The electrical potential profile follows the Poisson equation. If the electrical potential difference between the bulk solution and the OHP is defined as follows, the electrical potential, $\phi(\xi)$, has the relation of <3-45> in the diffuse layer.

$$\phi_0 = \phi_S - \phi_{OHP}$$

$$\tanh\{ze\phi(\xi)/4kT\} = -\tanh(ze\phi_0/4kT)e^{-\kappa\xi} \quad \text{<3-45>}$$

Here, $\xi = x - x_2$ and ξ is distance from OHP to bulk solution. κ is expressed by <3-46>, where n^0 is the concentration of an ion in the bulk solution, z is the charge of the ion, e is the unit charge, ε is the dielectric constant of the medium, and ε_0 is the vacuum permittivity.

$$\kappa = \left(\frac{2n^o z^2 e^2}{\varepsilon\varepsilon_o kT}\right)^{1/2} \quad \text{<3-46>}$$

If ϕ_0 is small enough in <3-45>, $(ze\phi_0/4kT) < 0.5$, $\tanh\{ze\phi(\xi)/4kT\} \approx -ze\phi(\xi)/4kT$ and thus <3-47> is derived.

$$\phi(\xi) = -\phi_0 e^{-\kappa\xi} \quad \text{<3-47>}$$

According to <3-47>, the electrical potential changes according to the exponential function of the distance from OHP. Here, ϕ_0 is called the **zeta potential**. At $\xi = 1/\kappa$, $\phi(\xi) = -\phi_0 e^{-1}$, which represents the thickness of the diffuse layer. The thickness of diffuse layer is determined by the ionic strength in the solution. The smaller the ionic strength, the smaller the κ, so the diffuse layer develops wider.

The electric double-layer can be formed not only on electrode surfaces but also on electrically non-conductive solid surfaces. For example, when a solid particle itself has a charge on its surface or ions are adsorbed on the surface, it has a charge on its surface.

Counter ions should be positioned in the solution side for electro-neutrality. The counter ions form a compact layer and diffuse layer just like the ions in the electrolyte solution interfacing with the polarized electrode surface. It is common that anions are strongly adsorbed on the surface of colloidal particles. In this case, excess positive ions are positioned in the space charge region as shown in Fig. 3-22.

Self learning 3-7

Mercury is a liquid, so the larger the surface tension, the more it contracts. This is called electrocapillarity. Consider how the surface tension changes as the potential of a mercury electrode changes from the potential of zero charge (E_z) to the positive or negative direction. Also consider how the shape of mercury electrode changes.

Problems

01 If the ionic conductivity of 10^{-4} *M* and 5×10^{-4} *M* NaCl solution is 1.26×10^{-3} S/m and 6.2×10^{-3} S/m, respectively, what is the ionic conductivity at infinite dilution?

02 (a) If the ionic conductivity of aqueous 0.1 *M* KCl at 25 °C is 1.16 S/m and the resistance measured by conductivity meter is 25.0 Ω, what is the cell constant for this conductivity meter?

(b) The resistance of 0.2 *M* NaCl solution is 16.6 Ω when measured using the same conductivity meter. What is the ionic conductivity of the NaCl solution?

(c) The resistance of 0.01 *M* acetic acid is 2,000 Ω when measured using the same conductivity meter. What is the molar conductivity of the acetic acid solution?

(d) The molar conductivities of infinitely diluted HCl, sodium acetate, and NaCl solution are 42.6, 9.1, and 12.6 mS m^2/mol, respectively. What is the molar conductivity of acetic acid at infinite dilution?

(e) From the above data, calculate the degree of dissociation (α) of 0.01 *M* acetic acid (see Example 3-1).

03 Let us assume that the permeation of Na^+, Cl^- and H_2O to some plant cell membrane is possible, but the permeation of proteins is impossible. Further assume that 0.05 *M* NaCl solution is present outside the cell membrane and 0.001 *M* protein (*P*) is present inside. This protein is ionized to P^{z+} with z = 20 and the counter ion is Cl^-. From this, calculate the Donnan potential that develops between the cell membrane and the external solution at equilibrium. Assume unit activity coefficient for all ions.

04 The voltage measured by using a pH meter for 0.1 M NaCl is −0.45 V at pH = 8.0, −0.57 V at pH = 10.0, and −0.69 V at pH = 13.0. Using <3-44>, calculate the potentiometric selectivity constant of the glass membrane used for the pH meter.

05 The transport number can be measured by the Hittorf method described below.

L C R

(cathode) Ag | $AgNO_3$ (0.1 M) || $AgNO_3$ (0.1 M) || $AgNO_3$ (0.1 M) | Ag (anode)

Here, || is a separator that prevents the physical mixing of solutions, but ion migration is possible. The volume of $AgNO_3$ solution in all three chambers is 25 mL. An external power supply is connected to pass electric charges of 96.5 C to induce both Ag electrodeposition at the cathode and Ag dissolution at the anode.

(a) What is the weight (in grams) of Ag that is electrodeposited on the cathode?

(b) If the transport number of Ag^+ is 1.0, what is the concentration of Ag^+ in the three chambers after the electrolysis?

(c) If the transport number of Ag^+ is assumed to be 0.0, what is the Ag^+ concentration in the three chambers after the electrolysis?

(d) The concentration of Ag^+ in the chamber R is 0.121 M after the real electrolysis. Calculate the transport number of Ag^+.

06 AgBr is a mixed-conducting solid; the electrical conduction by both electrons and ion (Ag^+) is possible. A cell is fabricated with Ag/AgBr/Ag configuration, in which the AgBr solid film is inserted between two Ag electrodes (each weighing 1.00 g). After flowing a current of 200 mA for 10 min, the weight of cathode is found to be 1.12 g. From this, calculate how much charges are passed by electron conduction.

07 Quinhydrone is a mixture of quinone (Q) and hydroquinone (H_2Q) (1:1 molar ratio), which reacts reversibly on Pt electrode as follows.

$$Q + 2H^+ + 2e = H_2Q \quad E^0 = 0.458 \text{ V } (vs. \text{ SCE})$$

In order to measure the pH of an aqueous solution, a cell is fabricated with a SCE reference electrode and a Pt indicator electrode.

SCE | quinhydrone (saturated), H^+ (unknown) | Pt

(a) The voltage between the Pt and SCE is measured according to the pH of aqueous solutions. What is the slope in the measured voltage *vs.* pH plot?
(b) If the measured voltage is 0.3 V, what is the solution pH? Ignore the liquid junction potentials.

08 D. C. Grahame *et al* (*Chem. Rev.*, **41**, 441, 1947) reported the change of C_d according to NaF concentration and electrode polarization for mercury electrode. The capacitance value (C_d) in the electric double-layer shown in Fig. 3-24 is the sum of the serially connected C_{OHP} and $C_{diffuse}$, thus the following is given:

$$1/C_d = 1/C_{OHP} + 1/C_{diffuse}$$

In general, the capacitance (in F/cm^2) is inversely proportional to the distance d between two planes as

$$C_d = \varepsilon\varepsilon_0/d$$

Here, ε is the dielectric constant of the medium and ε_0 is the vacuum permittivity. Assuming no specifically adsorbed anions, C_{OHP} is independent on ion concentration in the electrolyte, but has different values depending on the electrode polarization; when the electrode is negatively polarized, the cation forms an OHP, whereas the anion forms an OHP upon a positive polarization.

Meanwhile, the $C_{diffuse}$ depends on the ion concentration (Debye-Hűckel theory) and also on the electrode potential. Fig. 3-24 shows the change of C_d according to NaF concentration and electrode potential. Explain the following observations shown in Fig. 3-24.

(a) When NaF concentration is very small (0.001 *M*), C_d shows the minimum value when the electrode potential approaches E_z: the polarization is small.

(b) As NaF concentration increases, the feature of the minimum value of C_d near E_z disappears.

(c) When NaF concentration is as large as 0.1 *M*, C_d is larger upon a positive polarization compared to the negative polarization. Hint; the size of hydrated Na^+ is larger than that of hydrated F^-.

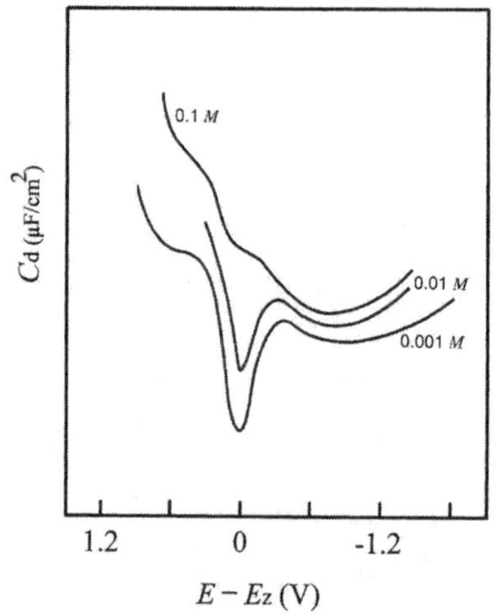

Fig. 3-24 Change of C_d according to NaF concentration and electrode polarization.

CHAPTER 04

Electrochemical kinetics

When an electroactive compound O in solution is reduced through an electrochemical reaction, O must first move from bulk solution to electrode surface. This is called mass transfer. Mass transfer can be classified into three types: diffusion, migration and convection. O arriving at the electrode surface is reduced to R by heterogeneous charge transfer reaction. In an electrochemical reaction consisting of two steps of mass transfer and charge transfer, the current (reaction rate) is determined by the slower step (rate-determining step) of the two. When charge transfer is slower, the rate of overall reaction (current) is determined by the charge transfer rate and, conversely, by the mass transfer rate when mass transfer is slower. The charge transfer rate varies with an exponential function of applied electrode potential, such that the current, which is the rate of whole electrochemical reaction, also varies by the exponential function of applied electrode potential.

In order for an electrochemical reaction $O + ne \rightarrow R$ to proceed, n electrons per formula unit of O must be transferred from the electrode to O in the charge transfer step. Here, n electrons can not be transferred all at once in one step. Rather, electrons are considered to be transferred one by one through continued elementary steps. Therefore, in order for n electrons to be transferred, n steps in which one electron is involved must be continued.

In Section 4-1, it will be discussed how the charge transfer rate varies according to the applied electrode potential in one-electron charge transfer step (4-1-2 ~ 4-1-6). Then, the discussion will be extended to how the charge transfer rate varies according to the applied electrode potential for the systems involving n charge transfer steps for n-electron reactions (4-1-7). Section 4-2 describes the relationship between mass transfer rate and current when mass transfer determines the overall rate for electrochemical reactions. The role of supporting electrolytes is also discussed.

4-1 Charge transfer rate

4-1-1 Chemical kinetics

The rate of forward and backward (or reverse) reaction for a general chemical reaction ($A + B \rightleftarrows C + D$) is expressed as

$$v_f = k_f C_A C_B,\ v_b = k_b C_C C_D,\ v_{net} = k_f C_A C_B - k_b C_C C_D$$

In equilibrium state, $v_{net} = 0$ because the reaction rate in both directions is identical. Here, k_f and k_b are the reaction rate constants for the forward and backward reaction, respectively. According to the activated complex theory (or transition state theory), the rate constants have the following relation with $\Delta G^{\ddagger}$ that is related with the activation energy of the reactions (Fig. 4-1).

$$k_f = A_f\, e^{-\Delta G_f^{\ddagger}/RT},\ k_b = A_b\, e^{-\Delta G_b^{\ddagger}/RT} \qquad \text{<4-1>}$$

In normal chemical reactions, the rate constants (k_f and k_b) can be increased by lowering the activation energy ($\Delta G^{\ddagger}$) by using a catalyst or by increasing the reaction temperature.

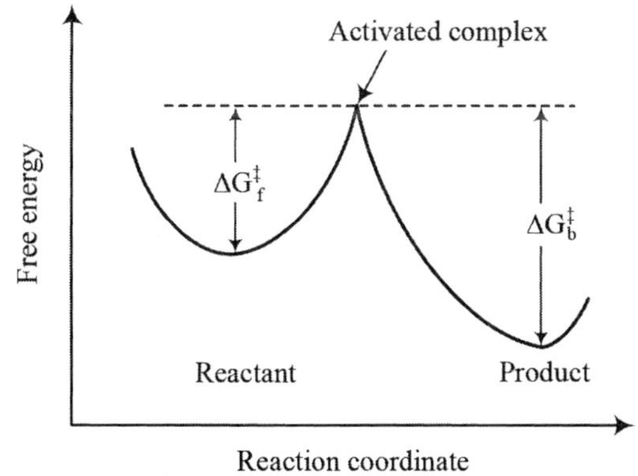

Fig. 4–1 Change of free energy according to reaction coordinate and activation free energy in chemical reaction.

4-1-2 Charge transfer kinetics for one-electron reaction

In the one-electron charge transfer reaction ($O_{surf} + e \rightarrow R_{surf}$), the charge transfer rate of the forward reaction (v_f) and the backward reaction (v_b) is expressed as below, from which the net current (i_{net}) is expressed as <4-2>.

$$v_f = k_f C_O(0,t) = \frac{i_c}{FA}, \quad v_b = k_b C_R(0,t) = \frac{i_a}{FA}, \quad v_{net} = v_f - v_b = \frac{i_{net}}{FA}$$

$$i_{net} = i_c - i_a = FA\,[k_f C_O(0,t) \; - \; k_b C_R(0,t)] \qquad \text{<4-2>}$$

Here, the $C_O(0,t)$ (= O_{surf}) means the surface concentration of O that is reduced to $C_R(0,t)$ (= R_{surf}) by one-electron charge transfer reaction. It can be assumed that the amount of electrons (e) in the electrode, which is one of the reactants in the charge transfer step, is sufficiently large so that its concentration does not change even if the charge transfer reaction progresses (refer to 1-1-6). Since the charge transfer step determines the overall rate, the current values (i_c, i_a, and i_{net}) correspond to the charge transfer rates. The concentration of O and R in solution changes with time (t) and position, such that they are expressed as $C_O(x,t)$ and $C_R(x,t)$ in one-dimensional space (x), where x is the distance from the electrode surface to the bulk solution. Thus, the $C_O(0,t)$ and $C_R(0,t)$ in <4-2> refer to the time-dependent concentration of O and R on the electrode surface (x = 0). In <4-2>, F is the Faraday constant, A is the electrode area, and k_f and k_b are the rate constant of the forward and backward charge transfer reaction, respectively, which is expressed as <4-3>.

$$k_f = k^0\, e^{-\alpha F(E-E^{0'})/RT}, \quad k_b = k^0\, e^{(1-\alpha)F(E-E^{0'})/RT} \qquad \text{<4-3>}$$

Here, k^0 is the ***standard rate constant***, α is the symmetry factor, E is the applied electrode potential, and $E^{0'}$ is the formal potential of the given half-cell. <4-3> reads that the charge transfer rate constants (k_f and k_b) vary by the exponential function of applied electrode potential. The electrode-potential-dependent rate constant for charge transfer reaction can be explained using the activated complex theory. The comparison of <4-1> and <4-3> reveals that the activation free energy ($\Delta G^{\ddagger}$ in <4-1>) and the applied electrode potential

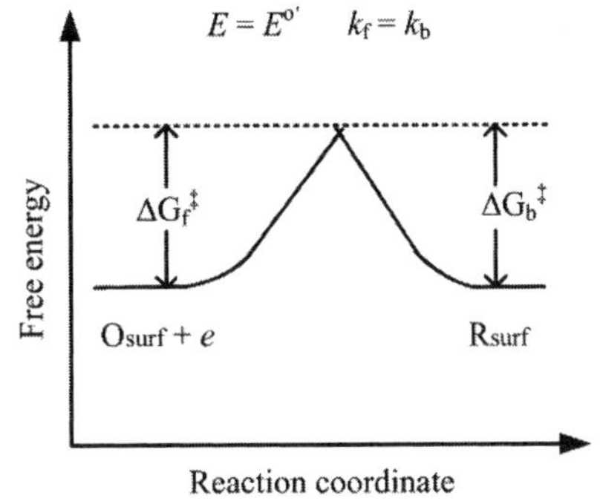

Fig. 4-2 Activation free energy for forward and backward charge transfer reaction when the applied electrode potential is identical with the formal potential of an electrochemical system ($E = E^{0'}$).

(E in <4-3> because $E^{0'}$ is constant in a given system) show one to one correspondence. This implies that the activation free energy can be changed by the applied electrode potential. This is true as discussed below.

Fig. 4-2 illustrates that the rate constant for both directions is the same ($k_f = k_b$) because the activation free energy for the forward and backward reaction is identical under the condition of $E = E^{0'}$. It can be seen from <4-3> that $k_f = k_b$ if $E = E^{0'}$.

Self learning 4-1

Prove that $k_f = k_b$ if $E = E^{0'}$ in <4-3>.

In Fig. 4-2, $k_f = k_b$ for $E = E^{0'}$. However, a change of applied electrode potential from $E^{0'}$ to a more positive value ($E > E^{0'}$) brings about the changes in k_f and k_b as illustrated in Fig. 4-3. Since the electrode potential is an expression of electron energy in the electrode (see 1-1-1), when the electrode potential is changed to a more positive value, the free energy of electrons in the electrode becomes lower while there is no change in the free energy of the reactant O_{surf}. The decrease in the electron free energy amounts to $F(E - E^{0'})$. Note here that the electric energy = Q x V; the electric charge $Q = n$F (n = 1) and $V = (E - E^{0'})$. Due to the decrease of electron free energy by $F(E - E^{0'})$, the sum of free energy for two reactants ($O_{surf} + e$) also decreases as shown in the dotted line in Fig. 4-3. As a result, the activation energy for the forward reaction ($\Delta G_c^\ddagger$) increases

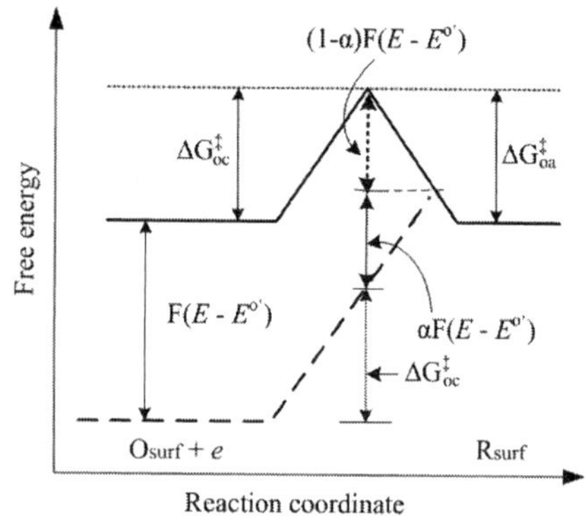

Fig. 4–3 Change of activation free energy for charge transfer reaction according to applied electrode potential.

whereas the activation energy for the backward reaction ($\Delta G_a^\ddagger$) decreases as <4-4>. k_f and k_b are then expressed by <4-5> and <4-6>, respectively. Here, if the standard rate constant (k^0) is defined as <4-7>, <4-5> and <4-6> become identical with <4-3>. <4-3> means that when the electrode potential is changed to a more positive value, k_b increases but k_f decreases. Furthermore, due to the change in the rate constants, the charge transfer rates also change to eventually lead to a change in the overall rates (current); the oxidation current increases at the expense of reduction current when the electrode potential is pushed to a more positive value.

$$\Delta G_c^\ddagger = \Delta G_{0c}^\ddagger + \alpha F(E - E^{0'}),\ \ \Delta G_a^\ddagger = \Delta G_{0a}^\ddagger - (1-\alpha)F(E - E^{0'}) \qquad \text{<4-4>}$$

$$k_f = A_f e^{-\Delta G_c^\ddagger/RT} = A_f e^{-\Delta G_{0c}^\ddagger/RT} e^{-\alpha F(E - E^{0'})/RT} \qquad \text{<4-5>}$$

$$k_b = A_b e^{-\Delta G_a^\ddagger/RT} = A_b e^{-\Delta G_{0a}^\ddagger/RT} e^{(1-\alpha)F(E - E^{0'})/RT} \qquad \text{<4-6>}$$

$$k^0 = A_f e^{-\Delta G_{0c}^\ddagger/RT} = A_b e^{-\Delta G_{0a}^\ddagger/RT} \qquad \text{<4-7>}$$

Example 4-1

In Fig. 4-3, the free energy of electron, which is one of two reactants, is affected by the applied electrode potential. Is there any way to change the free energy of O_{surf}, which is another reactant for the charge transfer reaction, to induce a change in the oxidation and reduction current?

Solution

In order to improve the quality of electroplated layers and uniformity of plating thickness (referred to a throwing power), the electroplating rate (that is, plating current) is controlled to be low enough. As a method for lowering the plating current, a chelating agent capable of forming a strong complex with to-be-plated metal ion is added to the plating solution. For example, the copper plating solutions include $Cu(CN)_2$; 40~50%, KCN; 20~30%, K_2CO_3; 10%. The cyanide (CN^-) forms a strong complex with Cu^{2+}. The strong complex formation leads to a decrease in the free energy of Cu^{2+} due to stabilization; the sum of free energy for two reactants (O_{surf} + e) decreases as shown in Fig. 4-3; O_{surf} is the Cu^{2+} complex in this case. As a result, the activation free energy for the forward charge transfer reaction (plating) increases to slow down the plating current; both the quality of plating layer and the throwing power can be improved (see Section 11-5).

From <4-2> and <4-3>, i_{net} is derived as follows.

$$\begin{aligned} i_{net} = i_c - i_a &= FA\,[k_f C_O(0,t) - k_b C_R(0,t)] \\ &= FAk^0[C_O(0,t)\, e^{-\alpha F(E-E^{0'})/RT} - C_R(0,t)\, e^{(1-\alpha)F(E-E^{0'})/RT}] \end{aligned} \qquad \text{<4-8>}$$

The first term on the right side of <4-8> is the charge transfer rate for the reduction reaction; it is the reduction current (i_c) because the charge transfer determines the overall rate, and the second term is the charge transfer rate for the oxidation reaction. Since i_{net} = 0 under an equilibrium, i_c = $|i_a|$. However, i_{net} = 0 does not mean that there is no current flow. In real, the current of the same magnitude flows in both directions under equilibrium (it is called a dynamic equilibrium), thereby the observed net current (i_{net}) appears to be zero. The current flowing for both directions at equilibrium is called an ***exchange current*** (i_0). Since i_0 = i_c = $|i_a|$ in equilibrium state, i_0 can be derived as follows. Using the condition of i_0 = i_c at E = E_{eq} for <4-8>, <4-9> is derived.

$$i_0 = i_c = FAk^0 C_O^* e^{-\alpha F(E_{eq}-E^{0'})/RT} \qquad \text{<4-9>}$$

In equilibrium state, the surface concentration is identical with the bulk concentration. Substitution of <4-10> into <4-11> and rearrangement yields <4-12>.

$$C_O(0,t) = C_O^*, \quad C_R(0,t) = C_R^* \qquad \text{<4-10>}$$

$$E_{eq} = E^{0'} + \frac{RT}{F} \ln \frac{C_O^*}{C_R^*}$$ <4-11>

$$\frac{C_O^*}{C_R^*} = e^{F(E_{eq}-E^{0'})/RT}, \quad \left(\frac{C_O^*}{C_R^*}\right)^{-\alpha} = e^{-\alpha F(E_{eq}-E^{0'})/RT}$$

$$i_0 = FAk^0 C_O^{*(1-\alpha)} C_R^{*\alpha}$$ <4-12>

In <4-12>, it is seen that the exchange current (i_0) is directly proportional to the ***standard rate constant*** (k^0) and electrode area (A). <4-13> is derived from <4-12>. The symmetry factor (α) can be obtained experimentally by using <4-13>.

$$\left(\frac{\partial \ln i_0}{\partial \ln C_O^*}\right)_{C_R^*} = 1-\alpha, \quad \left(\frac{\partial \ln i_0}{\partial \ln C_R^*}\right)_{C_O^*} = \alpha$$ <4-13>

Meanwhile, the standard rate constant (k^0) is not a function of electrode potential as shown in <4-7>, so that it is a value representing the intrinsic charge transfer property irrespective of applied electrode potential. The larger k^0 means a smaller $\Delta G_0^\ddagger$, which in turn indicates an intrinsically fast charge transfer kinetics. In Section 1-2, the R_{ct} of hydrogen evolution reaction on mercury electrode is so large that hydrogen evolution current is negligibly small over a wide overpotential range. This indicates that k^0 is very small for the charge transfer for H_2 evolution on Hg electrode. On the other hand, the R_{ct} of hydrogen evolution on Pt electrode is very small, indicative of a large k^0 for the charge transfer for H_2 evolution on Pt electrode. Electrocatalysts are frequently used in electrochemical reactions to increase the reaction rate. The electrocatalyst lowers $\Delta G_0^\ddagger$ to accelerate the charge transfer reaction; it plays a role of increasing k^0 by decreasing $\Delta G_0^\ddagger$ of a given system.

The observed current (i_{net}) can be expressed as <4-14> by combining <4-8> and <4-12>. In <4-14>, the current is expressed as an exponential function of overpotential ($\eta = E - E_{eq}$).

$$i_{net} = i_0 \left[\frac{C_O(0,t)}{C_O^*} e^{-\alpha F(E-E_{eq})/RT} - \frac{C_R(0,t)}{C_R^*} e^{(1-\alpha)F(E-E_{eq})/RT} \right]$$ <4-14>

In <4-14>, the first term in the bracket refers to the reduction current (i_c) and the second the oxidation current (i_a). The term expressed by the concentration ratio indicates the mass transfer rate, whereas the term given by the exponential function of overpotential corresponds to the charge transfer rate.

Self learning 4-2

The intrinsic property for charge transfer reaction for a given electrode/electrochemical reactions can be described by several parameters such as activation overpotential, i_0, R_{ct} and k^0. Among these, k^0 is most appropriate to represent the intrinsic charge transfer characteristics. Give an explanation on this.

4-1-3 Butler-Volmer equation

If mass transfer rate in an electrochemical reaction is sufficiently high to determine the overall rate by charge transfer, $C_O(0, t) \approx C_O^*$ and $C_R(0, t) \approx C_R^*$ prevail in <4-14>. That is, since the mass transfer rate is sufficiently high, the surface concentration, $C_{O/R}(0,t)$, can be assumed to be identical with the bulk concentration ($C_{O/R}^*$). Then, <4-14> is rearranged into <4-15> that is called ***Butler-Volmer equation***.

$$i_{net} = i_0[e^{-\alpha F\eta/RT} - e^{(1-\alpha)F\eta/RT}] \quad \text{<4-15>}$$

The Butler-Volmer equation illustrates how the current varies according to the overpotential when charge transfer step determines the overall rate. Note that the overpotential is defined as $\eta = E - E_{eq}$, in which E is the applied electrode potential and E_{eq} is the equilibrium potential for the given system. Since E_{eq} is constant for a given electrochemical system, the applied electrode potential (E) is only the variable to affect the current values in Butler-Volmer equation. Namely, when the applied electrode potential is pushed to a more negative value, the overpotential ($\eta = E - E_{eq}$) becomes more negative. Then, the first term in the bracket (the charge transfer rate for the reduction reaction and thus the reduction current) increases by the exponential function of applied potential.

The oxidation current, however, exponentially decreases as is seen in <4-15>. In opposite, when the electrode potential is pushed to a more positive value, η becomes more positive to exponentially decrease the reduction current, but to exponentially increase the oxidation current. Here, the overpotential (η) is called an ***activation overpotential*** (or charge transfer overpotential) because it relates to the charge transfer rate.

Example 4-2

Assuming that the charge transfer determines the overall rate for an electrochemical reaction (O + e = R), sketch the current-potential (i *vs.* V) profile using the Butler-Volmer equation. It is given that A = 0.1 cm^2, $E^{0'}$ = −0.15 V (*vs.* SCE), k^0 = 10^{-4} cm/sec, T = 298 K, α = 0.5, C_O^*= 5.0 mM, and C_R^* = 1.0 mM.

Solution

To simulate the current-potential profile using the Butler-Volmer equation, we first need to know the equilibrium potential (E_{eq}) of the given system (given half-cell) and the exchange current (i_0). The equilibrium potential (E_{eq}) is calculated as follows.

$$E_{eq} = E^{0'} + \frac{RT}{F}\ \ln \frac{C_O^*}{C_R^*}$$

$$= (-0.15\ \text{V}) + \frac{(8.314\ \text{J/mol}\cdot\text{K})(298\text{K})}{(9.65\times 10^4\ \text{C/mol})}\ \ln \frac{(5\times10^{-3}\text{M})}{(10^{-3}\text{M})} = -0.11\ \text{V}$$

The exchange current (i_0) is calculated as follows.

$$i_0 = FAk^0 C_O^{*(1-\alpha)} C_R^{*\alpha}$$

$$= (9.65\times10^4\ \text{C/mol})\times(0.1\ \text{cm}^2)\times(10^{-4}\ \text{cm/s})$$

$$\times(5\times10^{-6}\ \text{mol/cm}^3)^{0.5}\times(1\times10^{-6}\ \text{mol/cm}^3)^{0.5}$$

$$= 2.2\ \mu\text{A}$$

In this calculation, each variable is expressed in cgs unit and *the concentration must be used in mol/cm^3* to have amperes as the unit for current.

$$\left(\frac{\text{C}}{\text{mol}}\right)\left(\text{cm}^2\right)\left(\frac{\text{cm}}{\text{s}}\right)\left(\frac{\text{mol}}{\text{cm}^3}\right)^{0.5}\left(\frac{\text{mol}}{\text{cm}^3}\right)^{0.5} = \frac{\text{C}}{\text{s}} = \text{Amperes}$$

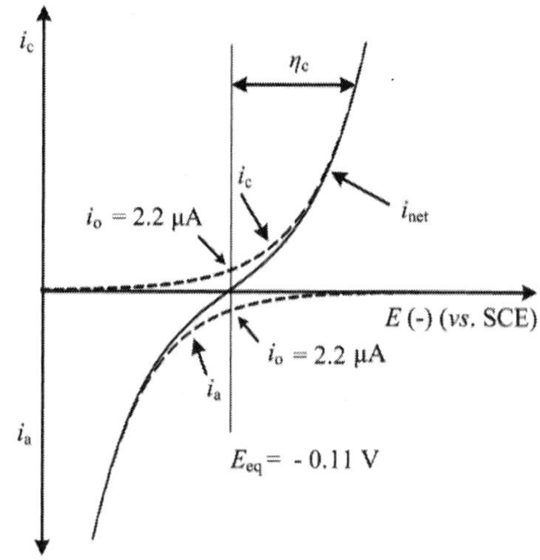

Fig. 4-4 The i_c, i_a and i_{net} according to Butler-Volmer equation. Note that the currents represent the charge transfer rates.

Fig. 4-4 sketches the i_c and i_a that are obtained as a function of applied electrode potential after substituting the as-calculated E_{eq} and i_0 values in <4-15>. As the electrode potential moves to the negative direction, the reduction current (i_c) increases by the exponential function (upper dotted line) and the oxidation current (i_a) decreases by the exponential function (lower dotted line). In opposite, as the electrode potential changes into a more positive value, the oxidation current (i_a) exponentially increases but the reduction current (i_c) decreases exponentially. At the equilibrium potential (E_{eq}), $i_c = |i_a| = i_0$. The i_{net} *vs.* electrode potential is represented by a solid line, which is the actually observed current. The i_c and i_a represented by the dotted lines are invisible and only the difference between them is detected as i_{net}. Therefore, E_{eq} is the reference for judging whether reduction or oxidation current is actually observed. In Fig. 4-4, a pure reduction current (the upper right-side solid line) is observed when the electrode potential is pushed to a more negative value than the equilibrium potential ($E < E_{eq}$), whereas a pure oxidation current (the lower left-side solid line) is observed $E > E_{eq}$. At E_{eq}, $i_{net} = i_c - |i_a| = 0$ because $i_c = |i_a|$. Therefore, in order to practically obtain a reduction current, the electrode potential should be pushed to a more negative value than E_{eq}. The difference between the pushed electrode potential and E_{eq} is expressed as an overpotential ($\eta = E - E_{eq}$). Here, "over" means that the electrode potential should be pushed to a more negative (or positive) value than E_{eq} to actually obtain the reduction (or oxidation) current. The overpotential is regarded as the driving force required to push the charge transfer reaction to proceed at the rate needed to obtain a targeted current.

Self learning 4-3

Apply the E_{eq} and i_0 values obtained in Example 4-2 to <4-15> to calculate i_c, i_a, and i_{net} when the applied electrode potential is -0.3 V (*vs.* SCE). What is the observed current?

4-1-4 Electrochemical reversibility

Reversibility in chemical reactions dictates that the reaction is reversible; the backward reaction is possible for a forward reaction. Electrochemical reversibility adds another concept to the chemical reversibility, which is that the standard rate constant (k^0) is very large.

Assuming that electrode area (A), symmetry factor (α), and bulk concentration of O and R are fixed in <4-12>, i_0 is proportional to k^0. According to <4-15>, i_c and i_a are proportional to i_0 and thus proportional to k^0. In Fig. 4-5, the current-potential profiles are compared for two cases. If k^0 is very large (Fig. 4-5-a), both i_c and i_a are substantially

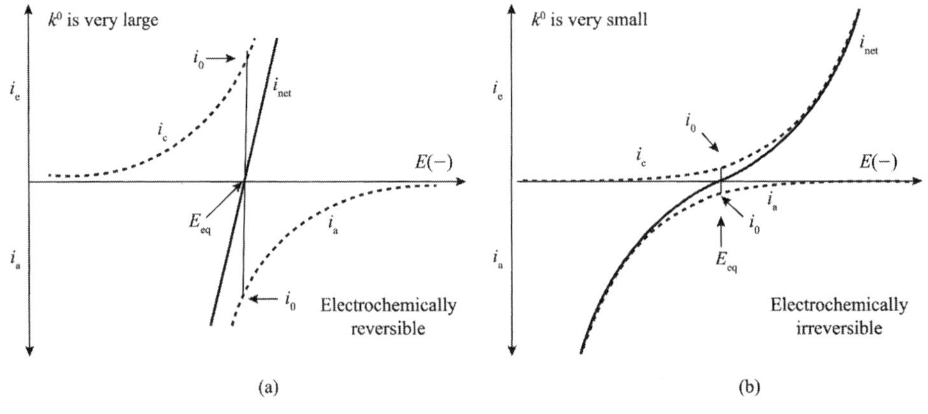

Fig. 4-5 The i_c, i_a and i_{net} representing the charge transfer rates for: (a); electrochemically reversible systems and (b); electrochemically irreversible systems.

large in a wide potential range and the observed current (i_{net} indicated by the solid line) is determined by the difference between two currents ($i_{net} = i_c - |i_a|$) in a wide potential range. It is said to be ***electrochemically reversible*** because both the reduction current (i_c) and oxidation current (i_a) flow 'reversibly' in a wide potential range. In ***electrochemically irreversible*** systems with very small k^0, i_c and i_a are notably small, such that the electrochemical reversibility (current flow for both directions and the net current is decided by the difference of two currents) appears only in a very small overpotential region. In the large negative overpotential region ($E << E_{eq}$), i_a is negligibly small to give $i_{net} \cong i_c$. In the large positive overpotential region ($E >> E_{eq}$), i_c is negligibly small to give $i_{net} \cong |i_a|$. The meaning of 'irreversible' is that current flows to only one direction; the reverse current is absent for a given forward current. Furthermore, with a decrease in k^0, the overpotential range where the electrochemical reversibility appears becomes narrower; the overpotential range where electrochemical irreversibility appears becomes wider. The i/V profile in Fig. 4-5-a is closer to the electrochemical reversibility; both i_c and i_a flow in a wide overpotential range. The i/V profile in Fig. 4-5-b is rather closer to the electrochemical irreversibility; only one-direction current flows except for a very small overpotential range.

In the electrochemical reversible system, k^0 is very large in <4-8>, then <4-16> is satisfied, which illustrates that the value in the bracket in <4-8> tends to zero. From this, <4-17> is derived, which is the ***Nernst equation***. Hence, the electrochemically reversible systems are called a ***nernstian*** because they satisfy the Nernst equation <4-17>.

$$\frac{i_{net}}{FAk^0} \approx 0 \qquad \text{<4-16>}$$

$$E = E^{0'} + \frac{RT}{F} \ln \frac{C_O(0,t)}{C_R(0,t)} \qquad \text{<4-17>}$$

From Fig. 4-4 and Fig. 4-5, it is seen that a larger activation overpotential is required to obtain a current when k^0 is smaller. Often, the activation overpotential is used to compare the intrinsic charge transfer characteristics. However, since the overpotential depends on the current size, it is not possible to quantitatively compare the overpotential values if the current is not specified.

Table 4-1 Surface concentration of *O* and *R* according to the change of applied electrode potential*

$E - E^{0'}$(mV)	$C_O(0, t) / C_R(0, t)$
0	1
59	10^1
118	10^2
236	10^4
−59	10^{-1}
−177	10^{-3}
−236	10^{-4}

[a]Temp. = 298 K, R = 8.314 J/mol K

In general, if $k^0 > 2\times10^{-2}$ cm/sec, it is said to be electrochemically reversible. It is irreversible if $k^0 < 5\times10^{-3}$ cm/sec. However, the limit is relative, and one should consider the kinetic response upon a potential perturbation. Namely, when the electrode potential (E) changes slowly, the electrochemical system may be able to maintain the condition of <4-17>; the surface concentrations, $C_O(0, t)$ and $C_R(0, t)$, are the values that are predicted (calculated) from <4-17>. Then, it appears to be a nernstian. At a very fast potential change, however, the charge transfer kinetics are not fast enough to adjust the surface concentrations to the values calculated from the Nernst equation. Then, the system appears to be electrochemical irreversible. In short, a single electrochemical system can appear as electrochemically reversible or irreversible depending on the rate of potential perturbation (change). Table 4-1 shows how the value of $C_O(0,t)/C_R(0,t)$ varies as the electrode potential changes. It can be seen that the ratio changes by 10^8 times upon an electrode potential change by 472 mV. Therefore, the nernstian refers to the case where the charge transfer kinetics is sufficiently fast to allow a huge change in surface concentrations even if the electrode potential changes fast. This means that the charge transfer kinetics are so fast that the surface concentrations adjust themselves to the equilibrium values regardless of how fast the electrode potential is perturbed.

Self learning 4-4

As can be seen in <4-12> and <4-15>, the current is determined by at least three variables: (1) k^0; the intrinsic charge transfer characteristic for a given electrochemical system, (2) electrode area (A), and (3) the applied electrode potential (E). Hydrogen evolution and oxidation ($2H^+ + 2e \rightleftarrows H_2$) currents appear to be very small on Hg electrode, but very large on Pt electrode. Which variable plays in deciding the currents? Which one can be said to be closer to the electrochemically reversible system? Sketch the hydrogen evolution/oxidation current that evolves on a Hg electrode in a wide positive and negative overpotential range.

4-1-5 Charge transfer and mass transfer

The Butler-Volmer equation describes the current-potential behavior for the electrochemical systems in which the charge transfer rate is slower than mass transfer. However, since the charge transfer rate increases by an exponential function of applied electrode potential, the charge transfer rate can exceed the mass transfer rate in large overpotential region. Then, the overall rate (current) is determined by the mass transfer rate. This point is illustrates in Fig. 4-6, in which the mass transfer rate is assumed to be constant. Even if the current is determined by the charge transfer rate in the small overpotential region, thereby the current changes according to the Butler-Volmer equation, the current appears to be constant due to a constant mass transfer rate at the large overpotential region.

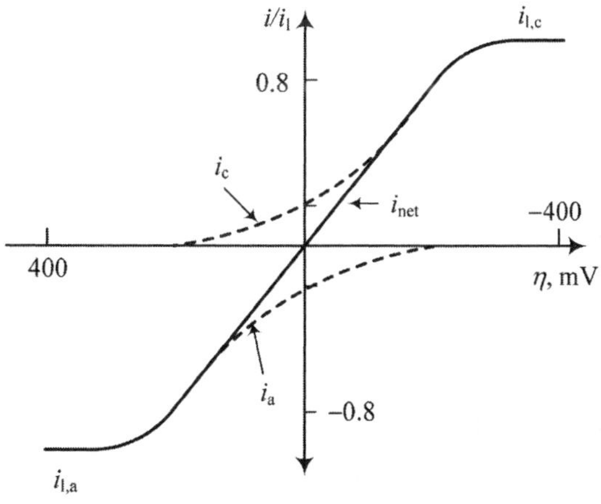

Fig. 4–6 Determination of current by charge transfer rate at a small overpotential region, but by mass transfer rate at a large overpotential region.

If k^0 is very small, one may simply assume that the electrochemical system is electrochemically irreversible and the charge transfer rate is slow to determine the overall rate (current); charge transfer limited. This is not always true; there is no direct correlation between electrochemical reversibility and rate-determining step (either mass transfer or charge transfer): the charge transfer is not always rate-determining even for the electrochemically irreversible systems. Namely, even if k^0 is very small, the charge transfer rate increases by the exponential function of applied potential. That is, even if the electrochemical system is electrochemically irreversible, the mass transfer can determine the overall rate at the large overpotential region, where the charge transfer rate exceeds the mass transfer rate. In opposite, even if the electrochemical system is electrochemically reversible with a large k^0, the charge transfer rate at the very small overpotential region can still be slower than the mass transfer rate; the charge transfer can determine the overall rate at a very small overpotential region even for electrochemically reversible systems. In short, one can judge the electrochemical reversibility on the basis of k^0, but the rate-determining step by considering both k^0 and applied electrode potential.

4-1-6 Tafel plot

When an electrochemical system is charge-transfer limited (i/V follows the Butler-Volmer equation), i_{net} observed in small overpotential region is linearly proportional to the applied potential as shown in Fig. 4-4 and Fig. 4-7. If $|\eta| < 118/n$ mV, the Butler-Volmer equation is converted into <4-18>.

$$e^x = 1 + x + \frac{x^2}{2} + \frac{x^3}{6} + \ldots, \quad e^x \cong 1 + x$$

$$i = i_0 \frac{F}{RT}(-\eta) \qquad \text{<4-18>}$$

In <4-18>, the observed current and the activation overpotential are linearly proportional. The slope between the current and potential in Fig. 4-7 corresponds to the inverse of resistance; $i = (1/R) \times V$. Here, since the Ohm's law ($V = iR$) holds between the current and electrode potential, the small overpotential region is called an ohmic region. Also,

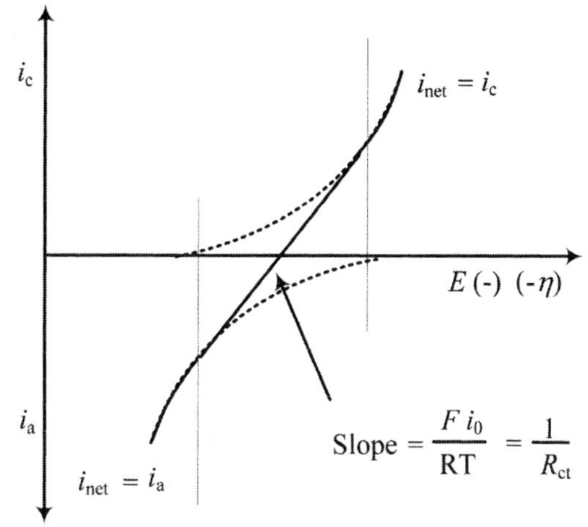

Fig. 4-7 Linear relationship between the activation overpotential and observed current (i_{net}) in a small overpotential region (ohmic region). Note that the currents represent the charge transfer rates.

since charge transfer determines the overall rate in this region; in most cases this assumption is true, the resistance is called a ***charge transfer resistance*** (R_{ct}), which is expressed as <4-19>. Note that R_{ct} and i_0 have an inverse relation. As discussed in Section 1-2, the electrochemical systems close to an ideally non-polarizable electrodes show $R_{ct} \rightarrow 0$. Since $i_0 \rightarrow \infty$ and also $k^0 \rightarrow \infty$ from <4-19>, one can say that the electrochemical systems showing ideally non-polarizable electrode behavior carry an electrochemical reversibility, and belong to nernstian. In opposite, the systems close to an ideally polarizable electrodes show an electrochemical irreversibility with $R_{ct} \rightarrow \infty$ and i_0 (k^0) $\rightarrow 0$. Also, as can be expected from Fig. 4-7, the slope in the i/V relation is steeper with a smaller R_{ct}. As can be deduced from Fig. 4-4, the activation overpotential is smaller if the slope is steeper with a smaller R_{ct} (larger i_0 and k^0).

$$R_{ct} = \frac{RT}{i_0 F} \qquad \text{<4-19>}$$

It is seen on Fig. 4-7 that $i_{net} \approx i_a$ and $i_{net} \approx i_c$ at the large overpotential regions. If $\eta < -(118/n)$ mV, $i_{net} \approx i_c$ is derived from the Butler-Volmer equation. <4-21> is obtained from <4-20>.

$$i_{net} \cong i_c = i_0 e^{-\alpha F\eta/RT} \qquad \text{<4-20>}$$

$$\log |i_{net}| = \log i_0 - \frac{\alpha F \eta}{2.303RT} \quad \text{<4-21>}$$

Similarly, <4-22> is obtained for η > (118/n) mV.

$$\log |i_{net}| = \log i_0 + \frac{(1-\alpha)F \eta}{2.303RT} \quad \text{<4-22>}$$

In Fig. 4-8, the log|i_{net}| is plotted against the activation overpotential. This is called a ***Tafel plot***. As predicted from <4-21> and <4-22>, log |i_{net}| is linearly proportional to η, which is described as a ***Tafel behavior***. In addition, the overpotential region where log|i_{net}| is linearly proportional to η ($i_{net} \approx i_a$ or $i_{net} \approx i_c$) is distinguished from the ohmic region by the ***Tafel region***. One can obtain i_0 from the intercept in the Tapel plot, from which k^0 is calculated by using <4-12>, and R_{ct} by using <4-19>. The symmetry factor (α) can be obtained from the slope.

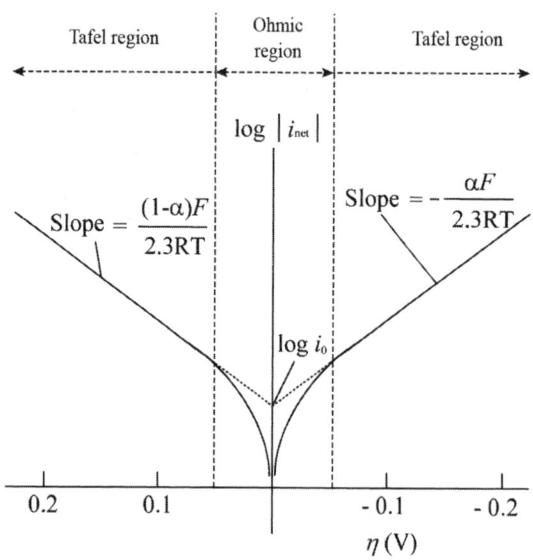

Fig. 4-8 The linear relation between the activation overpotential and log|i_{net}| (Tafel plot) in the large overpotential region (Tafel region).

Self learning 4-5

As explained in Chap. 1, the current changes according to the exponential function of electrode potential under the charge-transfer limited condition. Indicate the potential region in which the observed current changes exponentially. How big is the reverse current in this region?

4-1-7 Charge transfer kinetics for multi-electron reaction

In the charge transfer step for an electrochemical reaction $O + ne = R$, n electrons are not transferred at the same time. Rather, it is generally accepted that one electron is transferred step by step. For example, the electrochemical reaction involving two electrons, $Fe^{2+} + 2e = Fe$, proceeds through a chemical reaction (C) and two one-electron electrochemical reactions (E) as follows.

$$Fe^{2+}{}_{(aq)} + H_2O = FeOH^{+}{}_{(aq)} + H^{+}{}_{(aq)} \ (C)$$
$$FeOH^{+}{}_{(aq)} + e = FeOH_{(surface)} \ (E)$$
$$FeOH_{(surface)} + H^{+}{}_{(aq)} + e = Fe_{(s)} + H_2O \ (E)$$

The Butler-Volmer equation <4-23> is derived for ***multi-electron charge transfer reaction*** with some assumptions; the overall reaction is $A + ne = Z$ and n elementary steps for one-electron reaction are continued. Among these, there is a rate-determining step ($R + e \rightarrow S$), which can be repeated ν times.

$$
\begin{array}{c}
A + e = B \text{ [step 1]} \\
B + e = C \text{ [step 2]} \\
\bullet \quad \bullet \\
\bullet \quad \bullet \\
P + e = R \text{ [step } \gamma_c \text{]} \\
\nu (R + e \rightarrow S) \text{ [rds repeated } \nu \text{ times]} \\
S + e = T \\
\bullet \quad \bullet \\
\bullet \quad \bullet \\
Y + e = Z \text{ [step n]}
\end{array}
$$

$$i = i_0 [e^{-\alpha_c F\eta/RT} - e^{\alpha_a F\eta/RT}] \qquad \text{<4-23>}$$

In <4-23>, α_c and α_a have the following relationship.

$$\alpha_c = \frac{\gamma_c}{\nu} + \alpha, \quad \alpha_a = \frac{n - \gamma_c}{\nu} - \alpha, \quad \alpha_c + \alpha_a = \frac{n}{\nu}$$

Here, γ_c is the number of electrons transferred prior to the rate-determining step (rds), α is the symmetry factor in the rds and ν is the ***stoichiometric number*** that indicates how many times the rate determining step is repeated. The Tafel plots ($\log|i_{net}|$ *vs.* η) can be drawn by using <4-23>, from which the slopes for the reduction reaction and the oxidation reaction are obtained as follows.

Slope of reduction reaction: $\dfrac{-\alpha_c F}{2.3RT}$

Slope of oxidation reaction: $\dfrac{\alpha_a F}{2.3RT}$

Self learning 4-6

Prove that <4-23> is identical with <4-15> if $n = 1$.

Self learning 4-7

H_2 forms by combining two hydrogen atoms adsorbed on a metal electrode surface as follows in an acidic solution.

$$M_{(s)} + H_3O^+_{(aq)} + e \rightarrow M\text{-}H_{(surface)} + H_2O \quad \text{(slow)}$$

$$2M\text{-}H_{(surface)} \rightarrow 2M_{(s)} + H_2 \quad \text{(fast)}$$

Here, $M_{(s)}$ is the metal atom on electrode surface and $M\text{-}H_{(surface)}$ is the adsorbed hydrogen atom. Confirm that the stoichiometric number is two ($\nu = 2$).

Self learning 4-8

A Tafel plot was obtained for two-electron reaction, $Fe^{2+} + 2e = Fe$. The slope for the reduction and oxidation reaction was $-(40\ mV)^{-1}$ and $(120\ mV)^{-1}$, respectively. Identify the rate-determining step. Assume $\alpha = 0.5$.

4-2 Mass transfer rate

Under mass-transfer limited condition, the overall rate (current) is determined by the rate of mass transfer. The equality between mass transfer rate and current can be explained from <4-24>. The left term represents the mass transfer rate (J_i, flux), of which the unit is mol cm^{-2} s^{-1}, and the right term represents the current. The unit matches to each other, indicative of the equality.

$$J_O = \frac{i_c}{nFA}, \quad \left(\frac{\text{mol}}{\text{cm}^2\,\text{s}}\right) = \left(\frac{\text{C/s}}{\text{C/mol}\cdot\text{cm}^2}\right) \qquad \text{<4-24>}$$

The mass transfer rate of an ion i in an one-dimensional space is expressed as

$$J_i(x) = -D_i\frac{dC_i(x)}{dx} - \frac{z_iF}{RT}D_iC_i\frac{d\phi(x)}{dx} + C_iv(x) \qquad \text{<4-25>}$$

The first term on the right is the diffusion rate, the second the migration rate, and the third the convection rate. Using the Einstein relation <3-21>, the second term can also be expressed as a function of ionic mobility.

4-2-1 Diffusion

The diffusion rate in an one-dimensional space is expressed by <4-26>. The negative sign on the right side is due to the fact that diffusion proceeds to the direction opposite to the concentration gradient.

$$J_i(x) = -D_i\frac{dC_i(x)}{dx} \qquad \text{<4-26>}$$

The diffusion from higher to lower concentration is the phenomenological description. In a microscopic description, diffusion can be described by a ***random-walk model*** or a drunken sailor model, which illustrates that the probability for movement is identical for

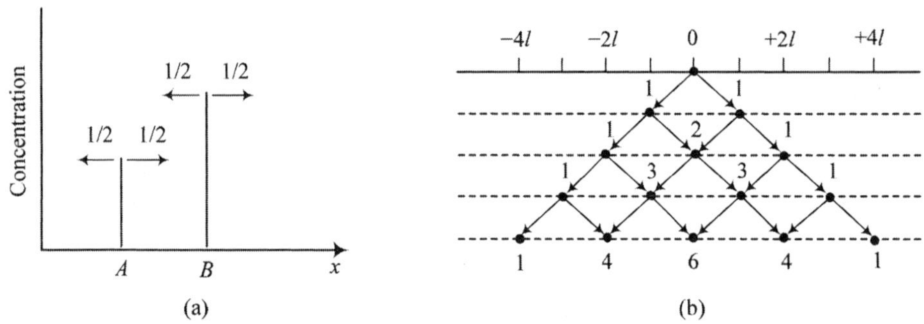

Fig. 4-9 (a) The microscopic point of view for diffusion and (b) Probability distribution by one-dimensional random-walk model.

both directions even if there is a concentration gradient for the to-be-diffused substance. Namely, even if there is a concentration difference at two points A and B in the one-dimensional space shown in Fig. 4-9-a, the substance has the same (50%) probability to move to the left and right. Even this, the absolute amount of substance moving from B to A is larger, so that the substance appears to move from B to A; from the higher to lower concentration region.

Now, consider where a drunken sailor (to-be-diffused substance) places after a certain time has elapsed. One assumption here is that the drunken sailor has the same probability to move back and forth (50% and 50%). Fig. 4-9-b shows the distribution of the position of the sailor as a function of time. When he walks four steps, the probability to find him is 1:4:6:4:1. This is a Gaussian distribution as shown in <4-27>.

$$\frac{N(x,t)}{N_0} = \frac{\Delta x}{2\sqrt{\pi Dt}} \exp\left(\frac{-x^2}{4Dt}\right) \qquad \text{<4-27>}$$

Now, how can his position be located? If one takes the average value, it is zero, which means that he does not move at all. However, the root-mean-square distance ($\overline{\Delta}$) is not zero with the unit of distance. Hence, $\overline{\Delta}$ is utilized to specify the diffusion distance. <4-28> is called the ***Einstein-Smoluchowski equation.*** It is seen that the larger the diffusion coefficient D is and the longer the time elapsed for diffusion (t), the longer the distance ($\overline{\Delta}$) is to be diffused.

$$\overline{\Delta} = \sqrt{2Dt} \qquad \text{<4-28>}$$

Example 4-3

The diffusion coefficient (D) of a substance in aqueous solution has a value of approximately $10^{-5} \sim 10^{-6}$ cm^2/s. Assuming $D = 10^{-5}$ cm^2/s, calculate the diffusion length for 1 s, 1 min, and 1 h using <4-28>. (a) From these results, consider what measures are needed in terms of mass transfer to increase the electrolysis rate when performing industrial electrolysis. (b) Voltammetry experiments performed in laboratory are usually carried out within 1 min. From the calculation, one-dimensional (linear) diffusion can be assumed for the substances participating in an electrochemical reaction. Explain why this is possible.

Solution

Diffusion length ($\overline{\Delta}$) = 4.5×10^{-3} cm for 1 s, 3.5×10^{-2} cm for 1 min, and 2.7×10^{-1} cm for 1 h.

(a) According to the above calculation, if there is no stirring during industrial electrolysis, the reactants can diffuse only 2.7 mm for 1 h. Therefore, mechanical convection such as stirring or pumping is necessary to increase the electrolysis rate.

(b) When a voltammetry experiment is conducted without stirring using a planar microelectrode, the experiment time is approximately 1 min or less. As an example, when performing the potential-step experiment described in Fig. 1-6, according to the above calculation, the diffusion layer expands 3.5×10^{-2} cm for 1 min. In other words, if the experiment is conducted for 1 min, only the reactants populated within 3.5×10^{-2} cm from the electrode surface can participate in the reaction. This feature has been explained in Fig. 1-7, which shows that the thickness of diffusion layer, which is the range where the reactants participating in the reaction are populated in the absence of stirring, is less than 0.1 cm. Since the diffusion layer expands to a very short distance from the electrode surface (since the ratio of the diffusion layer thickness/electrode surface area is very small when a planar microelectrode is used), one-dimensional (linear) diffusion rather than three-dimensional (radial) diffusion can be assumed.

4-2-2 Diffusion current and migration current

Let us assume that there is no convection in a mass-transfer limited electrochemical reaction. Then, the current, which corresponds to the mass transfer rate, is the sum of the current by diffusion and migration. The ***diffusion current*** (i_d) and ***migration current*** (i_m) are expressed by using the diffusion flux (J_d) and migration flux (J_m) as <4-29> by using the relation in <4-24>.

$$i_{total} = i_d + i_m$$

$$J_d = \frac{i_d}{nFA}, \quad J_m = \frac{i_m}{nFA} \qquad \text{<4-29>}$$

Consider that Cu plating is performed by reducing Cu^{2+}, $(Cu(CN)_4{}^{2-}$, or $Cu(CN)_2$ as shown in Fig. 4-10. Note that the copper species are cation, anion, and neutral one, respectively. For the Cu plating, the working electrode is polarized to a very negative value and thus the concentration of Cu species is zero on electrode surface. Under this condition, the Cu species tend to diffuse to the negative electrode because of the concentration difference between the bulk solution and electrode surface. That is, the three Cu species diffuse to the working (negative) electrode. However, the direction of ion migration depends on the Cu species as shown in Fig. 4-10. Note that the working electrode is negatively polarized while the counter electrode positively. Now, the ions in the solution migrate by the imposed electric field. The cation (Cu^{2+}) migrates to the negative electrode (working electrode), whereas the anion ($Cu(CN)_4{}^{2-}$) migrates to the positive (counter) electrode. Obviously, the neutral Cu species ($Cu(CN)_2$) can not migrate. As a

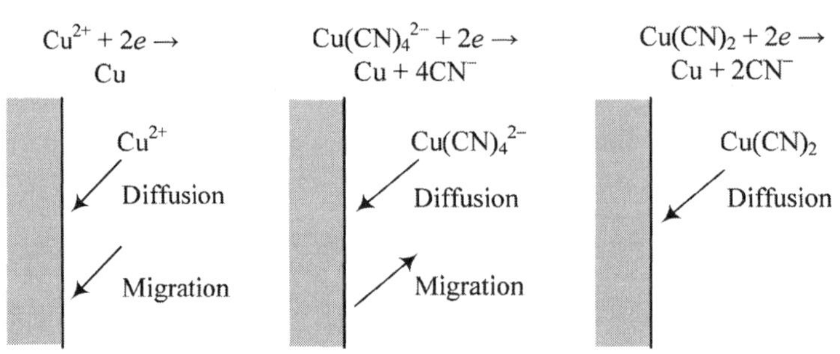

Fig. 4-10 Contribution of migration current according to the type of copper species (cation, anion and neutral one).

result, the total flux is the sum of diffusion and migration for the cation (Cu^{2+}), whereas it the difference between the two for the anion ($Cu(CN)_4^{2-}$).

As discussed, if the electroactive species (O) for $O + ne \rightarrow R$ is ions like Cu^{2+} and $Cu(CN)_4^{2-}$, their migration always prevails, such that the migration current can not be ignored under mass transfer limiting condition. However, if the contribution of migration can be neglected by any ways, which will be discussed shortly, only diffusion is possible; the overall rate (current) is decided by the diffusion rate. Then, the reduction current of O is expressed by the diffusion flux as shown in <4-30>.

$$\text{Rate} = -J_O(0,t) = \frac{i_d}{nFA} = D_O \left. \frac{dC_O(x,t)}{dx} \right|_{x=0}$$ <4-30>

Here, D_O is the diffusion coefficient of O and the other term is the concentration gradient of O at the electrode surface ($x = 0$). <4-30> illustrates that the diffusion current can be calculated (predicted) if one knows the concentration gradient near electrode surface (diffusion layer) under mass-transfer limited condition. Fig. 4-11 shows the movement of electrons and ions forming a closed loop in an electrolytic cell when the diffusion rate of O determines the overall rate. Since the reduction current (i_c) is determined by the diffusion rate of O, it corresponds to the diffusion current (i_d), and the same magnitude of oxidation current (i_a) flows in the opposite electrode to form a closed loop.

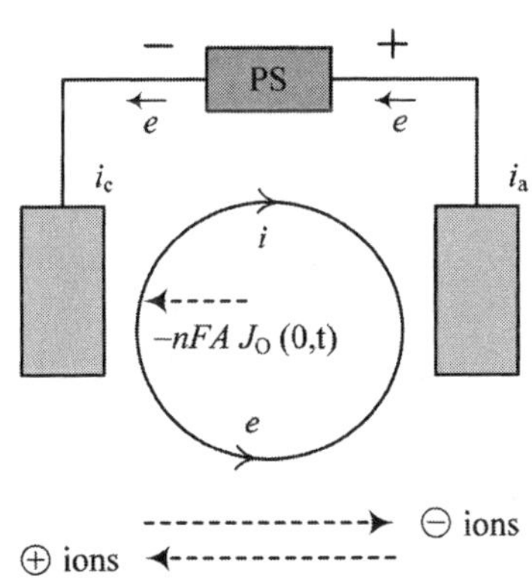

$-nFA\, J_O(0,t) = i_d = i_c = |i_a| = i$ (electrolyte)

Fig. 4-11 Formation of a closed loop when the diffusion of O determines the overall electrochemical reaction rate (current).

Then, a question is how we can discard the migration of ionic species. The answer is to add excessive amount of ions (as a supporting electrolyte) into electrolysis cell. This is explained by using a ***balance-sheet approach*** in Fig. 4-12 and Fig. 4-13. Let us consider that the reduction of H^+ and oxidation of Cl^- are performed in an electrolytic cell containing 10^{-3} M HCl solution. Let us assume that ten H^+ are electrochemically reduced per unit time to produce five H_2, then ten electrons are required. For a closed-loop formation, ten electric charges per unit time must move by ion migration in the solution. In the above electrolysis condition, only H^+ and Cl^- are present in the solution, thereby these two ions should participate in ion migration. Since the cation transport number (t_+) in aqueous HCl solutions is about 0.8, eight out of ten electric charges are transferred by migration of H^+ but two by Cl^- migration. Here, the movement of H^+ toward the negative electrode has the same effect as the movement of Cl^- toward the positive electrode, so that ten negative charges move counterclockwise in the solution to form a closed loop. Ten H^+ are required for the electrochemical reaction; eight H^+ are supplied by migration as discussed above, and thus two H^+ by diffusion. Similarly, two Cl^- are supplied by migration, and eight by diffusion. In short, the migration of electroactive species (H^+ and Cl^-) can not be neglected in this condition.

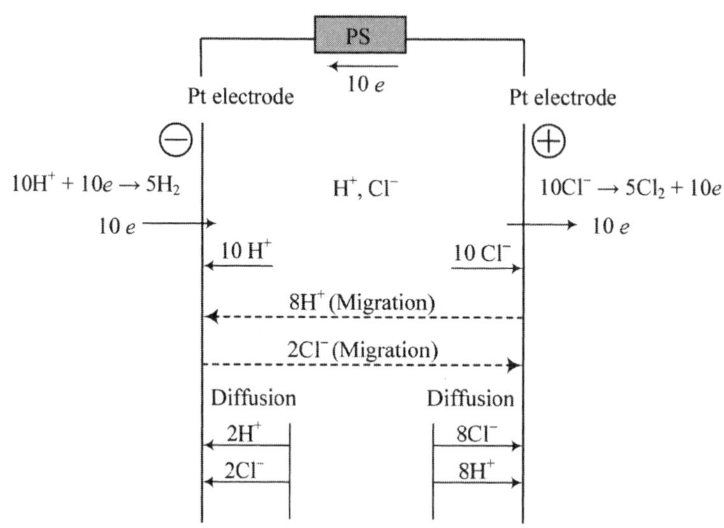

Fig. 4-12 Contribution to mass transfer by diffusion and migration for the reactants (H^+ and Cl^-) in the reduction of H^+/oxidation of Cl^-, which proceeds by forming a closed loop (in the absence of supporting electrolyte).

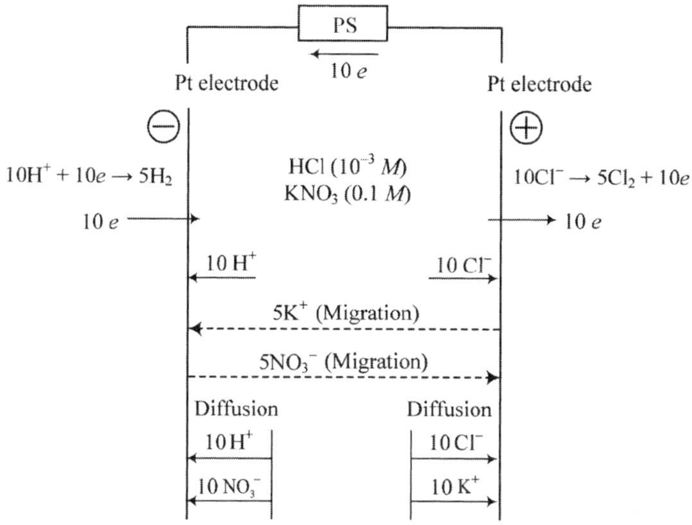

Fig. 4-13 Contribution to mass transfer by diffusion and migration for the reactants (H^+ and Cl^-) in the reduction of H^+/oxidation of Cl^-, which proceeds by forming a closed loop (excess KNO_3 is added as a supporting electrolyte).

Assume that excess amount of KNO_3 (0.1 *M*) is added into the above solution (Fig. 4-13). Note that the concentration of K^+ and NO_3^- is 100 times larger than the concentration of H^+ and Cl^-, such that most of charges (ten) are transferred by migration of K^+ and NO_3^- in the solution. Note here that all ions can migrate and the migration of K^+ and NO_3^- dominant because their concentration is 100 times larger than that for H^+ and Cl^-. Further, since $t_+ = t_- = 0.5$ in aqueous KNO_3 solution, five K^+ and five NO_3^- participate in the migration. Then, ten H^+ and ten Cl^- should be supplied by diffusion. That is, there is no migration for the electroactive species (H^+ and Cl^-) because the excessive amount of other ions is responsible for ion migration.

A solution containing ions in an electrochemical cell is called a ***supporting electrolyte*** (or simply an electrolyte), which plays two roles. One is to reduce the solution resistance (and separator resistance if any) in electrochemical cells. The other is to participate in ion migration in solution. In this case, the migration of *O*, even if it is an ionic species, can be discarded; The diffusion rate of *O* appears as a reduction current. Since the diffusion rate of *O* corresponds to the reduction current (as in <4-30>), the current can be calculated from the concentration gradient of *O* near electrode surface (diffusion layer). This principle will be utilized in Chap. 6~8 to predict the *i vs. V* behavior.

Problems

01 In Example 4-2 and Fig. 4-4, calculate the applied potential to have a reduction current of 10.0 μA.

02 With stirring, the reduction current according to the applied electrode potential is obtained as follows. The electrode area is 0.1 cm^2, the concentration of O and R is 0.01 M, and n = 1. Calculate i_0, α, k^0, and R_{ct}.

η (V)	−0.1	−0.12	−0.15	−0.5	−0.6
i (A)	4.6×10^{-5}	6.3×10^{-5}	1.1×10^{-4}	9.7×10^{-4}	9.7×10^{-4}

03 The exchange current for $M^+ + e = M$ is obtained as follows as a function of the concentration of M^+. Calculate α.

C_{M+}, 10^{-3} M	1.0	0.5	0.25	0.1
I_0, A/m^2	5.4	3.7	2.6	1.6

04 The exchange current (i_0) for $Fe(CN)_6^{3-} + e = Fe(CN)_6^{4-}$ is 2.0 mA in Pt/$Fe(CN)_6^{3-}$ (2.0 mM), $Fe(CN)_6^{4-}$ (2.0 mM), NaCl (1.0 M) cell. Calculate k^0 and R_{ct}. A = 1.0 cm^2 and α = 0.5.

05 Calculate the applied electrode potential for the Cu electroplating performed with a current of 20 A in a sulfuric acid solution containing 1.0 M Cu^+.

(a) $E^{0'}$ = 522 mV (*vs.* NHE) for Cu/Cu^+ half-cell, i_0 = 1.0 A and the slope of Tafel plot is (120 mV)$^{-1}$

(b) 5.0 M of CN^- is added into the sulfuric acid electrolyte to form $Cu(CN)_4^{3-}$

(the formation constant = 10^{20}), i_0 = 0.1 A, the slope of Tafel plot is (120 mV)$^{-1}$.

06 Assuming that copper dissolution/plating proceeds by two steps, determine the rate-determining step from the results shown below.

$$Cu^{2+} + e = Cu^{+}$$
$$Cu^{+} + e = Cu$$

η(V)	Reduction current density (A/m^2)	η(V)	Oxidation current density (A/m^2)
−0.10	6.8	0.01	−1.0
−0.09	5.6	0.03	−5.1
−0.07	3.8	0.05	−17.4
−0.05	2.6	0.07	−56.0
−0.03	1.6	0.09	−178.0
−0.01	0.7	0.10	−316.1

07 Consider that a molecule diffuses in a long cylindrical tube with a diameter of 2.0 mm. Calculate the diffusion coefficient of this molecule if it moves by 0.025×10^{-9} mol/s in a plane with a concentration gradient of 5.0 *M*/m.

08 The evolution and reduction of oxygen ($O_2 + 4H^+ + 4e = 2H_2O$) involve four electrons. The Tafel slope for the oxidation and reduction reaction is (38 mV)$^{-1}$ and −(118mV)$^{-1}$, respectively.

(a) Identify the rate-determining step.

(b) Calculate the stoichiometric number (v) with α = 0.5. Explain the meaning of v in your answer.

CHAPTER 05

Electrochemical instrumentations

The device that measures current by controlling electrode potential is called a potentiostat, whereas the device for measuring electrode potential by controlling current is called a galvanostat. In a two-electrode (working and counter electrode) cell, the potential of working electrode is controlled against the counter electrode using a potentiostat, and the current flowing between the working and counter electrode is traced. In a three-electrode (working, counter and reference electrode) cell, the potential of working electrode is controlled by using a potentiostat with respect to the reference electrode. The current flowing through the working and counter electrode is measured. Note that there is no current flow to the reference electrode. For controlled-current experiments, the galvanostat is used. A predetermined current is imposed between the working and counter electrode by using a galvanostat, and the potential between the working and reference electrode is measured.

Three-electrode cells are used in laboratory for voltammetry experiments, for which a precise control/measurement of potential (or current) is required. In industrial electrolysis, however, two-electrode cells are used. The control/measurement of potential/current is not precise compared to the three-electrode cell, but it is space-effective without reference electrode.

The operational amplifier (OP amp) is the key component in the electric circuits for potentiostat and galvanostat.

5-1 Working principle of OP amp

OP amp is fabricated in the form of a chip and consists of an integrated circuit combining a large number of resistors, capacitors, transistors, and other electronic components. The configuration of OP amp is displayed in Fig. 5-1. There are two inputs, into which the input signals are coming in; inverting input and non-inverting input. There is one output, through which the output signals after being processed by the OP amp are coming out. For the operation, +15 V and −15 V are required, which is called a power supply. The

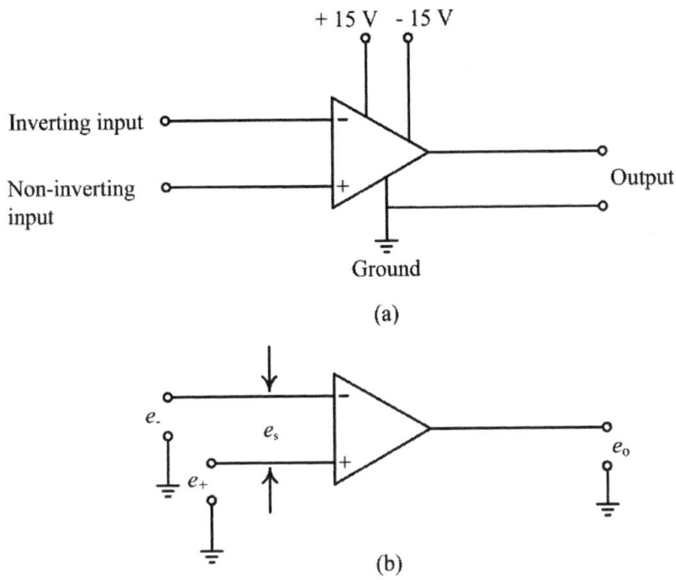

Fig. 5–1 Schematic diagram of operational amplifier (OP amp).

output voltage and waveform are determined by two input signals. The maximum output voltage (e_o) is the voltage deliverable by the power supply (practically ±13~14 V), such that e_o has the value anywhere between the voltage limits; from −13~14 V to +13~14 V.

The output voltage (e_o) is expressed as <5-2>, where A is the open-loop gain, and e_- and e_+ are the inverting and non-inverting input voltage, respectively. All voltages are referenced to the ground.

$$e_s = e_- - e_+ \qquad \text{<5-1>}$$

$$e_0 = -Ae_s = -A(e_- - e_+) \qquad \text{<5-2>}$$

The operational amplifier has the following operation characteristics.

(1) The open-loop gain (A) has a very large value of $10^5 \sim 10^7$. Assuming that the output voltage limit is +15 V and A is 10^6, it is calculate that 15 V = $-10^6 \times e_s$ and $e_s = -15 \times 10^{-6}$ V. It can be said that $e_s \approx 0.0$ V, which means that the two input terminals must be at the same voltage ($e_- \approx e_+$). In real, two input voltages are not identical ($e_- \neq e_+$). Hence, in order for OP amp to operate under the condition of $e_- \approx e_+$, the part of output should be fed back to the inverting input.

Two methods are employed; current feedback and voltage feedback.

(2) Input impedance is very high, $10^5 \sim 10^{13}$ Ω; the current flowing into the OP amp can be ignored.

(3) Output impedance is very small, ≈ 0 Ω; the output current going out of the OP amp is not zero.

(4) Not only DC but also AC signals can be supplied as the input.

5-1-1 Operation by current feedback

It is noted above that a very small voltage difference at the inputs (e_s ≈ 0.0 V) drives OP amp to its voltage limits, and the OP amp is stabilized by two feedbacks. In the ***current feedback*** operation, the current is routed from the output to the inverting input. An example of current feedback is shown in Fig. 5-2. When the input current (i_{in}) flows, the current at the point S must satisfy the ***Kirchhoff current rule*** <5-3>. Here, it is defined that the current input to S is positive (i_{in} and i_f), whereas the current output from S is negative (i). Since the input impedance is so large that the current (i) entering the OP amp is assumed to be zero. Then, $i_f = -i_{in}$ in <5-3> and <5-4> is derived. In <5-4>, the output voltage (e_o) is the product of the input current (i_{in}) and the feedback resistance (R_f). This circuit is called a ***current-to-voltage converter*** (i/V converter or current follower), which converts the current (i_{in}) into a voltage (e_o).

$$\sum i = 0$$

$$i_{in} - i + i_f = 0 \qquad \text{<5-3>}$$

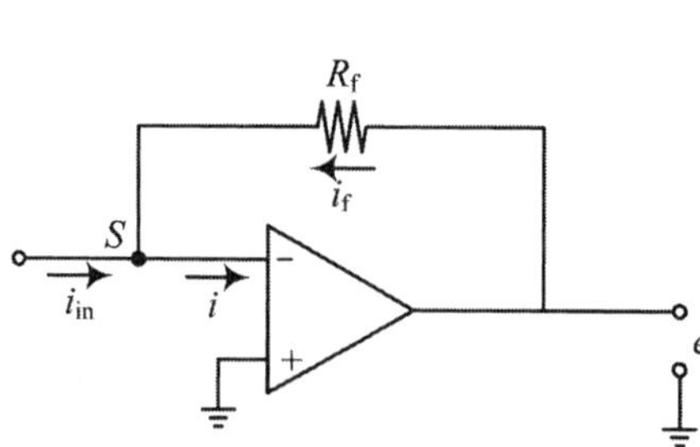

Fig. 5-2 A current-to-voltage converter (i/V converter or current follower).

$$\frac{e_o - e_S}{R_f} = -i_{in}$$

$$e_o\left(1+\frac{1}{A}\right) = -i_{in}R_f$$

$$e_o = -i_{in}R_f \qquad \text{<5-4>}$$

Fig. 5-3 shows the circuit for a scaler, wherein the voltage (e_i) is the signal to the inverting input. Here, <5-3> is also valid and thus <5-5> and <5-6> are derived. In <5-6>, the output voltage (e_o) is determined by the product of the input voltage (e_i) and R_f/R_i. If R_f/R_i = 100, the output voltage is 100 times larger than the input voltage with the sign being changed. This is called a ***scaler***. If R_f/R_i = 1, only the voltage sign is changed, thus called an inverter.

$$i_{in} = -i_f \qquad \text{<5-5>}$$

$$\frac{e_i - e_S}{R_i} = -\left(\frac{e_o - e_S}{R_f}\right)$$

$$e_o = -e_i\left(\frac{R_f}{R_i}\right) \qquad \text{<5-6>}$$

Meanwhile, the point S in Figs. 5-2 and 5-3 is called a ***summing point*** or virtual ground. In both circuits, the non-inverting input is grounded (e_+ = 0), so the inverting input (e_- and the voltage at the point S) must also be zero (e_- = e_+ = 0). That is why the summing point S is called the ***virtual ground***.

In Fig. 5-4, three voltage signals (e_1, e_2, and e_3) are provided to the inverting input.

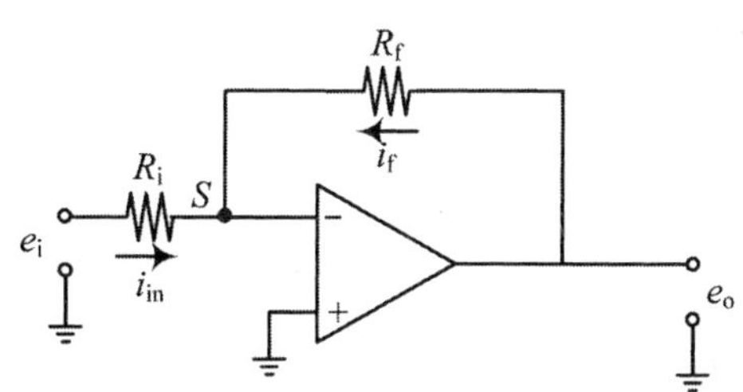

Fig. 5-3 A scaler.

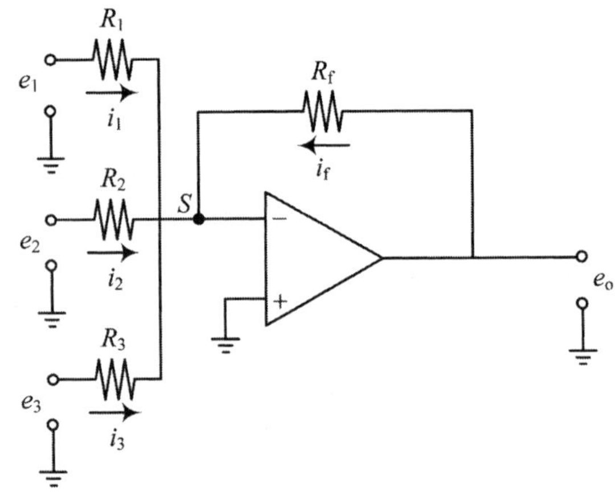

Fig. 5-4 An adder.

The output voltage (e_o) is given as the sum of the three input voltages as in <5-7>. Therefore, it is called an ***adder***.

$$i_f = -(i_1 + i_2 + i_3)$$

$$\frac{e_o}{R_f} = -\left(\frac{e_1}{R_1} + \frac{e_2}{R_2} + \frac{e_3}{R_3}\right)$$

$$e_o = -\left[e_1\left(\frac{R_f}{R_1}\right) + e_2\left(\frac{R_f}{R_2}\right) + e_3\left(\frac{R_f}{R_3}\right)\right] \quad \text{<5-7>}$$

OP amp can be utilized for analog-type mathematical operations. An example of multiplication is shown in Fig. 5-3, and an example of addition is shown in Fig. 5-4; if $R_f = R_1 = R_2 = R_3$ in <5-7>, the output voltage is the sum of three input voltages. In addition, subtraction, division, differentiation, and integration are possible. The circuit that integrates current is shown in Fig. 5-5. <5-8> is satisfied in this circuit, so that <5-9> is derived. This circuit integrates the input current for a given time and converts it into a voltage value.

$$C\frac{de_o}{dt} = -i_{in} \quad \text{<5-8>}$$

$$e_o = -\frac{1}{C}\int i_{in}\,dt \quad \text{<5-9>}$$

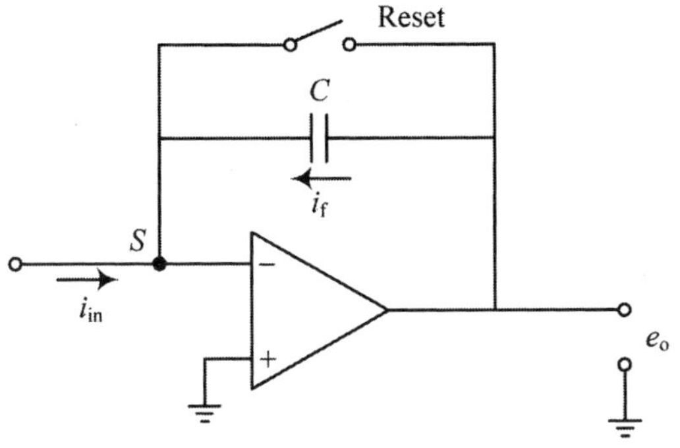

Fig. 5–5 A current integrator.

5-1-2 Operation by voltage feedback

The ***voltage follower*** shown in Fig. 5-6 is an example that operates by satisfying the condition of $e_- \cong e_+$ by voltage feedback. From <5-10>, <5-11> is derived. The output voltage is the input voltage. It is used to measure the voltage accurately without the interference of iR drop of the measuring device, or to perform mathematical operations after taking the voltage value (e_i) without errors.

$$e_o = -A(e_o - e_i)$$

$$e_o = \frac{e_i}{(1+1/A)} \qquad \text{<5-10>}$$

$$e_o \approx e_i \qquad \text{<5-11>}$$

Another example of voltage feedback is shown in Fig. 5-7. Here, since the voltage of the non-inverting input is grounded ($e_+ = 0$), $e_- = 0$. Therefore, $e_- = e_A + e_i = 0$,

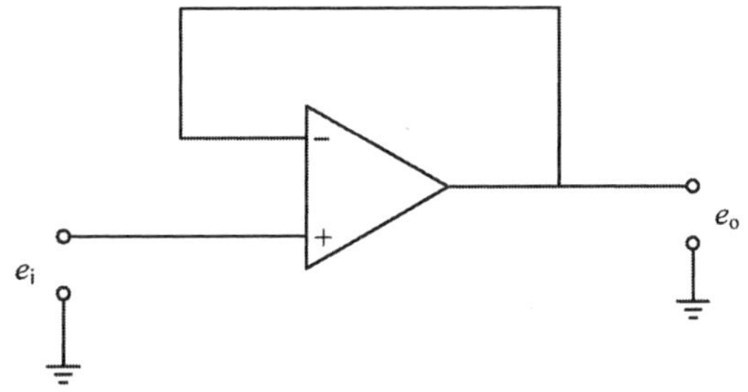

Fig. 5–6 A voltage follower.

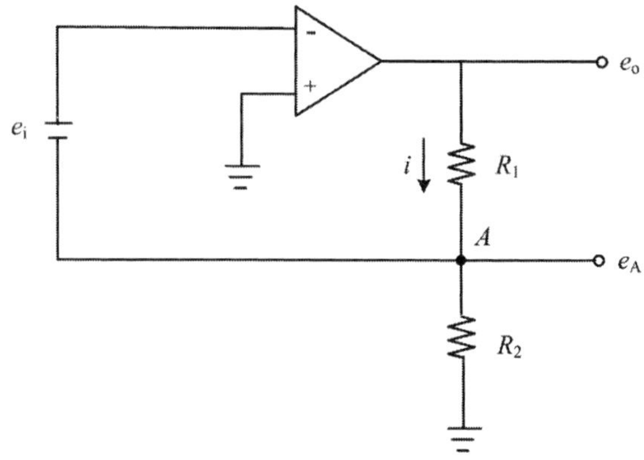

Fig. 5-7 Circuit that can control the voltage at the point A.

and $e_A = -e_i$. Thus, the voltage at A can be controlled by changing e_i. For example, if $e_i = 1.0$ V, $e_A = -1.0$ V. Since e_A is controlled by e_i, the current is determined by <5-12>.

$$i = -e_i/R_2$$ <5-12>

Self learning 5-1

The voltage (e_A) at point A can be controlled by changing e_i in the circuit in Fig. 5-7. Sketch the waveform of e_A with $e_i = V_p \sin 2\pi ft$ (f = 60 Hz, V_p = 2.0 V).

5-2 Potentiostat

The circuit in Fig. 5-7 can be applied to a 3-electrode cell, and the potential of reference electrode can be controlled to be $-e_i$ with respect to the working electrode being grounded as shown in Fig. 5-8. However, since the voltage is relative, the voltage of working electrode is controlled to be e_i with respect to the reference electrode.

$$e_{wk}(vs.\ \text{ref}) = e_i$$

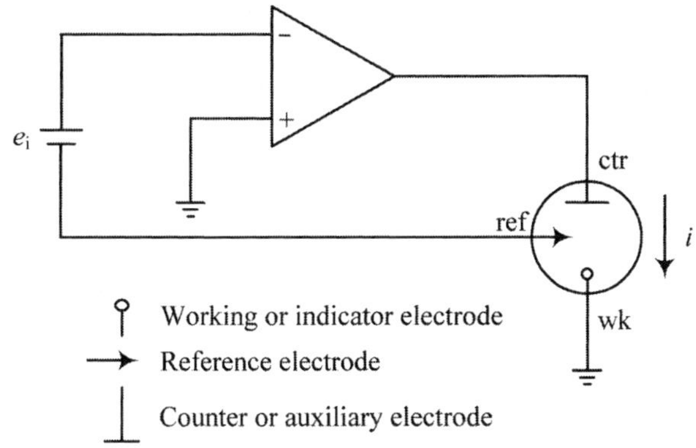

Fig. 5-8 A simple potentiostat based on the circuit shown in Fig. 5-7.

Fig. 5-9-a shows the equivalent circuit of a 3-electrode cell. The parallel connection of R_{ct} and C_{dl} shown in Fig. 1-14 is represented as Z_{ctr} and Z_{wk} for the electrochemical reactions on the counter and working electrode, respectively. The electrolyte resistance is divided into R_u and R_Ω because the reference electrode is placed between the two electrodes. This circuit can be further simplified as in Fig. 5-9-b. A careful inspection of Figs. 5-7, 5-8, and 5-9-b reveals that the circuits bear a strong analogy to each other; e_o in Fig. 5-7 corresponds to the voltage of counter electrode in Fig. 5-8, e_A is the voltage of reference electrode, and the working electrode is grounded. Also, R_1 and R_2 correspond to Z_1 and Z_2, respectively. Since the voltage of counter electrode is expressed by <5-13>

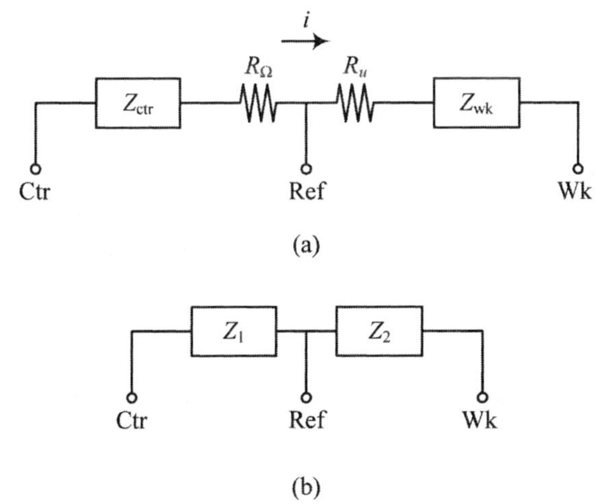

Fig. 5-9 The impedance network for 3-electrode cell.

and the current flowing from the counter to working electrode is given as <5-14>, the voltage of counter electrode (e_o) against the working electrode (grounded) is given by <5-15>.

$$e_o = i(Z_1 + Z_2) \qquad \text{<5-13>}$$

$$i = -e_i / Z_2 \qquad \text{<5-14>}$$

$$e_o = -e_i \left(\frac{Z_1 + Z_2}{Z_2} \right) \qquad \text{<5-15>}$$

The operating principles of three-electrode cells are as follows. Firstly, as described above, the potential of reference electrode is controlled against the grounded working electrode, but this has the same effect as controlling the potential of working electrode with respect to the reference electrode. Secondly, the current is determined by <5-14>, which manifests itself that only e_i and Z_2 decide the current while Z_1 is idling. This means that the current is determined only by the R_{ct} and C_{dl} of the working electrode, irrespective of the R_{ct} and C_{dl} of the counter electrode. The counter electrode has no influence on current size, but only plays an auxiliary role by giving or taking electrons through the oxidation and reduction reaction, which are the opposite reaction occurring on the working electrode. Therefore, the counter electrode is also referred to as an ***auxiliary electrode***. Thirdly, current flows only between working and auxiliary electrode. There is no current flow to reference electrode. Fourthly, the area of auxiliary electrode is designed to be larger than that of working electrode. If this condition is not met, frequently, the current flow through the auxiliary electrode may be smaller than that decided by e_i and Z_2, which are the characteristic between working and reference electrode. In another word, the counter electrode with a large surface area is used to avoid the current decision by the electrochemical characteristics of counter electrode (Z_1). Fifthly, the voltage of counter electrode is determined by <5-13>, and has a relation of <5-15> with the voltage (e_i) applied between reference and working electrode. Here, e_i is called the ***control voltage*** and e_o is called the ***compliance voltage***, which means a voltage difference between working and counter electrode. The commercial potentiostats have their own specifications on the control voltage and compliance voltage.

Self learning 5-2

If the electrolyte of an electrochemical cell has a large resistance or a large current flows in the cell, it is necessary to use a potentiostat that can provide a large compliance voltage. Explain why.

The most widely used ***potentiostat*** employs the adder circuit shown in Fig. 5-11. If the waveform shown in Fig. 5-10, which is the sum of three signals, is provided as an input, the three waveforms enter to e_1, e_2, and e_3 in the adder circuit. That is, if all the resistors have the same value in <5-16>, the applied voltage of working electrode with respect to a reference electrode is the sum of three signals as shown in <5-17>.

$$-i_{\mathrm{ref}} = i_1 + i_2 + i_3$$ <5-16>

$$-e_{\mathrm{ref}} = e_1\left(\frac{R_{\mathrm{ref}}}{R_1}\right) + e_2\left(\frac{R_{\mathrm{ref}}}{R_2}\right) + e_3\left(\frac{R_{\mathrm{ref}}}{R_3}\right)$$

$$e_{\mathrm{wk}}\,(vs.\ \mathrm{ref}) = e_1 + e_2 + e_3$$ <5-17>

In the circuit of Fig. 5-11, the voltage of the reference electrode is controlled against the working electrode that is grounded, but this has the same effect of a voltage control of the working electrode with respect to the reference electrode. Meanwhile, the voltage of reference electrode with respect to the working electrode (grounded) is read by using the voltage follower, and the current flowing between the counter and working electrode is read as a voltage value using the *i*/*V* converter.

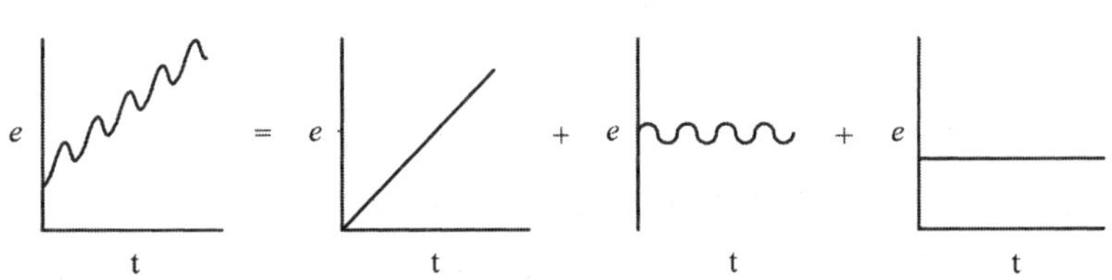

Fig. 5-10 Generation of a complex waveform by adding three input signals.

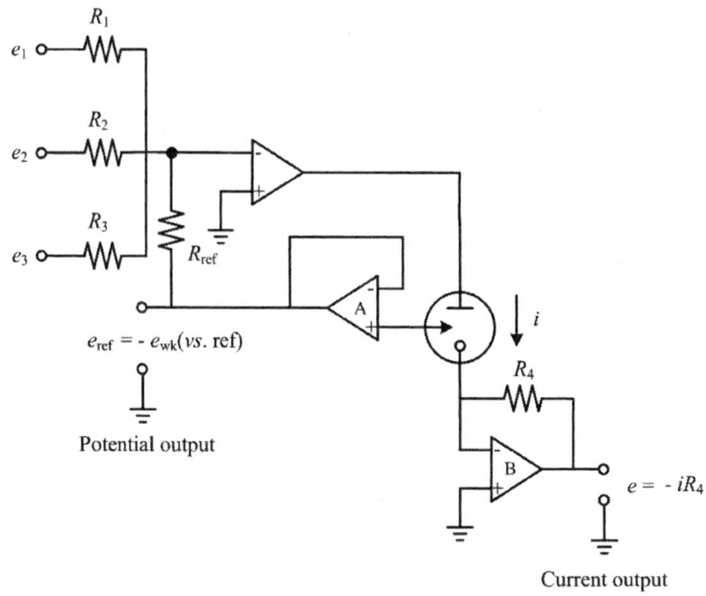

Fig. 5-11 An adder-type potentiostat.

Self learning 5-3

What is the role of OP amps *A* and *B* in the potentiostat shown in Fig. 5-11?

5-3 Galvanostat

Galvanostat is a device that reads the voltage change of a working electrode relative to a reference electrode by controlling the current flowing between a working and counter electrode. Fig. 5-12 shows the OP amp circuit for a galvanostat. Since the current has the relation of <5-18> at the summing point (*S*) by the Kirchhoff current rule, the current (i_{cell}) can be controlled by changing e_i. The voltage of reference electrode is read against the virtually grounded working electrode. This is the same as reading the voltage of work-

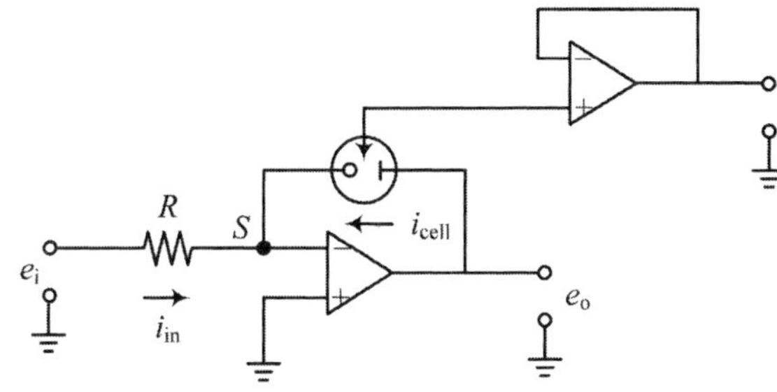

Fig. 5-12 A simple galvanostat.

ing electrode with respect to the reference electrode.

$$i_{cell} = -i_{in} = \frac{-e_i}{R} \qquad \text{<5-18>}$$

5-4 iR_u drop due to solution resistance

Fig. 5-13 shows the equivalent circuit for three-electrode cells. Since a reference electrode is placed in the electrolyte solution, the solution resistance is divided into R_u, which is the solution resistance between the reference and working electrode, and R_Ω, which is the solution resistance between the reference and counter electrode. If the working electrode is grounded and the voltage of the reference electrode is controlled as $e_1 + e_2 + e_3$, this gives the same effect as controlling the working electrode potential to $-(e_1 + e_2 + e_3)$ against the reference electrode. Here, the voltage (e_{true}) that is actually applied to the working electrode is smaller by iR_u due to the voltage drop. Therefore, e_{true} is given by

$$e_{true}\ (vs.\ \text{ref}) = |e_1 + e_2 + e_3| - |iR_u| \qquad \text{<5-19>}$$

In electrochemical reactions, the current changes sensitively by the applied voltage; when charge transfer determines the overall rate, the current varies by the exponential function of applied voltage. Therefore, if the difference is substantial between the apparent voltage

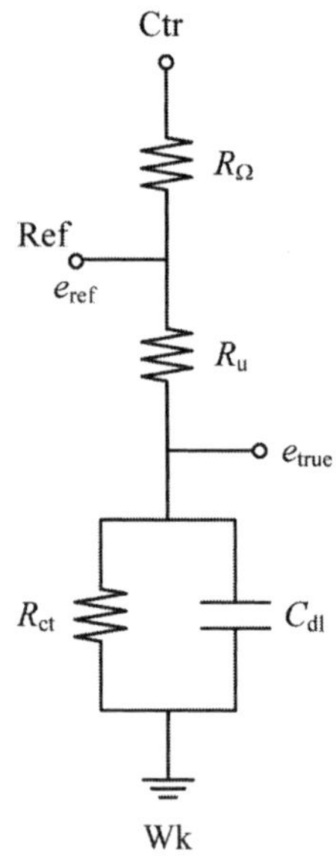

Fig. 5–13 Equivalent circuit for 3–electrode cell.

(e_{ref}) and actual voltage (e_{true}), it is difficult to obtain a correct current/voltage relationship. As shown in <5-19>, the larger the value of iR_u, the greater the difference between the two voltages. R_u is called ***uncompensated resistance***. To minimize R_u, the reference electrode is placed as close as possible to the working electrode by using a Luggin capillary. In addition, the modern potentiostats have the function of positive feedback compensation, in which iR_u is automatically compensated by using a specially designed electric circuit.

5-5 Two-electrode cell and three-electrode cell

Electrochemical cells are divided into ***two-electrode cell*** and ***three-electrode cell***. Fig. 5-14 shows the configuration of two-electrode cell. Here, the counter electrode also functions as a reference electrode. A potential is applied between two electrodes, and the current is also measured between two electrodes. Here, as the current flows to the reference electrode, the potential control is not accurate. However, as this configuration is space-effective, most commercial electrolysis cells adopt this, in which the potential control is not so demanding and the room is not so spacious to add reference electrode. On the other hand, three-electrode cells are used for laboratory experiments. Here, the potential of working electrode is precisely controlled against a reference electrode, and the current between the working and counter electrode is measured as shown in Fig. 5-15. Since no current flows through the reference electrode, the potential of reference electrode is not changed, such that it plays as a solid reference for potential control. Since the potential of working electrode is accurately controlled, one can obtain the accurate current-potential relation, from which one can obtain the electrochemical parameters (for example, R_{ct}, i_0 and D) and investigate reaction mechanisms.

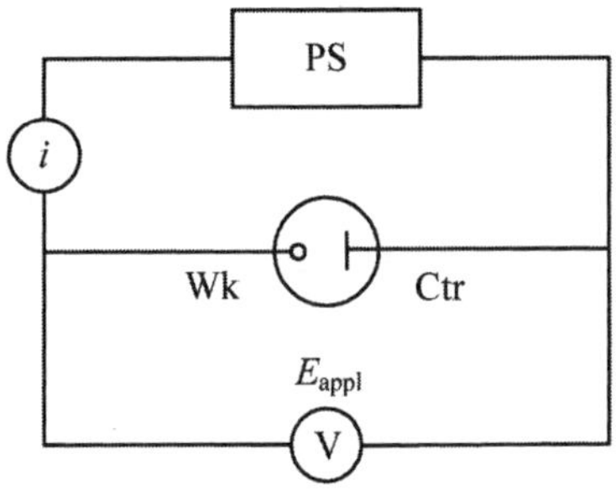

Fig. 5-14 Potential control and current measurement in two-electrode cell.

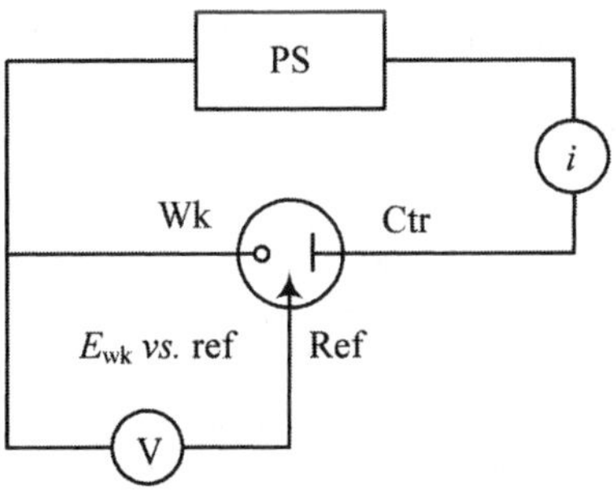

Fig. 5-15 Potential control and current measurement in three-electrode cell.

Problems

01 Draw a voltage integrator circuit using an OP amp. What is the input signal to obtain an output voltage of triangular shape, which is used for cyclic voltammetry?

02 In the next OP amp circuit: (a) Derive the relationship between the output voltage (e_o) and the input voltage (e_i), and (b) Compare the e_o values for $R_1 = R_2 = 0\ \Omega$ and $R_1 = R_2 \neq 0\ \Omega$, and explain why the given circuit is called "voltage follower with gain".

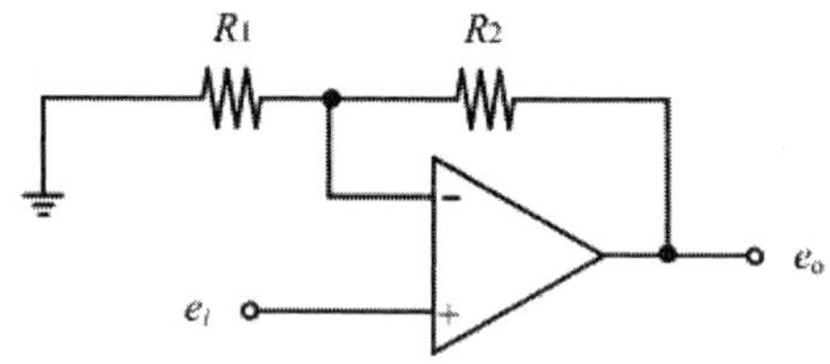

03 In the following circuit, the input (e_i) is the AC voltage ($e_i = V_p \sin 2\pi ft$), f = 60 Hz, V_p = 2.0 V), the power supply is +15 V/−15 V, and the open loop gain (A) is 10^6. Assuming that e_{ref} = 0.0 V, draw the output voltage (e_o).

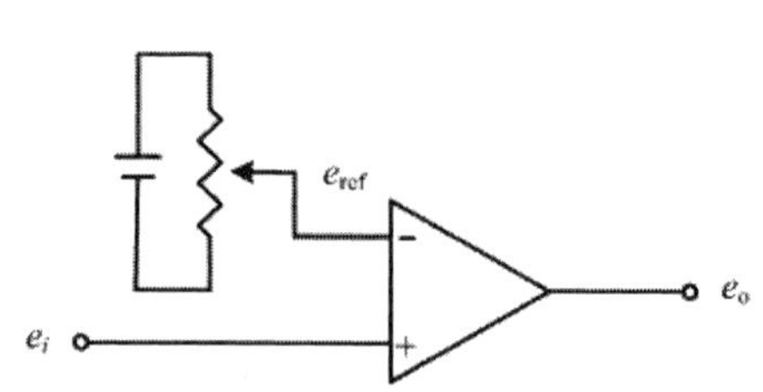

04 In Fig. 5-11, the positive feedback compensation can be used to solve the problem of uncompensated resistance (R_u). The idea is to add another input that corresponds to iR_u. Add this compensation circuit in Fig. 5-11.

05 Consider that E_{appl} = 3.0 V is applied for an electrolysis performed in a two-electrode cell. Calculate the current and the potential (E^a and E^c) applied to the anode and cathode using <1-6>. The equilibrium potentials for the cathode (E^c_{eq}) and anode (E^a_{eq}) are −0.80 V and 1.40 V (*vs.* NHE), respectively. The exchange current for the reduction and oxidation reaction is 8.0×10^{-8} A and 2.0×10^{-2} A, respectively. Assume that R_{total} = 4.0 Ω, n = 1, α = 0.5, and T = 298 K.

CHAPTER

06

Hydrodynamic methods

6-1 Rotating disk electrode (RDE) experiments
6-2 Rotating ring-disk electrode (RRDE) experiments
6-3 Concentration overpotential

A variety of electrochemical techniques is used to obtain electrochemical parameters related to charge transfer or mass transfer in electrochemical reactions, to analyze the concentration of unknown samples, or to analyze electrochemical reaction mechanisms. In these experiments, unlike the industrial electrolysis, it is common to use a very small area electrode and to use a three-electrode cell for precise potential or current control.

In hydrodynamic methods, a cylindrical working electrode is rotated to induce a mass transfer by mechanical convection. A rotating disk electrode (RDE) or a rotating ring-disk electrode (RRDE) are used as the working electrode. In both cases, the electrode area is small, such that the amount of reactants to be oxidized or reduced through electrolysis is not appreciable; it can be assumed that there is no change in the concentrations (C_O^* and C_R^*) of the electroactive species (O and R). Also, because the supporting electrolyte is always added into the solution by 100 times larger than the electroactive species, the mass transfer of electroactive species by migration can be neglected even if O or R are ionic species; only diffusion and convection are possible for mass transfer. Also, since the electrodes (RDE and RRDE) are rotated under the condition of maintaining a laminar flow, mechanical convection is possible only in the direction perpendicular to the working electrode surface; one-dimensional (linear) mass transfer is assumed.

In Section 6-1-2, the current-potential behaviors in RDE experiment are explained for the case of nernstian (electrochemically reversible) and mass transfer determines the overall rate in the whole potential range (including the extremely small overpotential region). This corresponds to the case where $k^0 = \infty$ ($R_{ct} = 0$), such that there is no electrode/electrochemical reaction having such a characteristic. However, when k^0 is very large, the current-potential behavior is close to that derived with such an extreme assumption.

Section 6-1-3 describes the current-potential behavior in all cases, including the case where the mass transfer and charge transfer rates are comparable to each other. Here, if $k^0 = \infty$, the current-potential characteristics obtained in this way should be identical with that obtained in Section 6-1-2. It will also explain how the current-potential behavior differs from that of nernstian when the electrochemical systems are far deviated from nernstian ($k^0 \rightarrow 0$).

6-1 Rotating disk electrode (RDE) experiments

6-1-1 Mass transfer in RDE experiments

Fig. 6-1 shows the structure of a ***rotating disk electrode***. A disk located at the bottom of cylinder serves as the working electrode. When the RDE is rotated in solution, the solution moves by mechanical convection towards the electrode surface, maintaining a laminar flow when the rotation speed is not too high. When the distance from the electrode surface is defined as x in the solution, the mechanical convection proceeds as shown in Fig. 6-2. Namely, the mass transfer takes place one-dimensionally in the vertical direction of the disk electrode surface by maintaining a laminar flow. Here, ω (= 2π × rotation

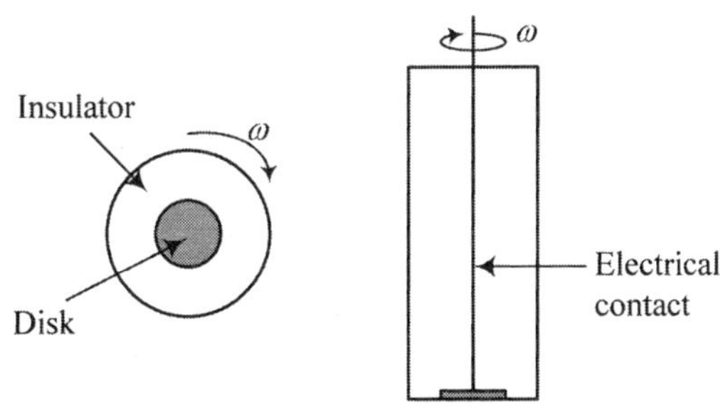

Fig. 6–1 Structure of a rotating disc electrode (RDE).

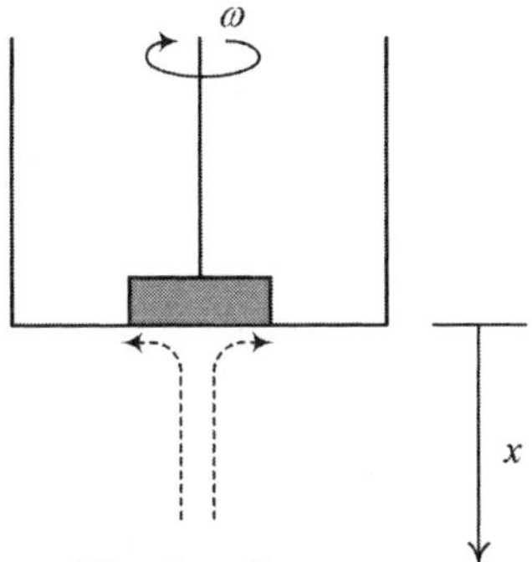

Fig. 6–2 Mechanical convection due to rotation of rotating disk electrode.

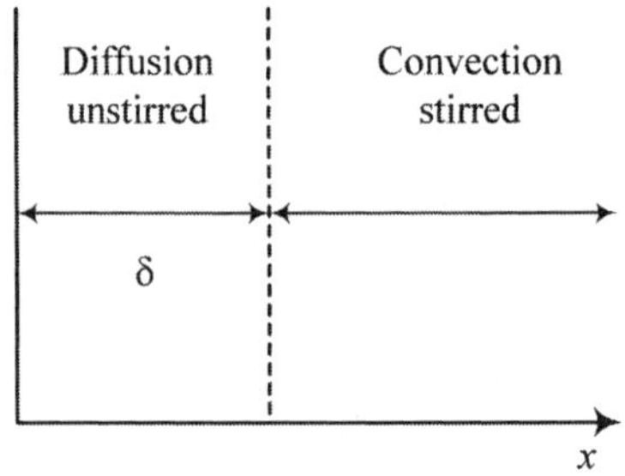

Fig. 6-3 Formation of Nernst diffusion layer by rotating an RDE electrode.

rate) is the angular frequency of rotation. Since there is no migration, mass transfer rate is the sum of diffusion and convection rate as shown in <6-1>.

$$J_i(x) = -D_i \frac{dC_i(x)}{dx} + C_i v(x) \qquad \text{<6-1>}$$

Here, $v(x)$ is the moving velocity of electroactive species i by mechanical convection. Since $v(x) \propto \omega^{3/2} x^2$, $v(x)$ can be neglected if x is very small, manifesting that the mechanical convection near electrode surface can be neglected; only diffusion is prevailing. However, $v(x)$ is large in the bulk solution with larger x, mass transfer by convection is dominant over diffusion (Fig. 6-3). Since $v(x) \to 0$ with $x \to 0$, it is like a condition without stirring (that is, convection). The solution near electrode surface where only diffusion contributes to mass transfer is called the ***Nernst diffusion layer***, and the thickness (δ) is given by <6-2>, wherein ν is the kinematic viscosity, which is approximately 0.01 cm^2/s in aqueous solutions. In short, diffusion is the only way for mass transfer near electrode surface, even if mass transfer by mechanical convection occurs in the solution far from the electrode surface.

$$\delta = \frac{1.61\, D_i^{1/3} \nu^{1/6}}{\omega^{1/2}} \qquad \text{<6-2>}$$

6-1-2 RDE experiments for electrochemically reversible systems

Assuming that an electrochemical system composed of an electrode/electrochemical reaction is a nernstian (electrochemically reversible because the charge transfer rate is high enough) and mass transfer determines the overall rate even in the very small overpotential region, the current in RDE experiments is determined by the diffusion rate at the electrode surface as shown in <6-3>; there is no migration because of excess amount of supporting electrolyte and the convection is negligible in the solution near the electrode surface. The current value can be obtained if the diffusion coefficient (D_O) and the concentration gradient of O near electrode surface ($x = 0$) are known. In <6-3>, the current (i) is referred to as a ***diffusion current*** (i_d) since the diffusion rate determines the current.

$$\text{Rate} = \frac{i}{nFA} = D_O \left. \frac{dC_O(x,t)}{dx} \right|_{x=0} \qquad \text{<6-3>}$$

The concentration gradient in the Nernst diffusion layer is given by <6-4>. From this, the current value can be obtained if δ and $C_O(0,t)$ are known because C_O^* is constant; this is valid because the electrochemical reaction proceeded on a small electrode surface is not so substantial; the change of C_O^* is negligible. Meanwhile, δ is constant because the rotation speed is kept constant, so only $C_O(0,t)$ is a variable.

$$\frac{C_O^* - C_O(0,t)}{\delta} \qquad \text{<6-4>}$$

$C_O(0,t)$ is determined by the applied potential, which can be obtained using the Nernst equation <4-17> because we have assumed a nernstian. The change of $C_O(0,t)$ and $C_R(0,t)$ according to the applied electrode potential is presented in Fig. 6-4, which was obtained with an assumption that $C_O^* = C_R^*$ for an electrochemical reaction $O + e = R$. The current predicted from the concentration gradient of O and R within the Nernst diffusion layer is displayed in Fig. 6-5. In Fig. 6-4, δ is fixed with an assumption that the rotation speed is fixed. C_O^* and C_R^* are invariant during the experiment, but C_O (x,t) (solid line) and C_R (x,t) (dotted line) are subjected to change according to the applied potential. When the applied potential $E = E^{0'}$ (E_{eq}), $C_O(0,t) = C_O^*$. This is shown in Fig. 6-4-c. Since the concentration gradient of O is zero at E_{eq}, the reduction current (i_c) is also zero according to <6-3>. Meanwhile, since $C_R(0,t) = C_R^*$ at E_{eq}, the concentration gradient of R is also zero to give $i_a = 0$. When the electrode potential moves to the negative direction

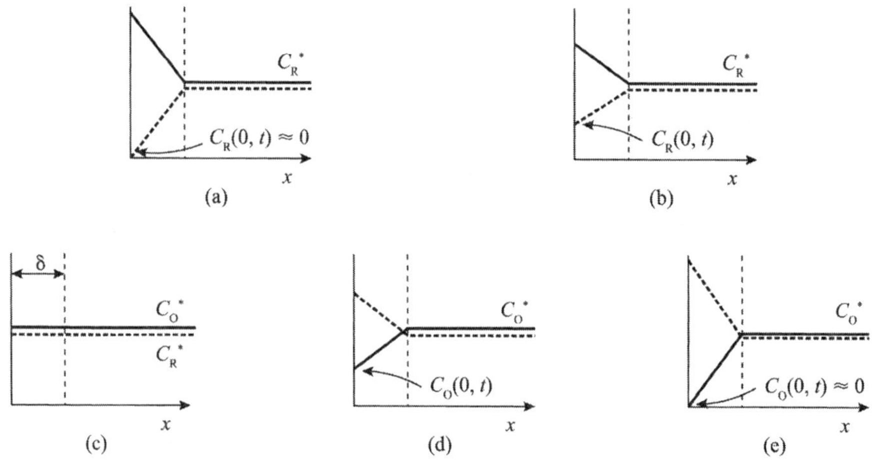

Fig. 6-4 Change of the concentration gradient of *O* (solid lines) and *R* (dotted lines) in the Nernst diffusion layer. The electrode potential is changed while the RDE rotation speed being fixed.

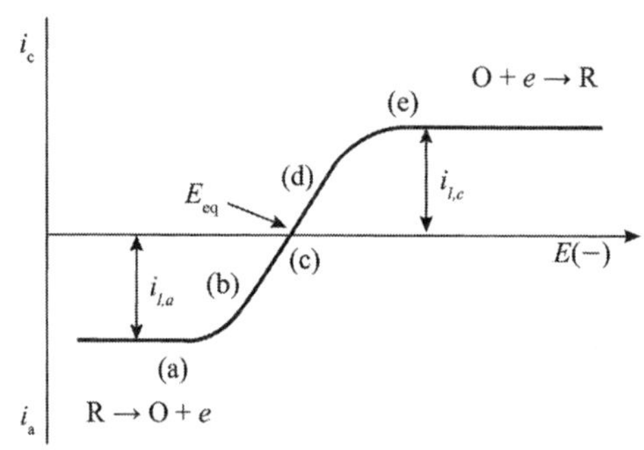

Fig. 6-5 The observed current according to the applied potential in RDE experiment. Mass-transfer limiting is assumed.

up to $E = E^{0'} - 28$ mV, $C_O(0,t)/C_R(0,t) = 1/3$ (d), and at $E = E^{0'} - 118$ mV, $C_O(0,t)/C_R(0,t) = 10^{-2}$, which means that $C_O(0,t) \approx 0$ (e). Even if the electrode potential moves to the more negative value ($E < E^{0'} - 118$ mV), the condition for $C_O(0,t) \approx 0$ is maintained. In short, as the electrode potential becomes more negative from E_{eq}, the concentration gradient of *O* increases to approach the invariant maximum value. Hence, as is seen in Fig. 6-5, the reduction current gradually increases to reach a constant value, which is called a ***limiting current***. The reason why the limiting current has a constant value is that there is no change in the concentration gradient once $C_O(0,t) = 0$ being reached, even when the electrode potential moves further to the negative direction.

The oxidation current (i_a) also shows a similar pattern. When $E = E^{0'} +28$ mV, $C_O(0,t)/C_R(0,t) = 3$ (b). At $E = E^{0'} +118$ mV, $C_O(0,t)/C_R(0,t) = 10^2$, indicative of $C_R(0,t) \approx 0$ (a). The condition for $C_R(0,t) \approx 0$ is maintained even though the electrode potential is pushed to a more positive value. As shown in Fig. 6-4, the concentration gradient of R increases in the Nernst diffusion layer as the electrode potential becomes more positive, but there is no change in the slope after $C_R(0,t)$ reaches zero. Consequently, the oxidation current (i_a) increases gradually to reach a limiting current.

Fig. 6-6-a shows how the concentration gradient of O in the Nernst diffusion layer changes as the rotational speed of RDE increases. When the rotational speed increases ($\omega_2 > \omega_1$), $\delta_2 < \delta_1$ according to <6-2>, so that the slope of concentration gradient becomes steeper to increase the diffusion current as shown in Fig. 6-6-b. Until now, the current has been decided only by the diffusion rate. Then, a question arises as to whether the mechanical convection performed in the hydrodynamic experiments does not affect the current at all. The answer is that it does affect. Since the Nernst diffusion layer becomes narrower with an increase in the rotation rate, the diffusion current becomes larger. This manifests itself that the mechanical convection affects the current as the thickness of Nernst diffusion layer is affected. In other word, the stirring reduces δ to increase the mass transfer rate by diffusion, which eventually leads to an increase in the diffusion

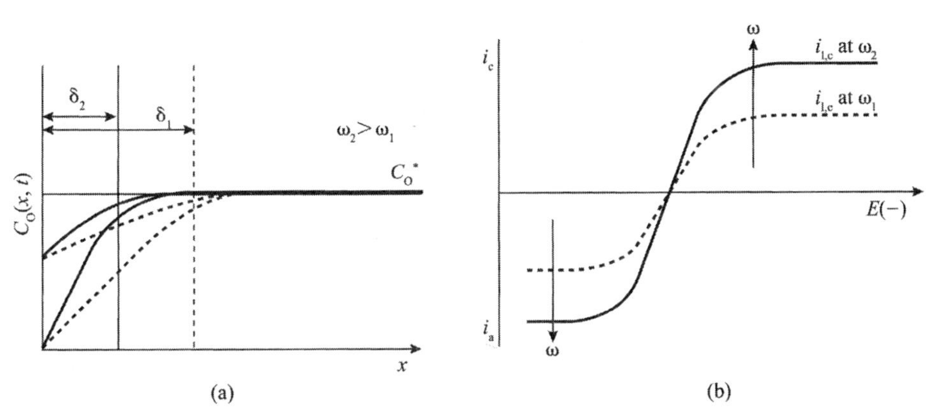

Fig. 6-6 (a); Thickness of Nernst diffusion layer and concentration gradient of O at two different rotation speeds of RDE, and (b); The current-potential profiles at two different rotation speeds.

current.

The current-potential behavior can be analyzed quantitatively. Since the current is determined by the diffusion rate, the current follows <6-5>. In the limiting current region, $C_O(0,t) \approx 0$. Hence, the limiting current can be expressed by <6-6>. Substituting <6-2> into <6-6> yields <6-7>, which is called ***Levich equation***.

$$\frac{i_c}{nFA} = D_O \left.\frac{dC_O(x,t)}{dx}\right|_{x=0} = D_O \frac{C_O^* - C_O(0,t)}{\delta} \qquad \text{<6-5>}$$

$$i_{l,c} = \frac{nFAD_O C_O^*}{\delta} \qquad \text{<6-6>}$$

$$i_{l,c} = \frac{0.62nFAD_O^{2/3} C_O^* \omega^{1/2}}{\nu^{1/6}} \qquad \text{<6-7>}$$

In Fig. 6-5, when the overpotential is not large, the current shows a sigmoid change. Here, the potential-dependent $C_O(0,t)$ should be known to quantify the current value.

$$i = nFAD_O \frac{C_O^* - C_O(0,t)}{\delta} \qquad \text{<6-8>}$$

If <6-9> is defined, <6-8> is rearranged into <6-10>.

$$k_O = \frac{nFAD_O}{\delta} \qquad \text{<6-9>}$$

$$i = k_O\,[C_O^* - C_O(0,t)] = i_{l,c} - k_O\,C_O(0,t) \qquad \text{<6-10>}$$

The first equation in <6-11> is derived from <6-10>. Similarly, one can derive the second equation in <6-11> after defining k_R. For nernstian, substituting $C_O(0,t)$ and $C_R(0,t)$ in <6-11> into <4-17> gives <6-12>.

$$C_O(0,t) = \frac{i_{l,c} - i}{k_O}, \quad C_R(0,t) = \frac{i - i_{l,a}}{k_R} \qquad \text{<6-11>}$$

$$E = E^{0'} + \frac{RT}{nF}\ln\frac{k_R}{k_O} + \frac{RT}{nF}\ln\left(\frac{i_{l,c} - i}{i - i_{l,a}}\right) = E_{1/2} + \frac{RT}{nF}\ln\left(\frac{i_{l,c} - i}{i - i_{l,a}}\right) \qquad \text{<6-12>}$$

Here, $E_{1/2}$ is called a ***half-wave potential***. This is the potential, at which the current is the average of two limiting currents.

$$i = \frac{i_{l,c} + |i_{l,a}|}{2}$$

From <6-12>, the slope is expected to be nF/RT when $\ln[(i_{l,c} - i)/(i - i_{l,a})]$ is plotted against $(E - E_{1/2})$. If the experimentally observed slope is identical with this, one can say that two assumptions made above are valid. Note that <6-12> is derived with two assumptions; nernstian and mass transfer limiting in the whole potential range. In practice, however, the experimentally observed slope frequently deviates from this theoretical one in small overpotential region, wherein frequently charge transfer is slower than mass transfer; charge transfer limited.

Self learning 6-1

In Fig. 6-5 and Fig. 4-6, the observed current shows a linear change in the small overpotential region, but has a completely different meaning. Explain how they are different.

*6-1-3 Current-potential characteristics for general cases in RDE experiments

In electrochemical reactions, the current is decided by charge transfer or mass transfer rate depending on which is slower than the other. The real world, of course, often contains intermediate cases where the rates of charge transfer and mass transfer are comparable. The current-potential behaviors for general cases including the comparable one will be explained.

Let us assume that an electrochemical reaction ($O + e \rightleftarrows R$) consists only of mass transfer and charge transfer without any complex processes such as adsorption and crystallization. Then, the following reaction scheme can be assumed.

$$C_O^* \underset{m_O}{\overset{m_O}{\rightleftarrows}} C_O(0,\ t) \underset{k_b}{\overset{k_f}{\rightleftarrows}} C_R(0,\ t) \underset{m_R}{\overset{m_R}{\rightleftarrows}} C_R^*$$

Here, the reactant O present in the bulk solution (C_O^*) moves in to the outside of Nernst diffusion layer by mechanical convection and then moves to the electrode surface by diffusion within the Nernst diffusion layer. The O approaching the electrode surface is reduced to R by charge transfer, and the as-generated R on the electrode surface diffuses out to the bulk solution. The reverse reaction (oxidation of R to O) also proceeds through a similar route. In the above scheme, m_O (= D_O/δ) and m_R (= D_R/δ) refer to the ***mass transfer coefficient***, and k_f and k_b are the rate constants for the charge transfer reactions. Let us assume a steady state. Namely, if the mass transfer rate of O (or R) is the same with the charge transfer rate at the electrode surface, there is no change in the surface concentration, $C_O(0,t)$ or $C_R(0,t)$, over time. The mass transfer rates are expressed by <6-13>. Further, when all O and R approaching the electrode surface is reduced or oxidized, the diffusion-limited cathodic and anodic limiting current are obtained as <6-14>.

$$J = \frac{i}{FA} = k_f C_O(0,\ t) - k_b C_R(0,\ t)$$
$$= m_O[C_O^* - C_O(0,\ t)]$$
$$= m_R[C_R(0,\ t) - C_R^*] \qquad \text{<6-13>}$$
$$i_{l,c} = FAm_O C_O^*, \quad i_{l,a} = -FAm_R C_R^* \qquad \text{<6-14>}$$

By making the appropriate substitutions, the concentration terms can be removed from <6-13> to yield <6-15>, where $p = m_O/m_R = D_O/D_R$.

$$i = \frac{k_f i_{l,c} + pk_b i_{l,a}}{m_O + k_f + pk_b} \qquad \text{<6-15>}$$

In general, $p = D_O/D_R \approx 1$, leading to <6-16>.

$$i = \frac{k_f i_{l,c} + k_b i_{l,a}}{m + k_f + k_b} \qquad \text{<6-16>}$$

The substitution of k_f and k_b into <6-16>, which are given below, yields <6-17>.

$$k_f(E) = k^0 e^{-\alpha F(E-E^{0'})/RT}$$
$$k_b(E) = k^0 e^{(1-\alpha)F(E-E^{0'})/RT}$$
$$f = \frac{F(E-E^{0'})}{RT}$$

$$i = \frac{i_{l,c}e^{-\alpha f} + i_{l,a}e^{(1-\alpha)f}}{\frac{m}{k^0} + e^{-\alpha f} + i_{l,a}e^{(1-\alpha)f}} \qquad \text{<6-17>}$$

In <6-17>, the mass transfer coefficient ($m = m_O = m_R$) and the standard rate constant (k^0) have the same unit of cm/s. Therefore, m/k^0 is a direct comparison between the mass transfer (m) and charge transfer rate (k^0). The following current-potential characteristics are derived according to the relative size of m and k^0, and the magnitude of f (*i.e.*, the applied potential from its definition as shown above).

1) $k^0 >> m$: This is the case for electrochemically reversible and mass transfer limited. <6-17> is rewritten as <6-18>.

$$i = \frac{i_{l,c} + i_{l,a}e^{f}}{1 + e^{f}} \qquad \text{<6-18>}$$

From <6-18>, <6-19> is derived and further <6-20> is obtained. <6-20> corresponds to the case for $D_O = D_R$ in <6-12> (assuming $p = D_O/D_R \approx 1$), and is identical with <6-12>, which was derived with an assumption of nernstian and mass transfer limiting. Note that $E_{1/2} \approx E^{0'}$. It can be seen that the current-potential characteristics derived for general cases are identical with those obtained in Section 6-1-2 under the condition of $k^0 >> m$.

$$e^{f} = \frac{i_{l,c} - i}{i - i_{l,a}} = e^{F(E-E^{0'})/RT} \qquad \text{<6-19>}$$

$$E = E^{0'} + \frac{RT}{F}\ln\frac{i_{l,c} - i}{i - i_{l,a}} \qquad \text{<6-20>}$$

2) $k^0 << m$: This corresponds to the electrochemically irreversible case. If the applied potential is negative, $i_{net} \approx i_c$. Therefore, $i_{l,a}$ in <6-17> can be discarded, and further $e^{(1-\alpha)f}$ in denominator can be deleted because $e^{(1-\alpha)f} << e^{-\alpha f}$. Finally, <6-21> is derived. From <6-21> and <6-22>, it is seen that $i = i_{l,c}$ at a very negative potential because $i_K > i_{l,c}$. That is, regardless of the magnitude of k^0, the limiting current is observed in large overpotential region. However, as k^0 becomes smaller (closer to the electrochemical irreversibility), $i_K > i_{l,c}$ prevails at the larger overpotential, such that a limiting current appears at a larger overpotential. This is shown in Fig. 6-7.

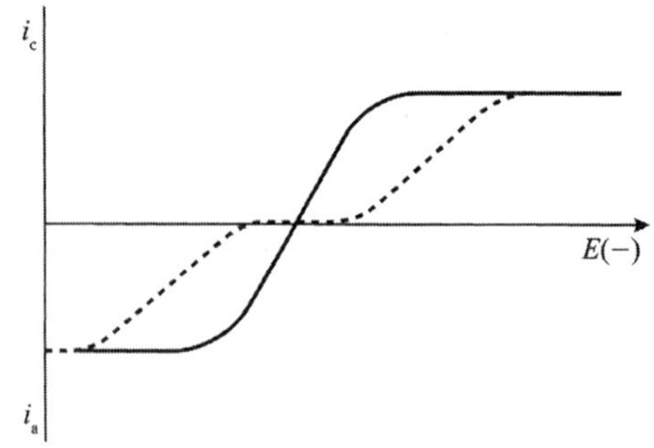

Fig. 6-7 Comparison of overpotential at which the limiting current appears according to k^0; k^0 is smaller for the dotted line.

$$\frac{1}{i} = \frac{1}{i_K} + \frac{1}{i_{l,c}} \qquad \text{<6-21>}$$

$$i_K = FA\,k_f(E)C_O^* = FAC_O^* k^0 e^{-\alpha F(E-E^{0'})/RT} \qquad \text{<6-22>}$$

Fig. 6-7 illustrates the following feature. Regardless of the magnitude of k^0, the limiting current always appears at some large overpotential; mass transfer determines the overall rate. However, as k^0 becomes smaller, a larger overpotential is required for the charge transfer rate to surpass the mass transfer rate, so that the overpotential at which mass transfer limitation begins to prevail also increases. On the other hand, as k^0 becomes larger (closer to nernstian), the overpotential at which the limiting current appears decreases, approaching what is shown in Fig. 6-5.

3) If f is very negative, <6-17> can be rewritten as <6-23>, where $i = i_{l,c}$ regardless of the size of m_O and k^0. If f is very positive, $i = i_{l,a}$. This means that, as described above, there always appears a limiting current in some large overpotential regardless of mass transfer and charge transfer rate.

$$i = \frac{i_{l,c}e^{-\alpha f}}{\frac{m_O}{k^0} + e^{-\alpha f}} \qquad \text{<6-23>}$$

4) If the overpotential is very small ($E \approx E^{0'}$), $f \rightarrow 0$ in <6-17>. Then, <6-24> is derived. In this case, regardless of the magnitude of m and k^0, the current is linearly proportional to the potential (see Fig. 6-5 and Fig. 4-6). This illustrates that the

current has a linear relation with applied potential in small overpotential region for both cases of mass transfer and charge transfer limitation. The slope in the current-potential profiles are different according to α, m, and k^0.

$$i = \frac{i_{l,c} + i_{l,a} - i_{l,c}\alpha f + i_{l,a}(1-\alpha)f}{\frac{m}{k^0} + 2} \qquad \text{<6-24>}$$

6-1-4 Investigation of reaction mechanism by RDE experiments

RDE experiments are frequently used to investigate the mechanism of electrochemical reactions. Fig. 6-8 shows an example for ECE-type electrochemical reaction. When the RDE rotation speed becomes high, the apparent n value approaches 1, but is closer to n = 2 as the rotation speed decreases. From this, it can be deduced that this reaction belongs to an ECE-type. Namely, compound A is reduced to B, which is converted into C by a homogeneous chemical reaction in solution, and C is further reduced to D. Here, if the rotation speed is sufficiently low, since a series of reactions from A to D occurs continuously on the disk electrode, it appears to be a two-electron reduction reaction. In opposite, if the rotation speed is very high, the residence time of B on the disk electrode surface is so short that the second electrochemical reaction ($C + e \rightarrow D$) can not take place. It appears an one-electron reduction.

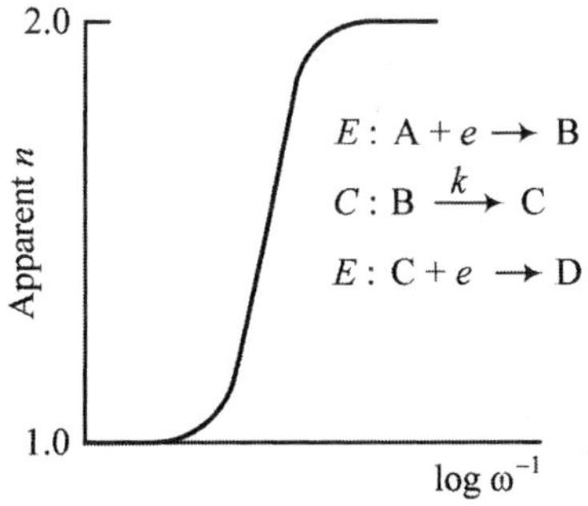

Fig. 6–8 Change of apparent n value in an ECE–type electrochemical reaction according to the rotation speed of RDE.

6-2 Rotating ring-disk electrode (RRDE) experiments

Fig. 6-9 shows the structure of an RRDE and the mass transfer pathway in solution when it is rotated. The difference from RDE is that a ring-shaped electrode is located outside a disk electrode. The potential of the disk and ring electrode is controlled separately using a bipotentiostat.

When RRDE is rotated, one-dimensional convection occurs in the direction perpendicular to the electrode surface. Here, a part of R generated through the electrochemical reaction on the disk reaches the ring electrode through the solution. The amount of R arriving at the ring electrode can be measured by oxidizing R on the ring electrode. This is called a ***collection experiment***. The ***collection efficiency*** is defined by <6-25>, which is the ratio of R generated on the disk (i_D) and re-oxidized on the ring ($-i_R$). The reason for the negative sign is that the reduction and oxidation current are expressed by positive and negative values, respectively.

$$N = \frac{-i_R}{i_D} \tag{6-25}$$

W. J. Albery and his coworkers (*Trans. Faraday Soc.*, **64**, 2831, 1968) obtained the

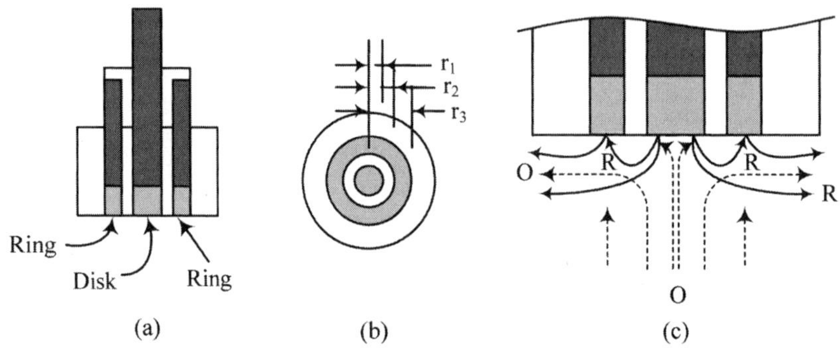

Fig. 6-9 Structure of an RRDE and the flow pattern in a collection experiment, where O is reduced to R at the disk and the as-generated R is collected at the ring to be oxidized. Note that some R escapes collection.

rate constant ($k = 3.2 \times 10^4/M$ s) for the reaction between Br_2 and anisole from the measurement of collection efficiency as a function of rotation speed.

$$Br^- \rightarrow 1/2\,Br_2 + e \quad (E)$$

$$Br_2 + \text{Anisole} \xrightarrow{k} \text{HBr} + \text{Brominated anisole} \; (C)$$

In the above EC-type electrochemical reaction, if the electrogenerated Br_2 on the disk reacts with anisole by chemical reaction in solution, the amount of Br_2 reduced back to Br^- on the ring is inevitably less than that generated. However, the higher the rotation speed of RRDE, the faster the mass transfer rate in the solution; the residence time of Br_2 in the solution decreases, indicating that the amount of Br_2 reacting with anisole decreases; the amount of Br_2 reaching the ring increases. This phenomenon is shown in Fig. 6-10. As the rotation speed (ω) increases, N increases as expected to approach N_L, which is called ***limiting collection efficiency***. N_L is the collection efficiency that can be obtained at an extremely high rotation speed. Under this condition, the residence time for Br_2 in solution is so short that the reaction with anisole can be ignored; no homogeneous chemical reactions during the journey from disk to ring. One may expect $N_L = 1.0$ for this case. In real, however, $N_L < 1.0$ because some of Br_2 escapes from collection as shown in Fig. 6-9-c. N_L is determined by the electrode configuration (r_1, r_2 and r_3 in Fig. 6-9-b). In Fig. 6-10, $N_L = 0.22$, manifesting itself that about 80 % of Br_2 escapes from collection due to the given electrode configuration; note that the reaction with ani-

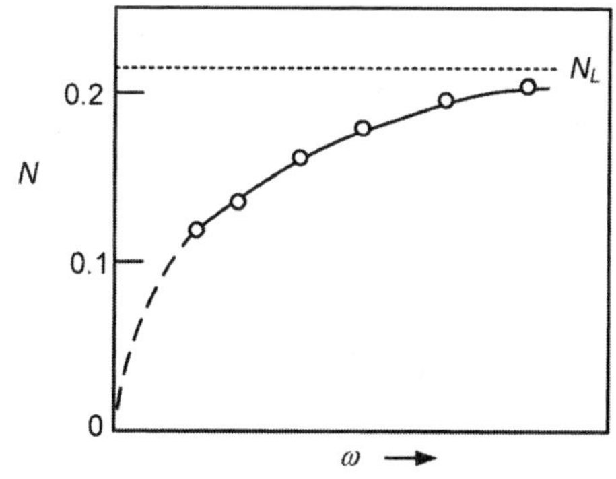

Fig. 6-10 Collection efficiency (N) of electrogenerated bromine in the presence of anisole as a function of rotation speed of RRDE. N_L corresponds to the collection efficiency at an extremely high rotation speed.

sole is not involved in the 80 % of escape.

Another example of collection experiment using RRDE is shown in Fig. 6-11 (R. N. Adams, *Electrochemistry at Solid Electrodes*, Marcel Dekker, New York, 1969). In a neutral or alkaline aqueous solution, oxygen is reduced either to hydrogen peroxide by two electrons, or to water by four electrons. In Fig. 6-11, the reduction current on the disk (i_{disk}) and the oxidation current on the ring (i_{ring}) are traced while the potential of disk electrode is swept to the negative direction. Note that the potential of ring electrode is fixed at a constant value, at which H_2O_2 can be electrochemically oxidized. In this setup, the H_2O_2 generated on the disk moves to the ring to be oxidized. As is seen, the H_2O_2 oxidation is observed when the disk potential has a certain range. It can be said that the generation of H_2O_2 by two-electron reduction of O_2 takes place only in the specified potential range. In fact, the O_2 reduction by four electrons proceeds at more negative potential.

In the RDE and RRDE experiments, the solution must maintain laminar flow characteristics so that linear mass transfer should be guaranteed. In general, laminar flow is maintained in the rotation speed range of 100 ~ 10,000 rpm in aqueous solution, but turbulence occurs at higher speed. When the turbulence occurs, the mechanical convection deviates from the linearity, so the current value changes. Also, if the surface of the disk is not

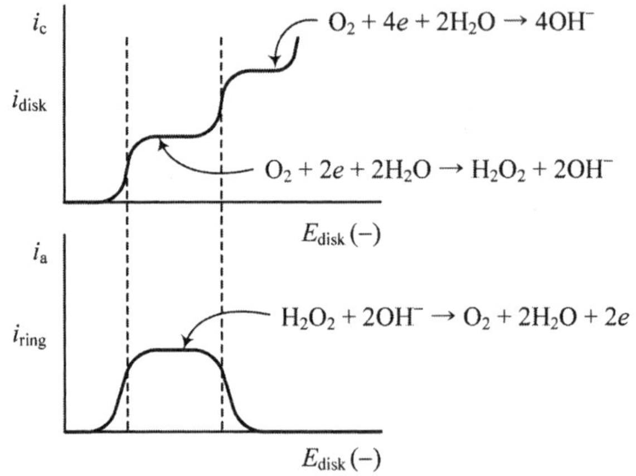

Fig. 6-11 The current at the disk (i_{disk}) and ring (i_{ring}) for the reduction of oxygen at an RRDE. Note that the two-electron reduction takes place within the potential range indicated by two dotted lines.

flat or not perpendicular to the axial direction, or precession is evolved during electrode rotation, turbulence may occur even at lower rotation speeds. In this case, an error occurs in the current measurement.

6-3 Concentration overpotential

It is a common observation that, in many electrochemical reactions, charge transfer determines the overall rate in small overpotential region, but mass transfer at large overpotential region (Fig. 4-6). However, if $k^0 = \infty$ ($R_{ct} = 0$), mass transfer determines the overall rate in the whole potential range. The *i-V* profile in Fig. 6-5 was obtained with this assumption. As is seen in Fig. 6-5, in order to obtain a current of a certain size, a potential greater than E_{eq} should be applied. The difference between the applied electrode potential and E_{eq} is called a ***concentration overpotential*** or mass transfer overpotential.

Concentration overpotential corresponds to the case where mass transfer determines the overall rate, so one can assume that the mass transfer rate, which appears as a current, can be enhanced by increasing the overpotential. With a simple mind, it is somewhat difficult to understand the potential-dependent mass transfer rate because mass transfer looks independent on the electrode potential as electroactive species moves in solution. In fact, the mass transfer rate is independent on the electrode potential when electroactive species is transferred by mechanical convection. However, since the convective moving velocity is $v(x) \propto \omega^{3/2} x^2$ and $v(x) \rightarrow 0$ near electrode surface ($x \rightarrow 0$), the mass transfer by convection is neglected near electrode surface; only diffusion is possible. As explained in Fig. 6-4, when the electrode potential moves from E_{eq} to the more negative direction (from (c) to (d) and (e)), the surface concentration of O decreases to give a steeper slope in the concentration gradient of O. As a result, the diffusion rate of O becomes faster to give a larger reduction current. In short, the mass transfer rate increases with an increase in overpotential; mass transfer rate is potential-dependent. The underlying reason for this is that diffusion is the only mass transfer route near electrode surface. In summary, the rate of mass transfer by convection in the bulk solution is independent on the electrode potential, but the surface concentration depends on electrode potential; the change in surface concentration by a potential change affects the diffusion rate and eventually current.

Concentration overpotential is explained for an electrochemical reaction $O + ne = R$, which is a nernstian. It is assumed that R is a metal and has a unit activity.

<6-26> is derived from <6-6> and <6-8>, and <6-27> is derived by substituting <6-26> into the Nernst equation.

$$\frac{C_O(0,t)}{C_O^*} = 1 - \frac{i}{i_{l,c}}, \quad C_O(0,t) = C_O^*\left(\frac{i_{l,c} - i}{i_{l,c}}\right) \tag{6-26}$$

$$E = E^{0'} + \frac{RT}{nF}\ln C_O^* + \frac{RT}{nF}\ln\left(\frac{i_{l,c} - i}{i_{l,c}}\right) \tag{6-27}$$

Since $i = 0$ at $E = E_{eq}$, the first two terms on the right side in <6-27> correspond to E_{eq}. If the concentration overpotential is defined as <6-28> to have a positive value, <6-29> is derived from <6-27>. Fig. 6-12 shows that η_{conc} increases as the current increases, and $\eta_{conc} \rightarrow \infty$ at $i = i_{l,c}$ as can be predicted from <6-29>.

$$\eta_{conc} = E_{eq} - E \tag{6-28}$$

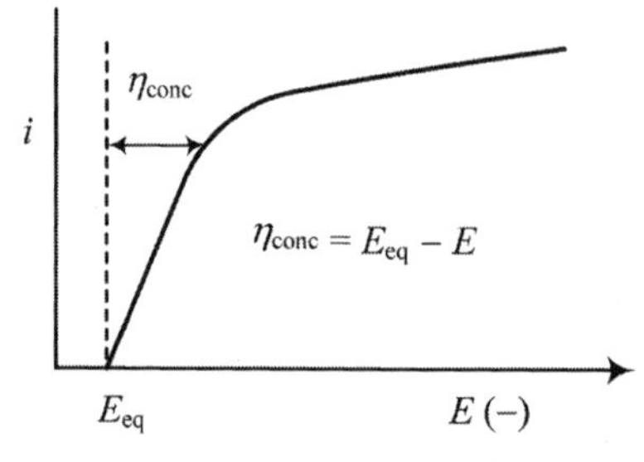

Fig. 6-12 Relationship between current and applied potential when $k^0 = \infty$ ($R_{ct} = 0$), which means that mass transfer limits the current in the whole potential region. Note that the limiting current increases with a slight slope due to electric double-layer charging current.

$$\eta_{\text{conc}} = \frac{RT}{nF} \ln\left(\frac{i_{\text{l,c}}}{i_{\text{l,c}} - i}\right) \qquad \text{<6-29>}$$

<6-29> is rewritten as <6-30>. After the exponential function is extended to power series and η_{conc} is assumed to be small, <6-31> is derived. <6-31> illustrates that η_{conc} is linearly proportional to the current (i) at small η_{conc} region, which is seen in Fig. 6-12. Meanwhile, from $V = i \times R$ relation in <6-31>, <6-32> is derived. The resistance is related to mass transfer, so it is called the ***mass transfer resistance*** (R_{mt}). This is contrasted with the charge transfer resistance (R_{ct}) shown in <4-19>.

$$1 - \frac{i}{i_{\text{l,c}}} = e^{-nF\eta_{\text{conc}}/RT} \qquad \text{<6-30>}$$

$$\eta_{\text{conc}} = \frac{RTi}{nFi_{\text{l,c}}} \qquad \text{<6-31>}$$

$$R_{\text{mt}} = \frac{RT}{nFi_{\text{l,c}}} \qquad \text{<6-32>}$$

<6-32> shows an inverse relation between the mass transfer resistance (R_{mt}) and limiting current ($i_{\text{l,c}}$). To minimize R_{mt}, $i_{\text{l,c}}$ should be maximized by increasing the concentration of O, enlarging the electrode area, increasing the rotation speed, and decreasing the solution viscosity as expected from <6-7>. Note that all these measures facilitate mass transfer rate.

R_{mt} is inversely proportional to $i_{\text{l,c}}$ as R_{ct} is inversely proportional to i_0 (refer to <4-19>); there is an one-to-one correspondence between R_{mt} and R_{ct}, and further between $i_{\text{l,c}}$ and i_0. The exchange current (i_0) represents the intrinsic charge transfer property; a larger i_0 indicates an intrinsically faster charge transfer. Similarly, a larger $i_{\text{l,c}}$ indicates an intrinsically faster mass transfer.

As is seen in <4-15>, the charge transfer rate is decided not only by the intrinsic charge transfer property (k^0 or i_0), but also by the activation overpoptetntial. Similarly, the mass transfer rate is affected by the intrinsic mass transfer parameter ($i_{\text{l,c}}$) and the concentration overpotential.

Example 6-1

In the cell polarization (cell polarization = $\eta_a + \eta_c + iR_{total}$) shown in <1-6> and <1-8>, the overpotential (η_a and η_c) at anode and cathode is the sum of all overpotentials associated with electrode reactions. Answer the following considering only the activation overpotential and concentration overpotential.

(1) Fig. 6-13 compares the i *vs.* V relationships to explain the activation overpotential, concentration overpotential, and iR_{total} voltage drop; all three contribute to cell polarization. How do the activation overpotential and concentration overpotential differ from the iR_{total} drop in the i *vs.* V relation?

(2) After overlapping the three pictures shown in Fig. 6-13, confirm that the dominant contributor for cell polarization differs according to the current; the activation overpotential is dominant over the other two at low current, the concentration overpotential at large current, and the iR_{total} term in the medium current region.

Solution

(1) iR_{total} term shows a linear i *vs.* V relationship over a wide potential (current) range. However, when charge transfer rate determines the overall rate, current (charge transfer rate) shows a linear relation with potential in small current region, but increases by an exponential function of applied electrode potential in large current region. As a result, the activation overpotential increases rapidly in small current region, but increasing rate becomes lower with an increase of current. The i *vs.* V profile shows another pattern when mass transfer limits the current; the current (mass transfer rate) shows a linear relationship with potential in small current region, but shows a limiting value in large current region; the concentration overpotential increases linearly at low current region, but becomes infinite (∞) in large current region. In short, the three show different i *vs.* V relationship to each other.

Meanwhile, since there is no linear i *vs.* V relationship in a wide range in (a) and (b), the concept of resistance can not be introduced. However, in both cases, a linear relationship is observed in the small overpotential (current) region, such that R_{ct} and R_{mt} can be defined and further be calculated from the slopes (indicated by the dotted lines). Moreover, the size of overpotential can be compared by using the resistance values. For example, in (a), if R_{ct} is smaller, the slope of the dotted line is steeper, indicating that the activation overpotential is smaller. The same is true in (b); a smaller R_{mt} gives a steeper slope to indicate a smaller concentration overpotential.

(2) Upon overlapping the three figures in Fig. 6-13, one can find that the activation overpotential is dominant over the other two in small current region, the concentration overpotential in large current region, whereas iR_{total} term is the largest in the medium current region. In short, the dominant contributor changes according to current; this behavior can be seen in Fig. 12-27.

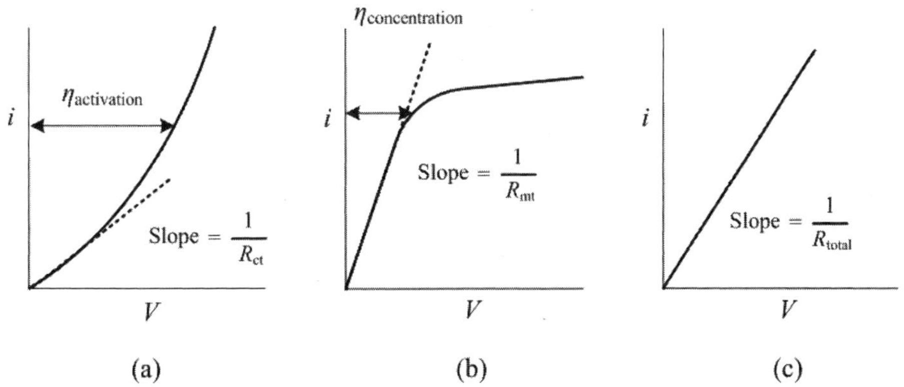

Fig. 6-13 The i–V relations illustrating the activation overpotential (a), concentration overpotential (b), and iR_{total} drop (c). The cell polarization is the sum of the three contributions. (a) and (b) are duplicated from Fig. 4-4 and Fig. 6-12, respectively.

Problems

01 Calculate ω (angular frequency of rotation, $\omega = 2\pi \times$ rotation rate) when an RDE is rotated in 10,000 rpm.

02 The reduction reaction of Ce^{4+} ($Ce^{4+} + e \rightarrow Ce^{3+}$, $E^{0'} = 1.61$ V *vs*. NHE) was performed on an RDE at 25 °C and the following current-potential relationship was obtained. The radius of the disk electrode was 4 mm, the rotation speed was 3000 rpm, and the kinematic viscosity (v) of the electrolyte solution was 0.95 cm^2/sec.

E/V	i/μA	E/V	i/μA
0.40	895	0.80	449
0.45	895	0.85	331
0.50	892	0.90	232
0.55	889	0.95	151
0.60	833	1.00	94.8
0.65	745	1.05	58.0
0.70	669	1.08	39.5
0.75	566	1.09	28.3

(a) Calculate the equilibrium potential with $[Ce^{4+}]$ = 5 mM and $[Ce^{3+}]$ = 0.1 mM.

(b) Calculate the diffusion coefficient of Ce^{4+}.

(c) Sketch the Tafel plot and calculate α and i_0.

CHAPTER

07

Potential step methods

7-1 Chronoamperometry and chronocoulometry with large-amplitude potential step

7-2 Limitations of potential step experiment

*7-3 Potential step experiment using planar ultra-microelectrode (UME)

*7-4 Small-amplitude potential step experiment

In potential-step experiments, the potential of working electrode is stepped from an initial value to a step potential. The current, which is the result of potential perturbation, is traced with time. This is called chronoamperometry. The electric charges that are obtained by integrating the current can also be traced with time, which is called chronocoulometry.

In potential-step experiments, the followings are assumed. Firstly, the initial bulk concentration (C_O^* and C_R^*) does not change during the experiment because the working electrode of small surface area is used. Secondly, the migration of electroactive ionic species is discarded since excess amount of supporting electrolyte is added. Thirdly, the solution is not stirred and thus mass transfer is possible only by diffusion. Fourthly, since a planar microelectrode is used and the experiment is performed within a few tens seconds, the diffusion proceeds in the vertical direction of electrode surface within a diffusion layer as shown in Fig. 7-1. That is, since the thickness of diffusion layer is much thinner compared to the electrode area (see Example 4-3), ***one-dimensional (linear) diffusion*** is assumed.

Let us say that an electroactive species O is reduced in a potential-step experiment and the current or electric charges are measured. Here, the step is given from an initial potential (E_1), which is sufficiently positive so that O can not be reduced, to the step potential (E_2), which is negative enough to allow the reduction of O (Fig. 7-2-a). Here, the factor (among the charge transfer or mass transfer) determining the overall rate is decided by k^0 (electrochemical reversibility) and E_2. For example, if $k^0 \rightarrow \infty$, mass transfer determines the rate even though E_2 is not so negative. If $k^0 \rightarrow 0$, however, E_2 should be sufficiently negative for the charge transfer rate to exceed the mass transfer rate. In the ***large-amplitude potential-step experiment***, $\Delta E = E_2 - E_1$ is set to be large; E_2 is

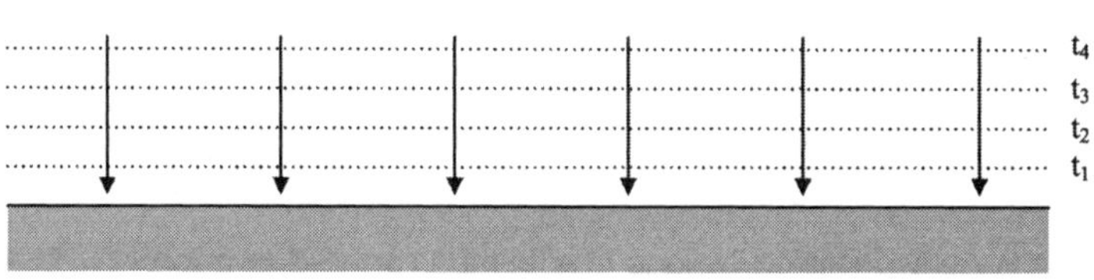

Fig. 7–1 One–dimensional expansion of diffusion layer with time (from t_1 to t_4) after a potential step on a planar microelectrode. The arrows indicate the direction of one–dimensional (linear) diffusion of electroactive species within a diffusion layer.

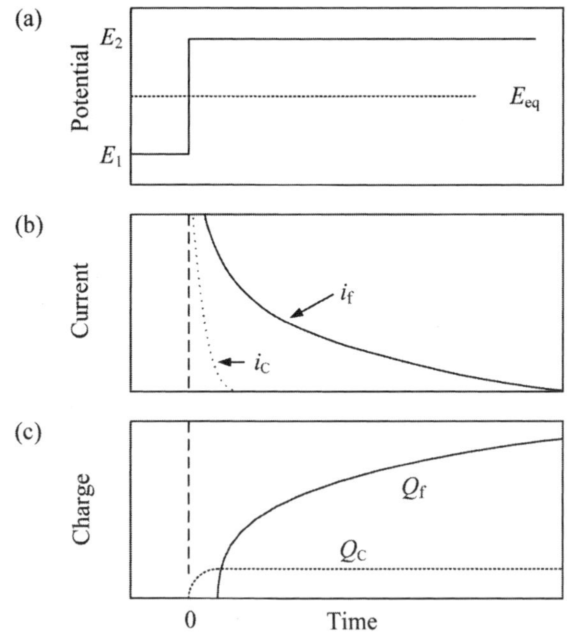

Fig. 7-2 (a); Change of the working electrode potential in potential-step experiment, (b); the charging current (i_c) immediately after the potential step and faradaic current (i_f), and (c); electric charges obtained by integrating the charging current and faradaic current.

controlled to be sufficiently negative. Two assumptions are possible. One is that the surface concentration of O is zero ($C_O(0,t) = 0$) because the reduction of O proceeds on the working electrode being imposed at E_2. The other is that mass transfer limits the overall rate regardless of the magnitude of k^0. The measured current is a diffusion current because diffusion is the way for mass transfer.

Section 7-1 describes the chronoamperometry and chronocoulometry obtained by the large-amplitude potential-step experiment. Section 7-2 explains the limitations of this technique. Section 7-3 explains the large-amplitude potential-step experiment performed with a planar ultra microelectrode (UME) where spherical diffusion is also prevailing. Section 7-4 describes the small-amplitude potential-step experiments, where E_2 is not so negative that $C_O(0,t) > 0$. It is also assumed that mass transfer limits the overall rate.

7-1 Chronoamperometry and chronocoulometry with large-amplitude potential step

Let us say that an electrochemical reduction, $O + ne \rightarrow R$, is performed. O can not be reduced at E_1. Hence, the surface concentration of O ($C_O\ (0,t)$) is equal to its bulk concentration (C_O^*) (see Fig. 1-6). At E_2, however, $C_O\ (0,t) = 0$ and this condition is maintained as far as the step is applied. Due to the concentration difference between the bulk solution and electrode surface, O diffuses toward the electrode and reduced. Since O continuously moves towards the electrode, the diffusion layer expands at a rate of $(2D_Ot)^{1/2}$ towards the bulk solution (Fig. 7-1). The current is determined by the concentration gradient of O in the diffusion layer according to <7-1>. The concentration gradient becomes less steep with time as shown in Fig. 1-6-b. Thus, the current steadily decreases with time (Fig. 1-12). Actually, the current decreases proportionally to $t^{-1/2}$ as shown in <7-2>, which is called the ***Cottrell equation***. Since the current is traced with time, it is called ***chronoamperometry***.

$$\text{Rate}= \frac{i_d}{nFA} = D_O \frac{dC_O(x,t)}{dx}\bigg|_{x=0} \qquad \text{<7-1>}$$

$$i_d = \frac{nFAD_O^{1/2}C_O^*}{\sqrt{\pi t}} \qquad \text{<7-2>}$$

In potential-step experiments, the potential of working electrode changes from E_1 to E_2 within a very short period of time, such that the double-layer charging current (i_C) flows immediately after the potential step as shown in Fig. 7-2-b. Also, a faradaic current (i_f) flows as the step potential is being imposed. The total current measured is the sum of the diffusion current (faradaic current) and the charging current (non-faradaic current). If the current is integrated over time, the electric charges can be calculated by <7-3>. This is called ***chronocoulometry***. When the electric charge Q is plotted against $t^{1/2}$ (referred to as an ***Anson plot***), a straight line is observed with a slope that is expressed by <7-4>. If n, A, and C_O^* are known, D_O can be calculated from the slope.

$$Q = \int_0^t i_d dt = \frac{2nFAD_O^{1/2}C_O^*}{\sqrt{\pi}} t^{1/2} \qquad \text{<7-3>}$$

$$\frac{2nFAD_O^{1/2}C_O^*}{\sqrt{\pi}} \qquad \text{<7-4>}$$

The electric charges are the sum of the integrated faradaic current (i_f) and charging current (i_C) as expressed by <7-5>.

$$Q = \int_0^t i_f dt + \int_0^t i_C dt = \frac{2nFAD_O^{1/2}C_O^*}{\sqrt{\pi}} t^{1/2} + Q_{dl} \qquad \text{<7-5>}$$

Therefore, the Anson plot (Q *vs*. $t^{1/2}$ plot) gives a straight line and an intercept as shown in Fig. 7-3-a. From the intercept, Q_{dl} can be calculated. Note that Anson plot gives a straight line only after a certain period of time. In the earlier period after potential step, Q *vs*. $t^{1/2}$ plot does not give a straight line because the charging current flows. In Fig. 7-3, the slope relevant to <7-4> is indicated by the solid lines. The dotted ones are just an extrapolation.

Sometimes, electroactive species are adsorbed on electrode surface. In this case, two faradaic currents are possible; one from the reduction of O that are diffused from the bulk solution and the other from the adsorbed O. In this case, the total electric charge is the sum of three components as indicated by <7-6>. From the Anson plot shown in

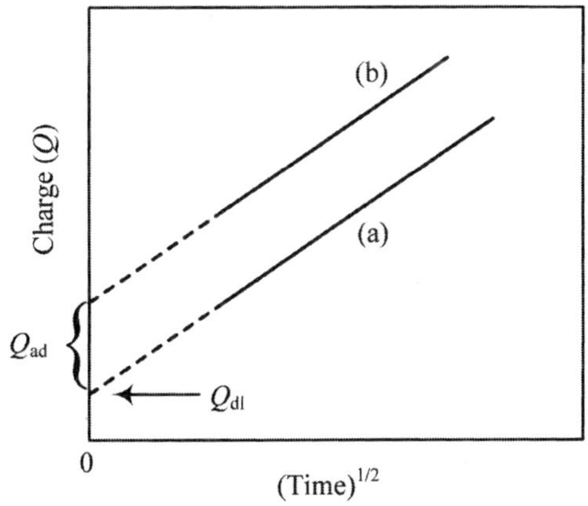

Fig. 7-3 Measurement of Q_{dl} and Q_{ad} from Anson plot.

Fig. 7-3-b, the amount of adsorbed O (Γ_O, in mol/cm^2) can be calculated using <7-7>. The value for Q_{ad} + Q_{dl} is obtained from the intercept, which is obtained by chronocoulometry with the electroactive O in the solution. Q_{dl} can be obtained by a separate potential-step experiment performed without O.

$$Q = Q_f + Q_{ad} + Q_{dl}$$ <7-6>

$$Q_{ad} = nFA\Gamma_O$$ <7-7>

Self learning 7-1

The slope in Anson plot for $O + e \rightarrow R$ was 1.2×10^{-5} C/sec$^{1/2}$. From this, calculate D_O: A = 0.02 cm^2 and C_O^* = 1.0 mM. Diffusion coefficients can also be determined by other electrochemical methods; for example, from the measured current ($i_{l,c}$ or i_{pc}) by using <6-7> or <8-11>. The D_O obtained from Anson plot is more accurate compared to others because it is obtained without the interference of electric double-layer charging current. Explain this.

Until now, it is shown that the faradaic current (diffusion current) decreases with time because the concentration gradient of electroactive species decreases within a diffusion layer. The time-dependent concentration gradient and the Cottrell equation can be quantitatively derived.

If one knows $C_O(x,t)$, which represents the concentration of O as a function of position (x; the distance from electrode surface) in the solution and time, the concentration gradient can be derived as a function of time. Note here that only x is variable (not y and z) since one-dimensional diffusion is assumed in potential-step experiments. In order to know $C_O(x,t)$, the Fick's second law <7-8> should be solved. However, since <7-8> is a partial differential equation (PDE), it should be converted into an ordinary differential equation (ODE) by ***Laplace transformation***, and then the solution is inversely transformed to obtain $C_O(x,t)$.

$$\frac{dC_O(x,t)}{dt} = D_O \frac{d^2C_O(x,t)}{dx^2}$$ <7-8>

The following boundary conditions are applied to solve <7-8>.

$$C_O(x,0) = C_O^*, \quad C_R(x,\ 0) = 0 \tag{7-9}$$

$$\lim_{x\to\infty} C_O(x,t) = C_O^*, \quad \lim_{x\to\infty} C_R(x,\ t) = 0 \tag{7-10}$$

$$C_O(0,t) = 0\ (t > 0) \tag{7-11}$$

<7-9> is the initial boundary condition, which illustrates that the concentration of O is constant irrespective of the position (x) and R is absent in the solution before the potential step (t = 0). <7-10> is the semi-infinite boundary condition, which explains that the concentration of O in the bulk solution ($x \rightarrow \infty$) is constant because only O locating within the diffusion layer participates in the electrochemical reaction, and the generated R accumulates within the diffusion layer. <7-11> is a surface condition, meaning that the surface concentration of O is zero during the period of potential step (t > 0) since E_2 is sufficiently negative (large-amplitude potential step).

The Laplace transform is defined by <7-12>.

$$L\{F(t)\} \equiv \int_0^\infty e^{-st}F(t)dt = f(s) = \overline{F}(s)\ (\text{for } s > 0) \tag{7-12}$$

The left term of <7-8> is Laplace transformed and <7-13> is derived by considering the relation of <7-9>.

$$L\left\{\frac{dF(t)}{dt}\right\} = \int_0^\infty e^{-st}\frac{dF(t)}{dt}dt = \left[e^{-st}F(t)\right]_0^\infty + s\int_0^\infty e^{-st}F(t)dt = -F(0) + sf(s)$$

$$L\left\{\frac{dC_O(x,t)}{dt}\right\} = -C_O(x,0) + s\overline{C}_O(x,s) = s\overline{C}_O(x,s) - C_O^* \tag{7-13}$$

Similarly, Laplace transformation of the right term of <7-8> gives <7-15> via <7-14>.

$$s\overline{C}_O(x,s) - C_O^* = D_O\frac{d^2\overline{C}_O(x,s)}{dx^2} \tag{7-14}$$

$$\frac{d^2\overline{C}_O(x,s)}{dx^2} - \frac{s}{D_O}\overline{C}_O(x,s) = -\frac{C_O^*}{D_O} \tag{7-15}$$

The solution of <7-15> is

Table 7–1 Laplace transformation of some representative functions

$F(t)$	$f(s)$
A (constant)	A/s
e^{-at}	$1/(s + a)$
$\sin at$	$a/(s^2 + a^2)$
$\cos at$	$s/(s^2 + a^2)$
t	$1/s^2$
$(\pi t)^{-1/2}$	$1/s^{1/2}$
$\mathrm{erfc}[x/2(kt)^{1/2}]$	$e^{-\beta x}/s$, where $\beta = (s/k)^{1/2}$

$$\overline{C}_O(x,s) = \frac{C_O^*}{s} + A'(s)\exp\left[-\left(\frac{s}{D_O}\right)^{1/2} x\right] + B'(s)\exp\left[\left(\frac{s}{D_O}\right)^{1/2} x\right]$$ <7-16>

Applying the Laplace transformation listed at the first row of Table 7-1 to the semi-infinite boundary condition <7-10> yields <7-17>, so <7-16> becomes <7-18>.

$$\lim_{x\to\infty} \overline{C}_O(x,s) = \frac{C_O^*}{s}$$ <7-17>

$$\overline{C}_O(x,s) = \frac{C_O^*}{s} + A(s)e^{-\sqrt{s/D_O}\,x}$$ <7-18>

Since <7-19> is obtained from the surface condition <7-11>, <7-18> becomes <7-20>, which is the final solution of the ODE.

$$\overline{C}_O(0,s) = 0, \quad A(s) = -\frac{C_O^*}{s}$$ <7-19>

$$\overline{C}_O(x,s) = \frac{C_O^*}{s} - \frac{C_O^*}{s}e^{-\sqrt{s/D_O}\,x}$$ <7-20>

The inverse transformation of <7-20> by using the relation in the last row of Table 7-1 yields <7-21> and <7-22>, wherein $C_O(x,t)$ is the final solution obtained from the Fick's second law. Here, *erf* is an ***error function***, which is defined by <7-23> (Fig. 7-4).

$$C_O(x,t) = C_O^*\left\{1 - erfc\left[\frac{x}{2(D_O t)^{1/2}}\right]\right\}$$ <7-21>

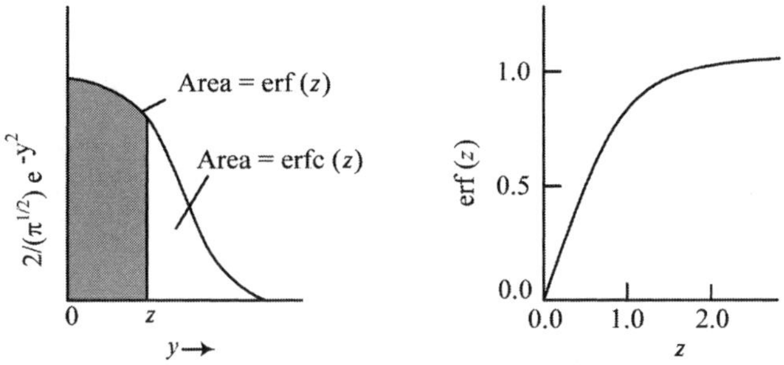

Fig. 7–4 Error function.

$$C_O(x,t) = C_O^* \ erf\left[\frac{x}{2(D_O t)^{1/2}}\right] \qquad \text{<7-22>}$$

$$erf(z) \equiv \frac{2}{\pi^{1/2}} \int_0^z e^{-y^2} dy \qquad \text{<7-23>}$$

In Fig. 7-5, $C_O(x,t)$ is simulated by using <7-22> as a function of x and time (t) assuming that $D_O = 10^{-5}$ cm^2/s. It is seen that the concentration profile of O according to x and t after the potential step is the same as that predicted qualitatively in Fig. 1-6-b.

Meanwhile, the Laplace transformation on the current equation <7-24> yields <7-25>.

$$-J_O(0,t) = \frac{i(t)}{nFA} = D_O\left[\frac{dC_O(x,t)}{dx}\right]_{x=0} \qquad \text{<7-24>}$$

$$\frac{\bar{i}(s)}{nFA} = D_O\left[\frac{d\bar{C}_O(x,s)}{dx}\right]_{x=0} \qquad \text{<7-25>}$$

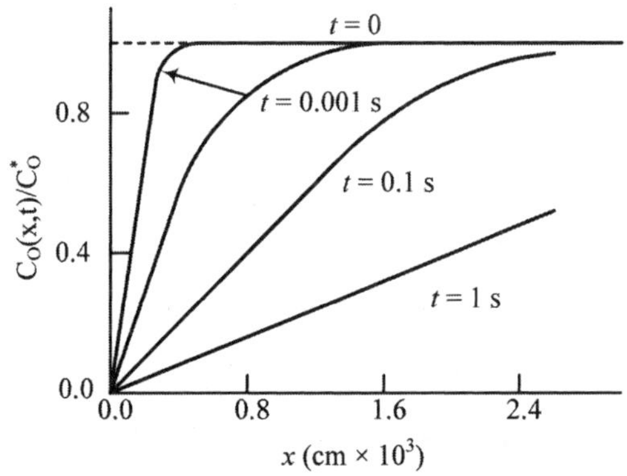

Fig. 7–5 Concentration profile of O in potential–step experiment, which was simulated using ⟨7–22⟩ with D_O = 10^{-5} cm^2/s.

<7-25> becomes <7-26> after differentiation of <7-20> with respect to x and considering $x = 0$. The inverse transformation of <7-26> by using the relationship in Table 7-1 yields <7-27>, which is the Cottrell equation.

$$\bar{i}(s) = \frac{nFAD_O^{1/2}C_O^*}{\sqrt{s}} \qquad \text{<7-26>}$$

$$i(t) = i_d(t) = \frac{nFAD_O^{1/2}C_O^*}{\sqrt{\pi t}} \qquad \text{<7-27>}$$

7-2 Limitations of potential step experiment

In potential-step experiment, the observed current is the sum of faradaic current (diffusion current) and non-faradaic current (electric double-layer charging current), and thus has the following limitations.

Firstly, the time required to charge the electric double-layer after a potential step is an important factor. Namely, the faradaic (diffusion) current is determined by the change in the concentration gradient as shown in Fig. 7-5. With an increase in the step-duration time, the natural convection, which has been ignored until now, can distort the concentration gradient shown in Fig. 7-5; the concentration gradient is disrupted by the natural convection, which usually results in currents larger than those predicted from the Cottrell equation. In order to eliminate the influence of natural convection, the step time should be generally limited to 10 to 20 seconds. This means that potential step experiments should be completed within 10 to 20 seconds. Here, an important parameter to be considered is the time required to charge the electric double-layer. As noted above, the charging current must flow right after a potential step. Hence, an appreciable non-faradaic current is superimposed with the faradaic current, such that it is difficult to identify the faradaic current precisely. Here, the longer the double-layer charging time, the narrower the time range, in which the diffusion current can be measured. Let us assume that a potential-step

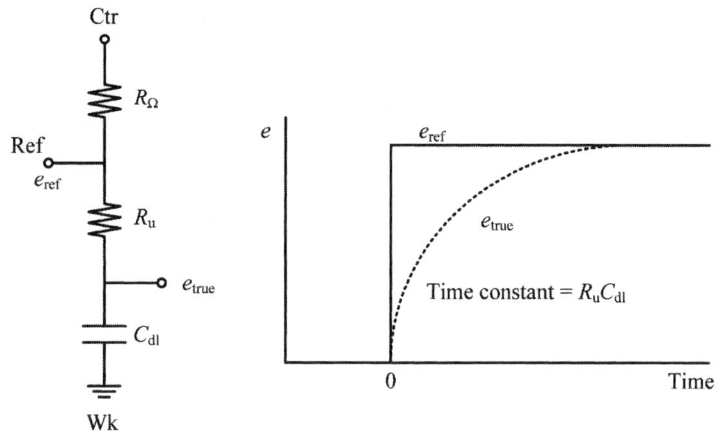

(a) (b)

Fig. 7-6 The equivalent circuit representing the charging current flow (a) and the change of e_{true} (b) immediately after a potential step.

experiment can be done within 20 seconds without the influence of natural convection, and 15 seconds are required to charge the electric double-layer. Then, the diffusion current can be measured only for the final 5 seconds without the interference of charging current. The diffusion current data obtained within a narrower time interval gives a less precise result. Obviously, the double-layer charging time should be minimized to enlarge the measuring time for diffusion current.

Fig. 7-6-a shows the equivalent circuit for the cell immediately after the potential step. Only the charging current for electric double-layer capacitor (C_{dl}) is represented while R_{ct} is removed from Fig. 5-13 because the charging current flows prior to the faradaic current. In this equivalent circuit, if a potential step is applied between the working and reference electrode, the charging current (i_C) flows according to <7-28>.

$$E = E_{R_u} + E_{C_{dl}} = i_C R_u + \frac{q}{C_{dl}}$$

$$i_C = \frac{E}{R_u} - \frac{q}{R_u C_{dl}} = \frac{dq}{dt}$$

$$q = EC_{dl}[1 - e^{-t/R_u C_{dl}}]$$

$$i_C = \frac{dq}{dt} = \frac{E}{R_u} e^{-t/R_u C_{dl}} \qquad \text{<7-28>}$$

Here, $\tau = R_u C_{dl}$, which is called a ***cell time-constant***. Note that R_u is the uncompensated resistance and C_{dl} is the electric double-layer capacitance, and their multiplication (τ) has a unit of time. <7-28> illustrates that the charging current decays exponentially with time, and further that the charging current decays to zero earlier when τ is smaller; the charging time is shorter, Therefore, in order to maximize the measuring time for faradaic current, the cell time-constant ($\tau = R_u C_{dl}$) should be minimized. To minimize the R_u, the distance between the reference and working electrode should be shortened. Meanwhile, to minimize C_{dl}, the working electrode of a small area should be used. However, since R_u increases with a decrease in electrode area, it is necessary to optimize the electrode area in consideration of two aspects (R_u and C_{dl}).

The second limitation is the problem of rising time. The potential profile in the potential step experiment should be that shown by the solid line in Fig. 7-6-b; the rectangular-shaped potential step. However, as is given in <7-29>, the actual potential (e_{true}) applied to the reference electrode against the grounded working electrode changes as the dotted line in Fig. 7-6-b; the potential reaches the step value after a certain period time, which is called a rising time. Considering that the electrochemical reaction rate (current) changes very sensitively to the applied potential, if the potential deviates from the ideal step form, the current is also forced to deviate from the expected value; there is an error in the current measurement. As can be seen from <7-29>, however, the applied potential (e_{true}) becomes closer to the ideal rectangular shape with a decrease in the cell time-constant (τ); the rising time becomes shorter. That is, the smaller the cell time-constant, the closer the measured current is to the actual value of the electrochemical system. Obviously, both R_u and C_{dl} should be minimized.

$$e_{R_u} = i_C R_u = E e^{-t/\tau}$$

$$e_{true} = e_{ref} - e_{R_u} = e_{ref}(1 - e^{-t/\tau}) \qquad \text{<7-29>}$$

Thirdly, immediately after a potential step, a large charging current flows between the working and counter electrode. Hence, the potentiostat providing a high compliance voltage should be used (refer to <5-15>). In particular, its use is more demanding when organic electrolytes having a high solution resistance are used in potential-step experiment.

*7-3 Potential step experiment using planar ultra-microelectrode (UME)

When a ***planar ultra-microelectrode*** (UME) is used as the working electrode in potential-step experiment, the pattern of diffusion layer expansion differs from that of a planar microelectrode. As shown in Fig. 7-1, with a planar microelectrode, the diffusion layer expands to the vertical direction (one-dimensional) because the diffusion layer (1 μm ~ 0.1 cm) is much thinner than the electrode area (See Example 4-3). In this case, as the step time elapses, the concentration gradient of reactant in the diffusion layer decreases to cause a decrease in the diffusion current (Cottrell equation).

Fig. 7-7 shows how the diffusion layer expands when a planar ultra-microelectrode (UME) is used. As is seen, the one-dimensional diffusion dominates immediately after the potential step, but the spherical (three-dimensional) diffusion becomes dominant in the later period of time. When the spherical diffusion prevails, two conflicting factors affect the diffusion rate (flux per unit area). One is the decrease of concentration gradient of the reactant due to the expansion of diffusion layer with time. The other is the facilitated diffusion rate due to three-dimensional diffusion; the reactant approaches the electrode surface from three-dimensional directions instead of one direction. The former causes a decrease in the diffusion rate, but the latter enlarges it; the flux decrease caused by the decrease of concentration gradient is compensated by the flux increase due to

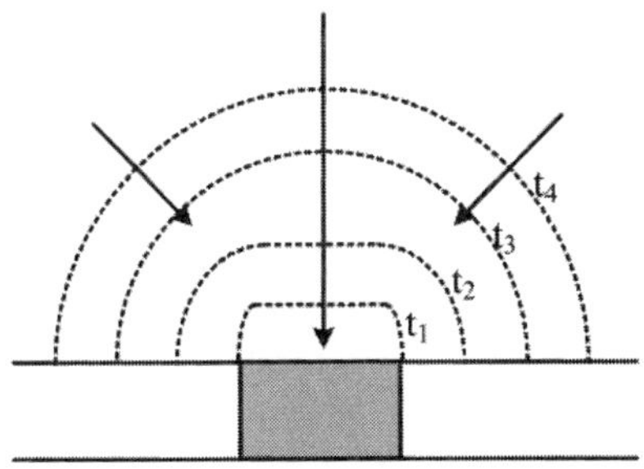

Fig. 7-7 Expansion of diffusion layer with time (from t_1 to t_4) after a potential step on a planar ultra microelectrode (UME); Change from initial linear diffusion to later spherical diffusion.

three-dimensional diffusion. As a result, when the spherical diffusion prevails, the diffusion current per unit area is larger compared to that observed with linear diffusion.

When a ***spherical electrode*** (e.g. a mercury-drop electrode) is used for the working electrode in potential-step experiments, the linear diffusion is dominant immediately after the step, but it is converted into ***spherical diffusion*** in the later period. Such a transition is also observed with planar UME. Hence, the boundary conditions relevant to spherical electrodes can be applied to planar UMEs. The Fick's second law for a spherical electrode is expressed by <7-30>, where r is the radial distance and r_0 is the radius of spherical electrode. By using the initial boundary condition <7-31>, semi-infinite boundary condition <7-32>, and the surface condition <7-33>, <7-30> can be solved to <7-34>.

$$\frac{\partial C_O(r,t)}{\partial t} = D_O \left\{ \frac{\partial^2 C_O(r,t)}{\partial r^2} + \frac{2}{r}\frac{\partial C_O(r,t)}{\partial r} \right\} \qquad \text{<7-30>}$$

$$C_O(r,0) = C_O^*, \quad (r > r_0) \qquad \text{<7-31>}$$

$$\lim_{r \to \infty} C_O(r,t) = C_O^* \qquad \text{<7-32>}$$

$$C_O(r_0,t) = 0 \ (t > 0) \qquad \text{<7-33>}$$

$$i_d(t) = nFAD_O C_O^* \left[\frac{1}{(\pi D_O t)^{1/2}} + \frac{1}{r_0} \right] \qquad \text{<7-34>}$$

<7-34> can be rearranged into <7-35> by using the Cottrell equation. An inspection of <7-35> and <7-27> reveals that the diffusion current with a spherical electrode, i_d (spherical electrode), is dominated by the linear diffusion current within a short step time. With an increase of step time, however, i_d (linear) decreases with time ($t^{-1/2}$) to finally approach zero (<7-36>). As a result, only the current associated with spherical diffusion remains with a constant value in the later period of time (<7-37>).

$$i_d \text{ (spherical electrode)} = i_d \text{ (linear)} + \frac{nFAD_O C_O^*}{r_0} \qquad \text{<7-35>}$$

$$\lim_{t \to \infty} i_d \text{ (linear)} = 0 \qquad \text{<7-36>}$$

$$\lim_{t \to \infty} i_d \text{ (spherical electrode)} = \frac{nFAD_O C_O^*}{r_0} = 4\pi nFD_O C_O^* r_0 = i_{ss} \qquad \text{<7-37>}$$

The diffusion current with a planar ultra-microelectrode (UME) can also be represented as in <7-38>, wherein r_o is the radius of planar UME.

$$i_d \text{ (UME)} = i_d \text{ (linear)} + \frac{nFAD_O C_O^*}{r_0} \qquad \text{<7-38>}$$

Fig. 7-8 compares the diffusion currents (current/electrode area, i_{ss}/A) that were obtained using disk electrodes of different radius (r_o). For the planar microelectrode (r_o = 10 mm), the current decreases according to the Cottrell equation ($\propto t^{-1/2}$). For the planar UME (r_o = 1 μm), however, the current shows the Cottrell behavior only in the very early period of step time, but reaches a constant value. As the radius decreases, the time required to reach the constant values becomes shorter as expected from <7-38>. In short, the use of UME not only alleviates the problems of charging time and rising time, but also allows a measurement of constant currents (the horizontal lines in Fig. 7-8) within a shorter period of time. From the constant current values, diffusion coefficient can be calculated.

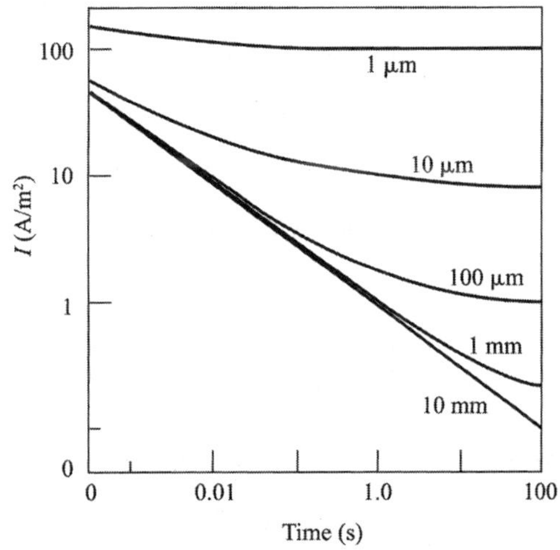

Fig. 7-8 The diffusion current (i_d) according to the radius of disk electrode.

Self learning 7-2

When a spherical electrode is used as the working electrode in the potential-step experiments, as in the case of UME, linear diffusion is dominant immediately after the potential step but is replaced by spherical diffusion in the later period. Confirm this transition by tracing the expansion of diffusion layer on a spherical electrode with step time.

Self learning 7-3

Using <7-38>, calculate how small the radius of a planar UME is in order to reach a constant current within 0.1 s. Assume that the diffusion coefficient is 10^{-5} cm^2/sec.

*7-4 Small-amplitude potential step experiment

An electrochemical reduction ($O + ne \rightarrow R$, which is a nernstian) is performed by applying potential step from E_1 to E_2. The used electrode is a planar microelectrode, in which only linear diffusion prevails. E_1 is fixed while E_2 is subject to a change. An assumption is that the electrochemical reduction at E_2 is mass transfer limited. The boundary conditions of <7-9> and <7-10> are still applicable to solve <7-8>. The flux balance is as follows.

$$D_O\left(\frac{dC_O(x,\ t)}{dx}\right)_{x=0} + D_R\left(\frac{dC_R(x,\ t)}{dx}\right)_{x=0} = 0 \qquad \text{<7-39>}$$

After rearranging <4-17> into <7-40>, and by performing Laplace transform and inverse transform, the diffusion current is derived as <7-41>.

$$\theta = \frac{C_O(0,\ t)}{C_R(0,\ t)} = e^{nF(E-E^{0'})/RT} \qquad \text{<7-40>}$$

$$i(t) = \frac{nFAD_O^{1/2}C_O^*}{\pi^{1/2}t^{1/2}(1+p\theta)} \quad \text{<7-41>}$$

$$p = \left(\frac{D_O}{D_R}\right)^{1/2} \quad \text{<7-42>}$$

In <7-40>, $\theta \rightarrow 0$ when E_2 becomes very negative. Then, <7-41> becomes <7-2>; the very negative E_2 is the condition for the large-amplitude potential-step experiment. Thus, the diffusion current should follow the Cottrell equation for one-dimensional diffusion. Meanwhile, if E_2 is not very negative, $\theta > 0$. <7-41> illustrates that, for $\theta > 0$, the shape of diffusion current is similar to that for $\theta = 0$ (Cottrell equation), but the current size decreases.

Problems

01 Calculate the time required to fill the capacitor by 95% when a potential-step is applied to the dummy cell shown in Fig. 7-6-a; R_u = 5 Ω, R_Ω = 20 Ω, e_{ref} = 1.0 V, and C_{dl} = 20 μF. Calculate the minimum compliance voltage of a potentiostat to perform this potential-step experiment.

02 Even if the electrode area, n, or C_O^* are not known, a diffusion coefficient can be calculated from the limiting current value obtained from RDE experiment and chronoamperometry data obtained using the same electrode. Rationalize this and suggest the causes of errors.

03 One wants to perform an electrochemical reduction ($O + e \rightarrow R$, which is a nernstian) by using a potential-step experiment. As shown in Fig. 7-9, five different E_2 were applied while E_1 (>> $E^{0'}$) was fixed at a constant value. Initially,

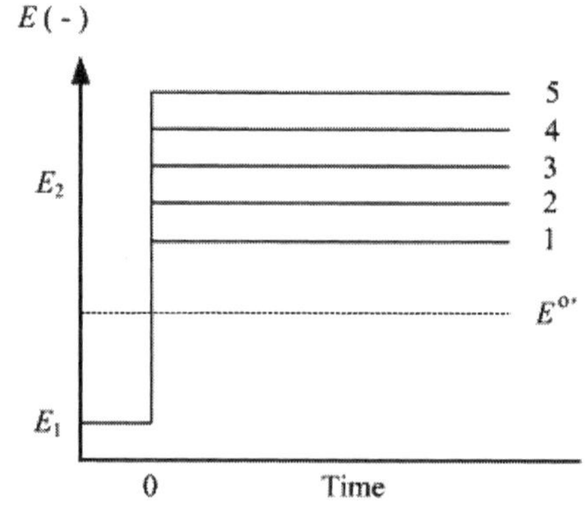

Fig. 7–9 Variation of electrode potential in a potential–step experiment, which was performed from E_1 to the five different E_2.

only O exists in the solution.

(a) Sketch the diffusion current [i(t)], assuming that E_2 (= 5) has a sufficiently large negative value (E_2 - $E^{0'}$<< −118 mV).

(b) Sketch the i(t), assuming that E_2 is more negative than E_2 (= 5), and compare it to that obtained in (a).

(c) Sketch i(t) for five E_2 values. Explain the reason for the change.

CHAPTER

08

Potential sweep methods

In potential sweep methods, the potential of working electrode is changed linearly with time and the current is measured. In linear sweep voltammetry (LSV), the electrode potential is swept from an initial value to a final one. In cyclic voltammetry (CV), however, the scan direction is changed at a switching potential to return back to the initial one.

In potential sweep methods, it is assumed that there is no change in the concentration (C_O^* and C_R^*) of electroactive species during the experiment due to the use of a small-area electrode. Also, the migration of ionic electroactive species is ignored because of the addition of supporting electrolytes. In addition, the experiments are performed without stirring in general case, such that diffusion is the way for mass transfer. Furthermore, since a planar microelectrode is used as the working electrode, only the linear diffusion is assumed to the vertical direction of electrode surface. This method is used to diagnose the electrochemical systems rather than the quantitative analysis. In diagnose, major concerns are the measurement of kinetic and thermodynamic parameters, and investigation of reaction mechanisms.

8-1 Linear sweep voltammetry (LSV)

In this experiment, the potential of working electrode is swept from an initial value (E_i) to a final one at a constant scan rate (v) ($E = E_i - vt$) and the resulting current is measured. The scan rate or sweep rate (v) is generally 1 mV/s ~ 1 V/s.

Electrochemical reactions can be divided into electrochemically reversible systems (nernstian), quasi-reversible and irreversible systems depending on the charge transfer rate. The current-potential characteristics are also different for the three cases.

8-1-1 Linear sweep voltammetry for nernstian

Let us assume that an electrochemical reaction ($O + ne = R$) is a nernstian, such that $C_O(0,t)$ and $C_R(0,t)$ are determined by the Nernst equation according to the electrode potential. Also let us assume that mass transfer determines the overall rate in the whole potential range ($R_{ct} = 0$). Note that any electrochemical systems do not meet this condition ($R_{ct} = 0$) but some are close to this. Hence, their current-potential behaviors are close to what is expected from $R_{ct} = 0$. The followings are expected from the above two assumptions. When the electrode potential is swept from a to e as shown in Fig. 8-1, the surface concentration $C_O(0,t)$ decreases with time as shown in Fig. 8-2. Meanwhile, the diffusion layer expands toward the bulk solution at a rate of $l = (2D_O\ t)^{1/2}$; this is

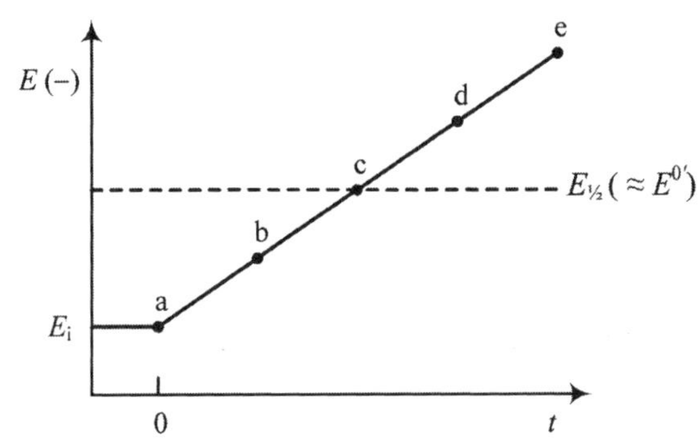

Fig. 8-1 Change of electrode potential with time in LSV experiment.

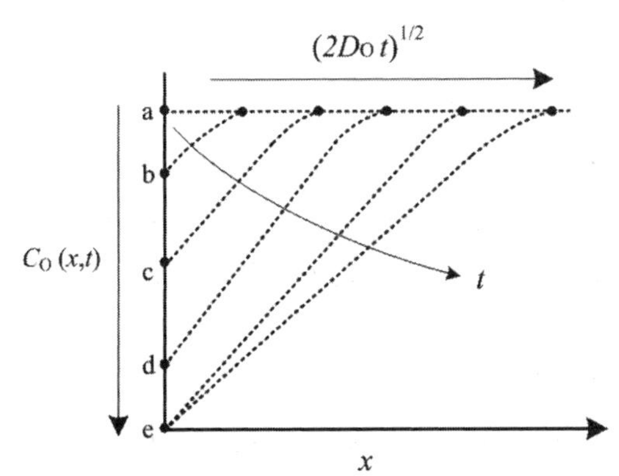

Fig. 8-2 The evolution of concentration gradient of *O* with time in LSV experiment.

also the case for the potential-step experiments (Chap. 7). Thus, the concentration gradient of O in the diffusion layer gradually increases (from a to c) to have a maximum slope (d), and then decreases (e and thereafter).

As the current is determined by the concentration gradient of O in the diffusion layer, the current also shows the same changing pattern. The peak current is observed at the peak potential (E_{pc}) (d in Fig. 8-2), at which the steepest slope is observed. $E_{pc} = E_{1/2} - (28.5/n)$ mV for nernstian at 25 °C from the numerical analysis as described below.

To obtain diffusion currents, the following Fick's second law should be solved.

$$\frac{dC_O(x,\ t)}{dt} = D_O \frac{d^2C_O(x,\ t)}{dx^2}, \quad \frac{dC_R(x,\ t)}{dt} = D_R \frac{d^2C_R(x,\ t)}{dx^2} \qquad \text{<8-1>}$$

The following initial and semi-infinite boundary conditions are derived, assuming that only O is initially present in the solution and the reduction reaction does not proceed at the initial potential (E_i).

$$C_O(x,\ 0) = C_O^*, \quad C_R(x,\ 0) = 0 \qquad \text{<8-2>}$$

$$\lim_{x \to \infty} C_O(x,\ t) = C_O^*, \quad \lim_{x \to \infty} C_R(x,\ t) = 0 \qquad \text{<8-3>}$$

In addition, the following flux balance should be satisfied.

$$D_O\left(\frac{dC_O(x,\ t)}{dx}\right)_{x=0} + D_R\left(\frac{dC_R(x,\ t)}{dx}\right)_{x=0} = 0 \qquad \text{<8-4>}$$

Meanwhile, θ is defined as follows since it is a nernstian.

$$\theta = \frac{C_O(0,\ t)}{C_R(0,\ t)} = e^{nF(E-E^{0'})/RT} \qquad \text{<8-5>}$$

As the electrode potential changes from the initial value (E_i) with a constant rate (v) until $t = \lambda$ as shown in <8-6>, <8-5> is rewritten as <8-7>.

$$E(t) = E_i - vt \ (0 < t \leq \lambda) \qquad \text{<8-6>}$$

$$\frac{C_O(0,\ t)}{C_R(0,\ t)} = e^{nF(E(t)-E^0)/RT} = e^{nF(E_i - vt - E^{0'})/RT} \qquad \text{<8-7>}$$

By applying the above boundary conditions and obtaining the solution through the Laplace transform and inverse transform, the diffusion current is obtained as <8-8>.

$$i = nFAC_O^*(\pi D_O \sigma)^{1/2}\chi(\sigma t) \qquad \text{<8-8>}$$

$$\sigma = \left(\frac{nF}{RT}\right)v \qquad \text{<8-9>}$$

$$\sigma t = \frac{nF}{RT}[E_i - E(t)] \qquad \text{<8-10>}$$

As shown in Fig. 8-3, the $\pi^{1/2}\chi(\sigma t)$ values obtained from numerical analysis gradually increase with a negative potential change, and then decrease after showing a peak value.

In Fig. 8-3, the peak value of $\pi^{1/2}\chi(\sigma t)$ is 0.4463 at $n(E - E_{1/2}) = -28.5$ mV, so that the peak current appears at the potential that is negative by $28.5/n$ mV than $E_{1/2}$ (Fig. 8-4-a). The current-potential profile can be explained by the change of concentration gradient shown in Fig. 8-2. The concentration gradient of O increases initially to reach the maximum value at $E_{pc} = E_{1/2} - (28.5/n)$ mV), such that the current also shows the peak value at this potential. The potential giving the peak current is called a ***peak potential*** (E_{pc}). Note that the half-wave potential ($E_{1/2}$) is very close to the formal potential ($E^{0'}$). Meanwhile, the concentration gradient decreases due to the expansion of diffusion layer after $C_O(0,t) = 0$ in Fig. 8-2; in the potential region more negative than e in Fig. 8-2.

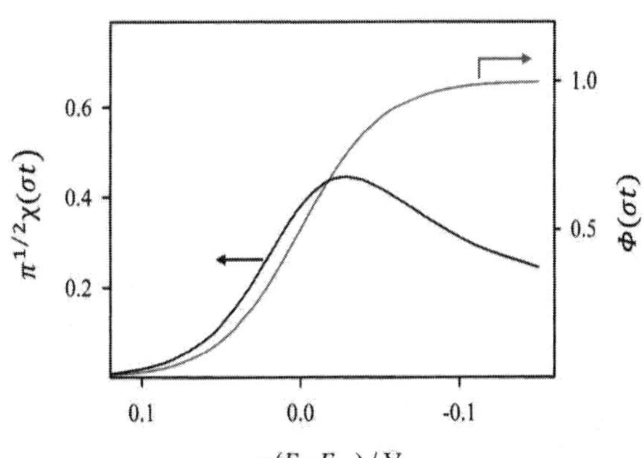

Fig. 8-3 Change of $\pi^{1/2}\chi(\sigma t)$ (planar electrode) and $\Phi(\sigma t)$ (spherical correction) in linear sweep voltammetry for reversible charge transfer reaction.

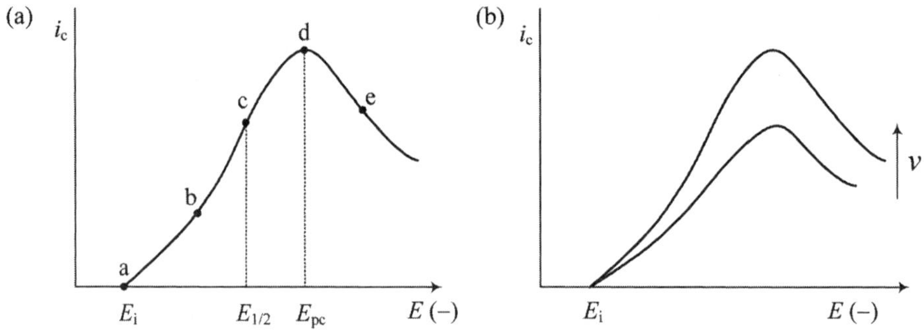

Fig. 8-4 (a); A linear sweep voltammogram obtained by substituting the potential-dependent $\pi^{1/2}\chi(\sigma t)$ values into ⟨8-8⟩. (b); The magnitude of diffusion current according to scan rate.

The condition is identical with the potential-step experiment in this potential range; the current decays by $t^{-1/2}$ according to the Cottrell equation.

The substitution of the maximum value of $\pi^{1/2}\chi(\sigma t)$ = 0.4463 into <8-8> gives the ***peak current*** (i_{pc}) as expressed in <8-11>. Here, the value of 2.69×10^5 is obtained by using R = 8.314 J/mol K and T = 298 K, and the units of the parameters are: A; cm^2, D_O; cm^2/s, C_O^*; mol/cm^3, v; V/s, and i_{pc}; amperes.

$$i_{pc} = 0.4463\, n^{3/2}\left(\frac{F^3}{RT}\right)^{1/2} A\, D_O^{1/2} C_O^* v^{1/2} = (2.69\times10^5) n^{3/2} A\, D_O^{1/2} C_O^*\, v^{1/2} \qquad \text{<8-11>}$$

<8-11> illustrates that $i_{pc} \propto v^{1/2}$, which is shown in Fig. 8-4-b. This feature can be accounted for by using the concentration profile shown in Fig. 8-2. At a faster scan rate, $C_O(0,t)$ decreases at a faster rate while the expansion rate of diffusion layer is constant. As a result, the concentration gradient of O is steeper to give a higher current.

In potential sweep experiments, the potential of working electrode changes, such that the ***electric double-layer charging current*** also flows. Thus, the measured current is the sum of a ***faradaic current*** (reduction current for O) and the double-layer charging current. The charging current is proportional to the scan rate (v) as shown in <8-12>, wherein C_{dl} (unit: F/cm^2) is the electric double-layer capacitance.

$$Q = AC_{dl}E$$

$$i_C\text{(charging current)} = \frac{dQ}{dt} = A\, C_{dl}\,\frac{dE}{dt} = AC_{dl}v \qquad \text{<8-12>}$$

If one knows n, A and C_O^*, D_O can be obtained using <8-11>. However, since the non-faradaic current (electric double-layer charging current) is superimposed with the faradaic one, the calculated D_O value is not accurate. To obtain a more accurate value, the contribution of charging current must be minimized. The ratio of two currents can be expressed as <8-13>, which illustrates that, to minimize the interference of charging current, the scan rate should be smaller and more concentrated O should be used for the linear sweep voltammerty.

$$\frac{i_C}{i_{pc}} = \frac{C_{dl} v^{1/2}}{(2.69 \times 10^5)\, n^{3/2} D_O^{1/2} C_O^*} \qquad \text{<8-13>}$$

Self learning 8-1

Using <8-13>, calculate i_C/i_{pc} under the following conditions; C_{dl} = 20 μF/cm^2, v = 200 mV/s, D_O = 10^{-5} cm^2/s, C_O^* = 0.1 mM, and n =1. What is the i_C/i_{pc} value when C_O^* is increased tenfold?

*8-1-2 Linear sweep voltammetry for electrochemically irreversible systems

When the charge transfer is irreversible, the current flows only in one direction; the reverse current can be neglected. Thus, <8-7> for a nernstian is replaced by <8-14>.

$$O + e \xrightarrow{k_f} R$$

$$\frac{i}{FA} = D_O \left[\frac{dC_O(x,\ t)}{dx}\right]_{x=0} = k_f(t) C_O(0,\ t) \qquad \text{<8-14>}$$

Since the charge transfer rate constant (k_f) is a function of electrode potential (<8-15>), which varies with time as <8-6>, <8-14> can be rewritten as <8-16>. Here, $f = F/RT$.

$$k_f(t) = k^0 e^{-\alpha f[E(t)\ -\ E^{0'}]} \qquad \text{<8-15>}$$

$$k_f(t) C_O(0,\ t) = k_{fi} C_O(0,\ t) e^{bt} \qquad \text{<8-16>}$$

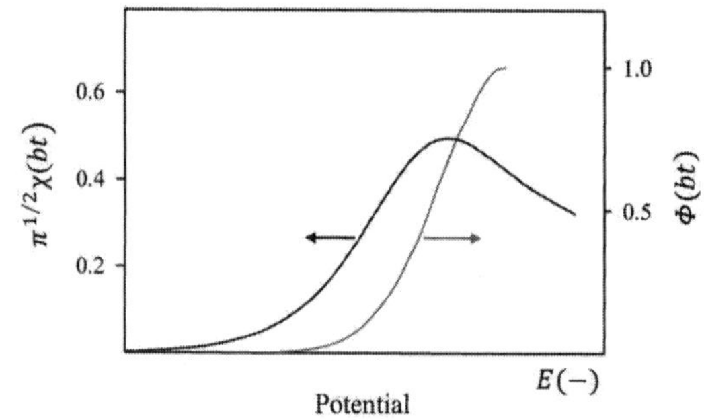

Fig. 8-5 Change of $\pi^{1/2}\chi(bt)$ (planar electrode) and $\Phi(bt)$ (spherical correction) in linear sweep voltammetry for irreversible charge transfer reaction.

k_{fi} and b are as follows.

$$k_{fi} = k^0 e^{-\alpha f(E_i - E^{0'})} \quad \text{<8-17>}$$

$$b = \alpha f v = \frac{\alpha F v}{RT} \quad \text{<8-18>}$$

The current is derived as <8-19> after a Laplace transform and inverse transform.

$$i = FAC_O^* D_O^{1/2} v^{1/2} \left(\frac{\alpha F}{RT}\right)^{1/2} \pi^{1/2} \chi(bt) \quad \text{<8-19>}$$

The current change can be plotted by substituting the value of $\pi^{1/2}\chi(bt)$ in Fig. 8-5 into <8-19>. Since the maximum value of $\pi^{1/2}\chi(bt)$ is 0.4958, the peak current is expressed by <8-20>.

$$i_{pc} = (2.99 \times 10^5)\alpha^{1/2} AC_O^* D_O^{1/2} v^{1/2} \quad \text{<8-20>}$$

The feature of $i_{pc} \propto v^{1/2}$ is similar to the nernstian (<8-11>). This illustrates that charge transfer rate increases exponentially with electrode potential even for electrochemically irreversible systems, such that mass transfer determines the overall rate when electrode potential is sufficiently negative. Obviously, however, the peak potential is more negative than that for the reversible case. As shown in <8-21>, when k^0 decreases (deviates more from nernstian), E_{pc} appears at a more negative value. <8-21> further explains that E_{pc} moves to a more negative value with an increase in scan rate (v), which is in contrast with nernstian. <8-22> is derived and the peak current (i_{pc}) is given as <8-23>.

$$E_{pc} = E^{0'} - \frac{RT}{\alpha F}\left[0.78 + \ln\left(\frac{D_O^{1/2}}{k^0}\right) + \ln\left(\frac{\alpha F v}{RT}\right)^{1/2}\right] \quad \text{<8-21>}$$

$$|E_{pc} - E_{pc/2}| = \frac{1.857RT}{\alpha F} = \frac{47.7}{\alpha} mV\ (25\ °C) \quad \text{<8-22>}$$

$$i_{pc} = 0.227\ FAC_O^{*}k^0 e^{-\alpha f(E_{pc}-E^{0'})} \quad \text{<8-23>}$$

*8-1-3 Linear sweep voltammetry for quasi-reversible systems

Only the current in one direction flows for electrochemically irreversible systems, but the reverse current should be considered for quasi-reversible systems. Therefore, as the boundary conditions to solve <8-1>, <8-24> should also be considered along with <8-2> ~ <8-4>.

$$\mathrm{O} + e \underset{k_b}{\overset{k_f}{\rightleftharpoons}} R$$

$$\frac{i}{FA} = D_O\left[\frac{dC_O(x,\ t)}{dx}\right]_{x=0} = k_f C_O(0,\ t) - k_b C_R(0,\ t) \quad \text{<8-24>}$$

For quasi-reversible systems, the current-potential profile depends on the symmetry factor (α) and Λ defined below. As shown in <8-25>, Λ represents the relative value of charge transfer and mass transfer rate. As is seen, the more the deviation from the electrochemical reversibility, the smaller Λ. If $D_O = D_R = D$, <8-25> is simplified as <8-26>.

$$\Lambda = \frac{k^0}{(D_O^{1-\alpha} D_R^{\alpha} f v)^{1/2}} \quad \text{<8-25>}$$

$$\Lambda = \frac{k^0}{(Dfv)^{1/2}} \quad \text{<8-26>}$$

The current is given by <8-27>. The maximum value of ψ(E) decreases and the peak potential moves to the negative direction with a decrease in Λ. This means that the peak potential moves to a more negative value and the peak current decreases as the systems deviate more from the electrochemical reversibility.

$$i = FAD_O^{1/2} C_O^{*} f^{1/2} v^{1/2} \psi(E) \quad \text{<8-27>}$$

Self learning 8-2

Sketch how the linear sweep voltammograms change with a decrease in Λ for quasi-reversible charge transfer. Assume α = 0.5.

8-2 Linear sweep voltammetry for multicomponent systems

If the electroactive species O and O' are reduced at similar potentials, their reduction currents are overlapped as shown by the solid line in Fig. 8-6. There is no problem in measuring the peak current (i_{pc}) of O because the baseline can be easily identified. The measurement of peak current (i'_{pc}) of O' is impossible because there is no baseline. The baseline for the second reduction current can be obtained by using the method shown in Fig. 8-7. Let us assume that the O' reduction reaction is negligible at a potential E^* and E^* corresponds to (e) in Fig. 8-2. When the potential is swept from the initial value to E^* and stopped at E^* for some period of time (experiment 1 in Fig. 8-7), the reduction current of O appears without the interference of O' reduction. The reduction current of O will show $t^{-1/2}$ dependence as the Cottrell equation suggests. This is because $C_O(0,t)$ = 0 at E^* and the diffusion layer expands at a rate of $l = (2D_O\ t)^{1/2}$ even if the electrode

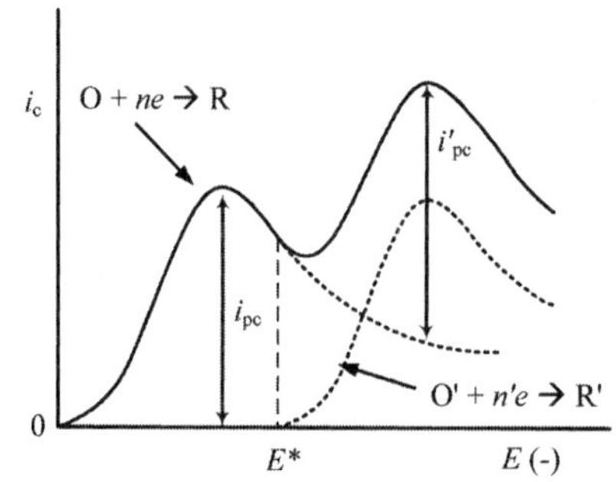

Fig. 8–6 The reduction current (dotted lines) for O and O', and the total reduction current (solid line) for the mixture of O and O'.

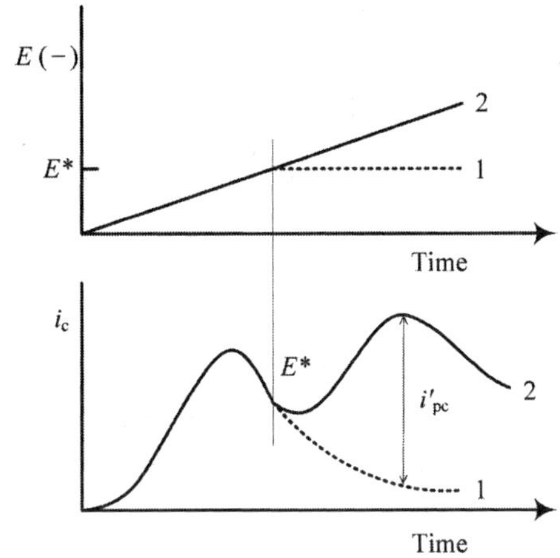

Fig. 8-7 Experimental method for obtaining the baseline for the measurement of i'_{pc}. The top; potential program and the bottom; voltammogram with potential stop at E^* (experiment 1) and potential scan continued (experiment 2).

potential is stopped at E^*. This feature is identical with the potential-step experiment, thereby the current shows the Cottrell behavior. The peak current (i'_{pc}) of the second wave can be obtained by using the as-obtained baseline after obtaining the total current (solid line) by the experiment 2 in Fig. 8-7.

8-3 Cyclic voltammetry (CV)

The ***cyclic voltammetry*** experiment is the same as LSV except that the scan direction is reversed at a switching potential (E_λ) (Fig. 8-8). The potential change according to time is expressed by <8-28>.

$$E(t) = E_i - v\lambda + v(t - \lambda) = E_i - 2v\lambda + vt \quad (t > \lambda) \qquad \text{<8-28>}$$

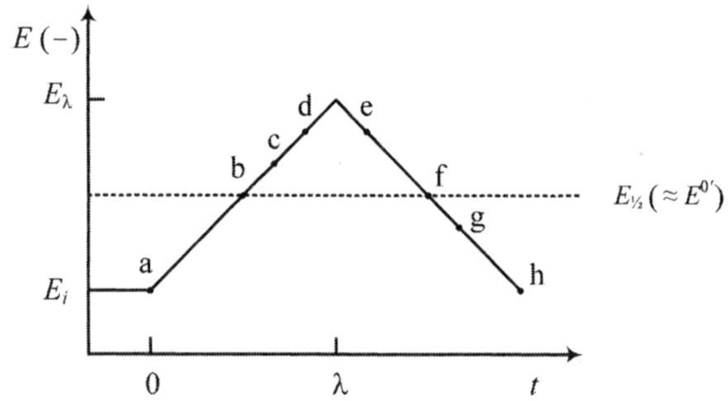

Fig. 8–8 Change of electrode potential with time in cyclic voltammetry. The potential sweep direction is reversed at the switching potential (E_λ).

8-3-1 Cyclic voltammetry for reversible charge transfer

The current *vs*. potential (time) profile can be predicted from the change of the concentration gradient of O or R in the diffusion layer. Here, it is assumed that the electrochemical reaction is a nernstian, mass transfer limited in the whole potential range, initially only O is present in the solution, and n = 1. At (a) in Fig. 8-8, the potential is sufficiently more positive than $E^{0'}$, such that $C_O\,(0,t) = C_O^*$ because the reduction of O is negligible (Fig. 8-9). At $E = E^{0'}(\approx E_{1/2})$ (b), $C_O\,(0,t)/C_R\,(0,t) = 1$ and $C_O\,(0,t) = 1/2C_O^*$. At $E = E^{0'} - 28.5$ mV (c), the concentration gradient of O shows the highest value. At (d), $C_O\,(0,t) = 0$ because $E \ll E^{0'}$, but the diffusion layer expands further to the bulk

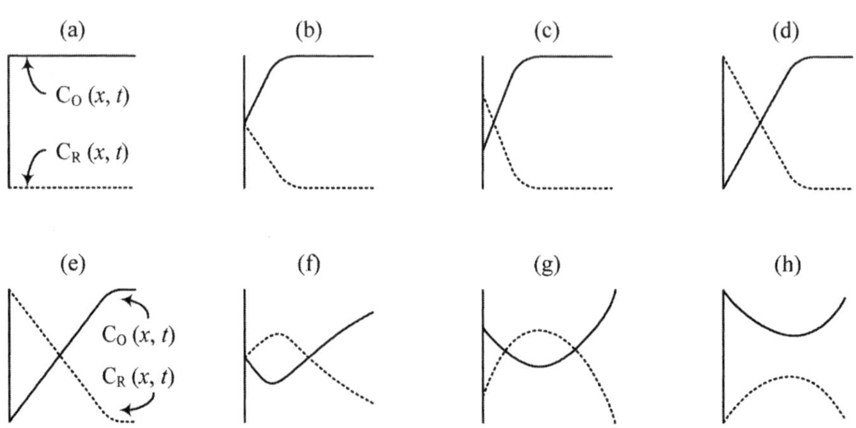

Fig. 8–9 Concentration profiles of O and R in solution according to the potential change shown in Fig. 8–8.

solution. At (e), the concentration gradient of O becomes smaller than at (d) as the diffusion layer expands further.

At (a), R does not exist, but R is generated from (b) to (e) and the diffusion layer of R expands since R tends to diffuse out to the bulk solution. The concentration profiles of R are indicated by the dotted lines in Fig. 8-9. As the potential moves to a more positive value than (e), the oxidation of R becomes more dominant than the reduction of O. At (f), $C_O(0,t)/C_R(0,t) = 1$. The concentration gradient of R shows a maximum slope at (g). The slope decreases further from (h) as the diffusion layer of R expands further toward the bulk solution.

Fig. 8-10 shows the current-potential profile that was drawn from the change of concentration gradient of O and R in Fig. 8-9. The current shape for the reverse scan may vary slightly depending on the switching potential, but it can be assumed that the current shape is identical with that for the forward scan if E_λ is more negative than E_{pc} by $35/n$ mV. In the forward scan, the cathodic peak current appears at (c) with the highest concentration gradient of O, and the anodic peak current appears at (g) with the highest concentration gradient of R. The potentials at which the peak currents (i_{pc} and i_{pa}) appear are called the ***peak potential***; E_{pc}; cathodic peak potential and E_{pa}; anodic peak potential.

In order to measure the ***peak current***, it is necessary to determine the baseline. It is not easy to determine the baseline for the reverse scan. The following method may be

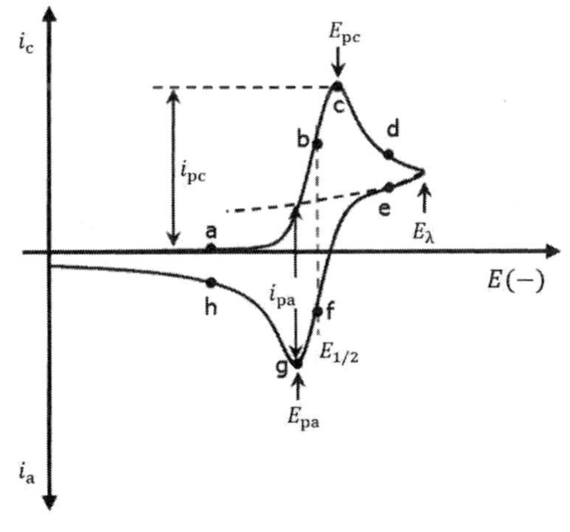

Fig. 8-10 A cyclic voltammogram; the evolution of diffusion current predicted from the concentration profiles of O and R shown in Fig. 8-9.

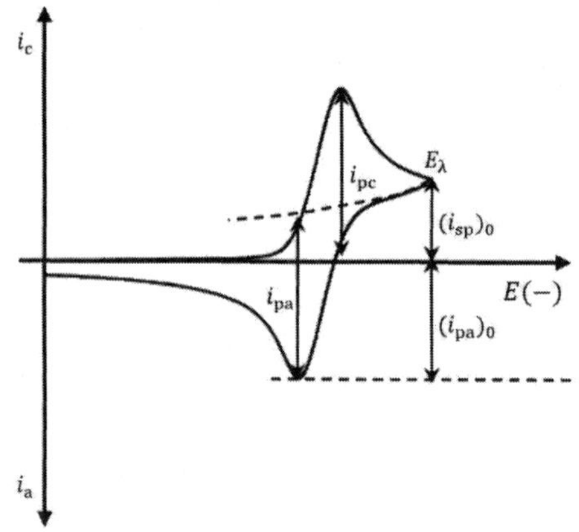

Fig. 8–11 Determination of the baseline for the reverse peak current (i_{pa}) in cyclic voltammetry.

used. As shown in Fig. 8-10, when the potential moves to the more negative value than (d), the current decreases by $t^{-1/2}$ by the Cottrell equation, which can be extended to the more negative potential than E_λ. Next, the mirror image of the current drawn at $E < E_\lambda$ is taken and attached at $E > E_\lambda$ as shown by the dotted line at $E > E_\lambda$ in Fig. 8-10. This can be used as the baseline for i_{pa} determination. Also, one can obtain i_{pa} using <8-29>, where $(i_{pa})_0$ and $(i_{sp})_0$ are defined in Fig. 8-11. In order for <8-29> to be applied, the electrochemical reaction must be a nernstian, mass transfer limited, and $E_\lambda \ll E_{pc} - 35/n$ mV.

As described above, the potential sweep experiment has limitations; the interference of charging current and uncertainty of baseline. Therefore, the measured faradaic current is imprecise. Consequently, the potential sweep experiments are not the ideal method for quantitative analysis that is based on the peak current values. However, this method is so powerful to be utilized for diagnostic purposes; e.g. identification of chemical species participating electrode reactions and analysis of reaction mechanisms (see Section 8-4).

$$\frac{i_{pa}}{i_{pc}} = \frac{(i_{pa})_0}{i_{pc}} + \frac{0.485(i_{sp})_0}{i_{pc}} + 0.086 \qquad \text{<8-29>}$$

In case of nernstian, the peak potential for the forward scan is given as $E_{pc} = E_{1/2} - (28.5/n)$. If $E_\lambda \ll E_{pc} - 35/n$ mV, the peak potential for the reverse scan is given

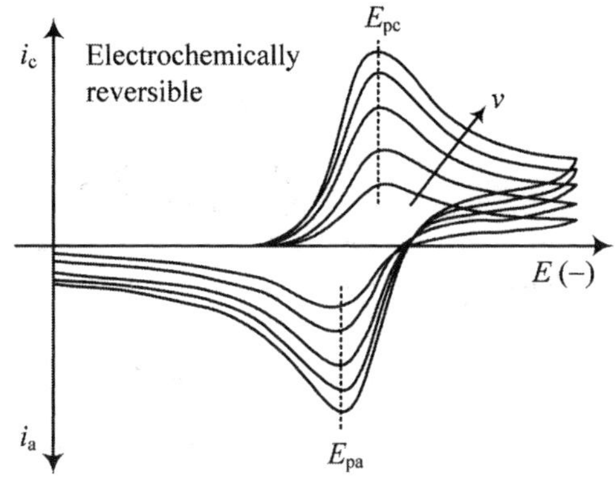

Fig. 8–12 Cyclic voltammograms as a function of scan rate for reversible charge transfer.

as <8-30>. Therefore, the difference of peak potentials is given as $\Delta E_p = E_{pc} - E_{pa} = 57/n$ mV.

$$E_{pa} = E_{1/2} + \frac{28.5}{n}\mathrm{mV} \qquad \text{<8-30>}$$

Fig. 8-12 displays the scan-rate-dependent cyclic voltammograms for a reversible charge transfer (nernstian). The electrochemically reversible systems satisfy the following conditions in cyclic voltammograms. (i) $i_{pc} = i_{pa}$ regardless of scan rate, (ii) E_{pc} and E_{pa} are constant irrespective of scan rate, thus ΔE_p (= $E_{pc} - E_{pa}$) does not change, and (iii) $i_p \propto v^{1/2}$.

During CV experiment, R generated by the reduction of O is accumulated in the diffusion layer. However, some of R will diffuse out to the bulk solution. Therefore, when the electric charge for the reverse scan (Q_r) is compared with that for the forward scan (Q_f), it is always true that $Q_r < Q_f$.

Self learning 8-3

Frequently, $\Delta E_p > (57/n)$ mV even for electrochemically reversible systems, and ΔE_p becomes larger with an increase in scan rate. Explain this using <5-19>.

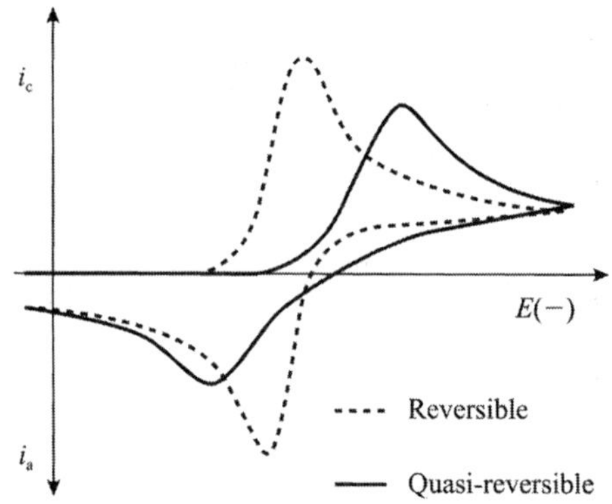

Fig. 8–13 The effect of increasing electrochemical irreversibility on the shape of cyclic voltammogram.

*8-3-2 Cyclic voltammetry for quasi-reversible charge transfer

<8-26> and <8-27> illustrate that the maximum value of $\psi(E)$ decreases and the peak potential moves to the negative direction as Λ decreases in LSV experiment. Based on this, the current-potential characteristics can be predicted for CV experiment. Firstly, when an electrochemical system deviates from nernstian, the forward peak potential moves to the negative direction together along with a decrease in the peak current. In the reverse scan, the peak potential moves to a more positive value along with a decrease in the peak current (solid line in Fig. 8-13). Therefore, $\Delta E_p > (57/n)$ mV. Secondly, as predicted from <8-26>, Λ decreases as the scan rate increases. Therefore, the peak potential moves to the negative direction in the forward scan while the reverse peak potential moves to the positive direction. As a result, ΔE_p becomes larger with an increase in scan rate.

8-4 Investigation of reaction mechanisms by cyclic voltammetry

An example of *EC*-type reaction is shown in Fig. 8-14. A feature for *EC*-type reaction is that the peak current of reverse scan is smaller than that for the forward scan, which

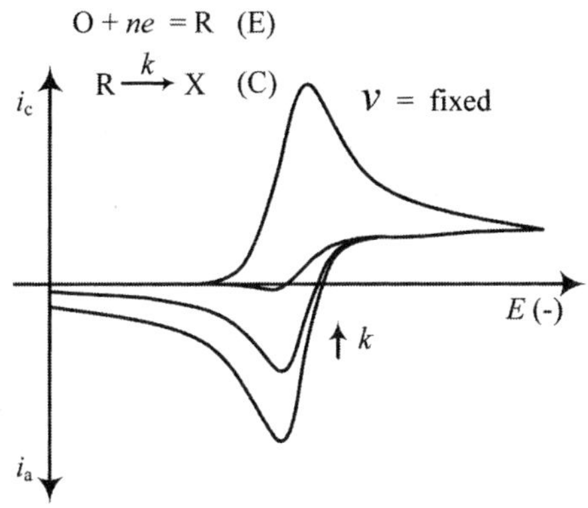

Fig. 8-14 The magnitude of reverse current according to the rate constant (*k*) of a coupled homogeneous chemical reaction in *EC*-type reaction.

becomes more prominent at slower scans. This is because the *R* generated by the reduction of *O* is converted into another compound by a homogeneous chemical reaction, and the chemical reaction proceeds to a larger degree at slower scan. Note that it takes more time at slower scan to reach the potential at which *R* is oxidized, such that more *R* is lost by the chemical reaction. Fig. 8-14 shows the change of reverse peak current according to the rate constant (*k*) of the homogeneous chemical reaction at a fixed scan rate (*v*). The larger the rate constant (*k*), the smaller the reverse current is.

J. Bacon and coworkers investigated the electrochemical oxidation mechanism of aniline (*J. Am. Chem. Soc.*, **90**, 6596, 1968). Fig. 8-15 represents the cyclic voltammograms. When the potential moves from 0.0 V to the positive value, the oxidation current of aniline appears at *ca.* 0.9 V (*vs.* SCE) (solid line indicated by (a)). Upon switching the scan direction at 1.2 V, there appears no reduction current corresponding to the oxidation current observed at 0.9 V, but instead two reduction currents are detected at points (b) and (c). When the second scan starts from 0.0 V to the positive direction (dotted line), oxidation currents are detected at (d) and (e), and the oxidation current is also detected at (a). In the second reverse scan, no reduction current is observed again near 0.9 V, but two reduction currents at (b) and (c). Based on these results, the mechanism can be inferred (Fig. 8-16). In the first forward scan, aniline is oxidized near 0.9 V, but the corresponding reduction current is not detected near 0.9 V because the oxidized product of

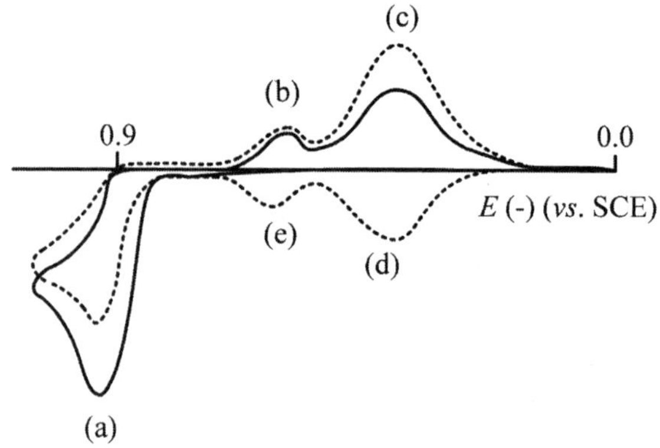

Fig. 8–15 Cyclic voltammograms for investigation of electrochemical oxidation mechanism of aniline. The electrode potential was swept from 0.0 V to 1.2 V and switched back to 0.0 V. The first scan is traced with the solid line and the second one by the dotted line.

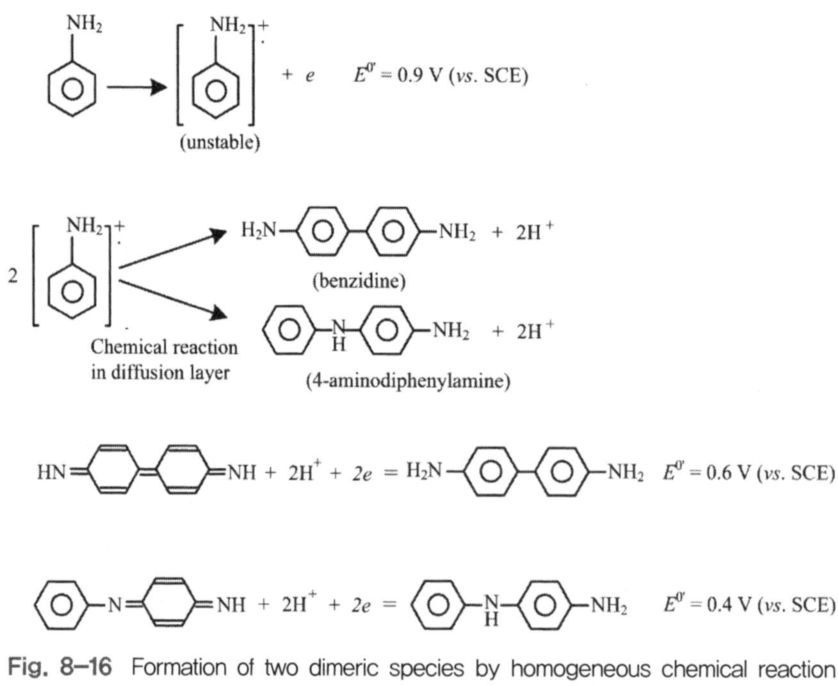

Fig. 8–16 Formation of two dimeric species by homogeneous chemical reaction of electrochemical oxidation product of aniline (the cation radical shown in the square bracket), and their redox reactions.

aniline is chemically unstable to be converted into another compound by homogeneous chemical reaction in the solution. The newly formed chemical species are then reduced at (b) and (c), and the as-reduced species are oxidized at (d) and (e) in the second forward scan. Since the electrochemical oxidation product of aniline is a cationic radical, its instability can be assumed; it dimerizes rapidly into benzidine or *N*-phenyl-*p*-phenylene diamine. To confirm this, two separate CV experiments were performed with benzidine and *N*-phenyl-*p*-phenylene diamine in solution. In one CV, benzidine gives an oxidation and reduction current at (e) and (b), respectively. In the other CV, *N*-phenyl-*p*-phenylene diamine shows a redox behavior at (d) and (c). From this, the nature of two dimeric species can be identified. Aniline is oxidized near 0.9 V in the first positive scan. The cation radical is dimerized into benzidine and *N*-phenyl-*p*-phenylene diamine. Here, the formal potential for two dimers (*ca.* 0.4 V and 0.6 V *vs.* SCE) is more negative than that for aniline (0.9 V); the dimers have a higher oxidation tendency. Consequently, the two dimers being generated near 0.9 V are immediately oxidized. The huge oxidation current near 0.9 V is resulted from the oxidation of aniline and two dimers just produced.

The type of electrochemical reactions occurring on electrode surface can be predicted from cyclic voltammograms. Fig. 8-17 shows the cyclic voltammogram taken with a Ni^{2+} solution on an inert (Pt) electrode. It is immediately apparent that this CV is far different from normal ones. Even if the reduction current for Ni^{2+} plating in the forward negative scan is similar to normal ones, the current shape in the reverse scan is totally different. Namely, the reduction current in the reverse scan at $E < E^{0'}$ is larger than that for the forward scan. This can be explained from the fact that Ni^{2+} reduction needs a larger acti-

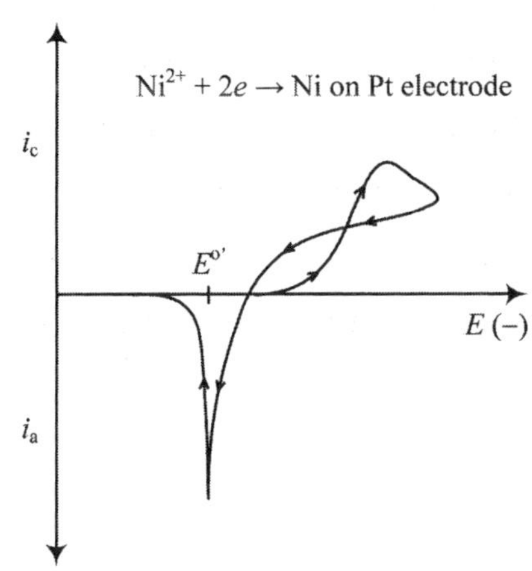

Fig. 8-17 Cyclic voltammogram showing the plating of Ni^{2+} on Pt electrode and the dissolution of the plated Ni.

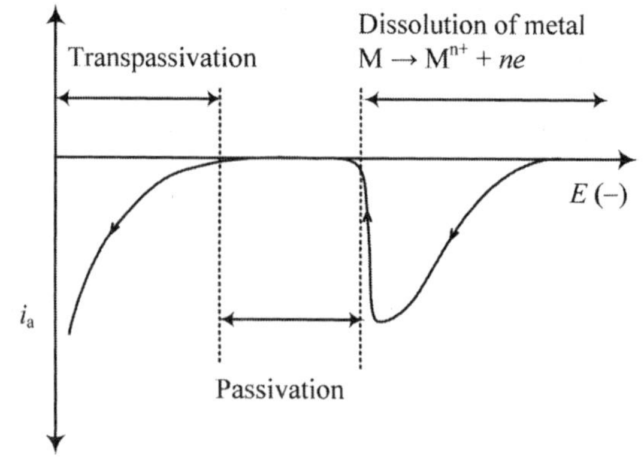

Fig. 8–18 Linear sweep voltammogram showing the passivation and transpassivation of a metal (for example, Cr).

vation energy on a different metal (Pt) surface as compared to that needed on the same metal (Ni) surface. In the forward scan, Ni is plated on Pt surface with a larger activation energy, but Ni plating takes place with a smaller activation energy on Ni surface in the reverse scan because Ni has already been plated on the Pt surface during the potential approach to the switching potential.

Fig. 8-18 shows the linear sweep voltammogram that explains the oxidation behavior of some metals. As the potential moves to the positive direction, the oxidation current flows due to metal dissolution. However, the oxidation current rapidly decreases, which is caused by the deposition of metal oxides or hydroxides that are the insoluble products of dissolved metal ions; e.g. metal ion reacts with OH^- to produce metal hydroxide. Due to the deposition of such insulating layers, the tunnelling rate decreases; the electrochemical oxidation (metal dissolution) rate becomes extremely slow when the deposition layer is thick enough, which is referred to as a ***passivation***. When the potential moves to the more positive value, the oxide (or hydroxide) layers covering the metal surface are destructed. Hence, the metal dissolution becomes serious, which is called a ***transpassivation***.

8-5 Cyclic voltammetry for adsorbed or fixed species on electrode surface

If an electroactive species O is adsorbed or fixed on an inert electrode surface, its mass transfer from bulk solution to electrode surface is not required. Then, the overall rate must be determined by the charge transfer rate even if it is a nernstian. The rate is expressed by <8-31>, which is the rate for charge transfer.

$$\text{Rate} = \frac{i}{nFA} = kC_O(0,t) = -\frac{dC_O(0,t)}{dt} \qquad \text{<8-31>}$$

Here, if the concentration of adsorbed O is C_O^*, $C_O^* = C_O(0,t) + C_R(0,t)$ should be satisfied during its reduction reaction. For a nernstian, <8-32> must be satisfied, from which <8-33> is derived.

$$E = E^{0'} + \frac{RT}{nF} \ln \frac{C_O(0,t)}{C_R(0,t)} \qquad \text{<8-32>}$$

$$C_R(0,t) = C_O^* - C_O(0,t)$$

$$E = E^{0'} + \frac{RT}{nF} \ln \left\{ \frac{C_O(0,t)}{C_O^* - C_O(0,t)} \right\} \qquad \text{<8-33>}$$

<8-33> is rearranged into <8-34>, which represents the potential-dependent $C_O(0,t)$.

$$\frac{C_O(0,t)}{C_O^* - C_O(0,t)} = \exp\left[\frac{nF(E - E^{0'})}{RT}\right]$$

$$C_O(0,t) = \frac{C_O^* \exp\left[\frac{nF(E - E^{0'})}{RT}\right]}{1 + \exp\left[\frac{nF(E - E^{0'})}{RT}\right]} \qquad \text{<8-34>}$$

Meanwhile, the reduction current is given by <8-35>, which is converted into <8-36> after considering that $v = -dE/dt$.

$$i = -nFA\frac{dC_O(0,t)}{dt} = -nFA\frac{dC_O(0,t)}{dE}\frac{dE}{dt} \qquad \text{<8-35>}$$

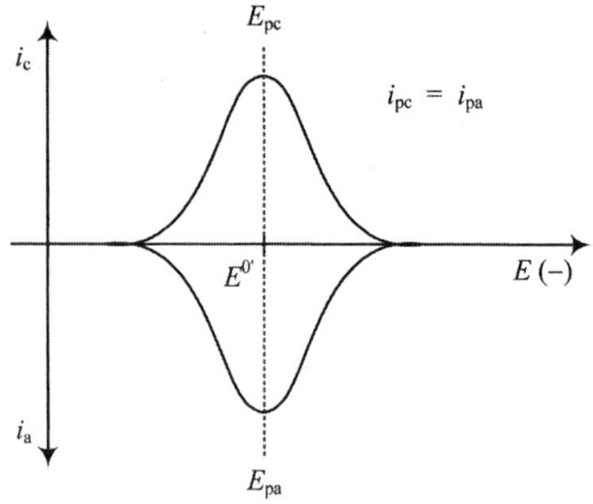

Fig. 8–19 Cyclic voltammogram for the electroactive species adsorbed or fixed (immobilized) on an inert electrode surface.

$$i = nFA\frac{dC_O(0,t)}{dE}v = \frac{n^2F^2AvC_O^*}{RT}\frac{\exp\left[\frac{nF(E-E^{0'})}{RT}\right]}{\left\{1+\exp\left[\frac{nF(E-E^{0'})}{RT}\right]\right\}^2} \quad \text{<8-36>}$$

In <8-36>, the peak current appears at $E = E^{0'}$, which is obtained as <8-37>. It is seen that the peak current is proportional to the scan rate ($i_{pc} \propto v$). This is contrasted by the situation where the electroactive species are not fixed on electrode surface, rather transferred by diffusion. Note that $i_{pc} \propto v^{1/2}$ in the latter case. From the relation between the peak current and scan rate, one can judge whether the electroactive species are confined on electrode surface ($i_{pc} \propto v$) or transferred by diffusion in solution ($i_{pc} \propto v^{1/2}$). As can be expected from <8-36>, the current is symmetrical with respect to the potential. Also, $E_{pc} = E_{pa} = E^{0'}$ and $i_{pc} = i_{pa}$ (Fig. 8-19).

$$i_{pc} = \frac{n^2F^2AvC_O^*}{4RT} \quad \text{<8-37>}$$

The ***surface-confined electroactive species*** show the behavior of $i_p \propto v$, which is also valid for the charging current of electric double-layer capacitor. Hence, the surface-confined electroactive species are called ***pseudo-capacitors***. The meaning of "pseudo" is that these are similar to electric double-layer capacitors ($i_p \propto v$) even if they are not capacitors; the observed current is a faradaic one in nature.

Fig. 8-20 shows the cyclic voltammograms that illustrate the electrochemical oxidation

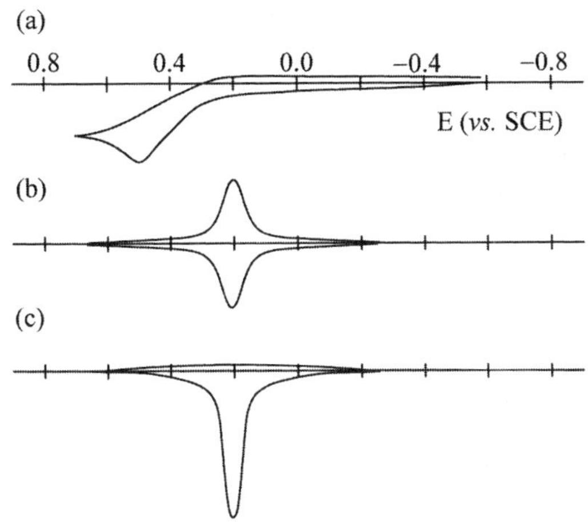

Fig. 8-20 (a); Cyclic voltammogram (CV) obtained with NADH on a glassy carbon electrode, (b); CV obtained with the glassy carbon electrode, on which a polymer containing an o-quinoline functional group is coated, and (c); CV obtained with the same glassy carbon electrode as for (b) with NADH being added into the solution.

behavior of NADH (nicotinamide adenine dinucleotide). For NADH/NAD$^+$ half-cell;

$$NAD^+ + H^+ + 2e = NADH \quad E^{0'} = -0.5 \text{ V } (vs. \text{ SCE})$$

Thermodynamics predict that the oxidation current for NADH is observed when the potential of an inert electrode is more positive negative than the equilibrium potential of the NADH/NAD$^+$ half-cell; here, $E_{eq} = -0.5$ V (*vs.* SCE) assuming a standard state. A sizable current is, however, observed at 0.5 V (Fig. 8-20-a); the overpotential amounts to 1.0 V to have this current. Such an overpotential must be due to a small k^0 for NADH oxidation on the glassy carbon electrode. Moreover, the corresponding reduction current does not appear in Fig. 8-20-a, which is the characteristic feature for irreversible charge transfer. Fig. 8-20-b presents the cyclic voltammogram taken with a glassy carbon electrode, onto which a polymer layer containing *o*-quinoline functional group is coated. The symmetrical voltammogram arising from the reduction/oxidation of *o*-quinoline/*o*-quinoline hydroxide is apparent at $E = E^{0'}$ (0.2 V *vs.* SCE). This is because *o*-quinoline functional groups are fixed on the electrode surface, and thus mass transfer is not required. A small amount of NADH was injected into the cell and cyclic voltammogram was obtained with the *o*-quinoline polymer-coated glassy carbon electrode. As shown in Fig.

8-20-c, the oxidation current at 0.2 V greatly increases at the expense of reduction current. This can be explained by the scenario that the *o*-quinoline functional group mediates the oxidation reaction of NADH. As suggested in Fig. 8-21-a, *o*-quinoline functional group reacts with NADH that diffuses into the polymer layer from the solution. By this reaction, the *o*-quinoline moiety (M_O) is reduced to *o*-quinoline hydroxide (M_R), while NADH is oxidized to NAD^+. There is an electron transfer from NADH to *o*-quinoline (M_O). Then, the as-generated *o*-quinoline hydroxide (M_R) is oxidized to *o*-quinoline (M_O) on the glassy carbon electrode at 0.2 V. The sequential process comprising the electron transfer from NADH to M_O (NADH oxidation and M_O reduction) and M_R oxidation on the glassy carbon electrode is repeated. Consequently, the oxidation current of M_R increases at 0.2 V, but the reduction current of M_O decreases. Note that the concentration of M_R in the polymer layer is much larger than that of M_O because M_R is produced while M_O is depleted by the reaction with NADH.

$$\text{NADH} + o\text{-quinoline} + H^+ \rightarrow \text{NAD}^+ + o\text{-quinoline hydroxide} \qquad \text{<8-38>}$$

Here, the *o*-quinoline (M_O)/*o*-quinoline hydroxide (M_R) redox pair is called an ***electron mediator*** because they provide an electron transfer channel from NADH to the glassy carbon electrode.

In order for any M_O/M_R redox pairs to serve as an electron mediator, at least three conditions should be met. Firstly, the electron transfer between NADH and M_O must be

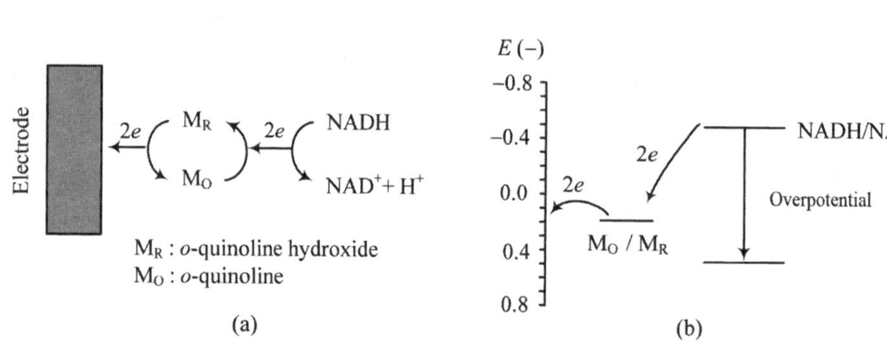

Fig. 8–21 (a); The electron flow from NADH in the solution to the *o*–quinoline functional group and finally to an inert electrode. (b); Electron transfer in the mediated process of NADH oxidation by *o*–quinoline (M_O)/*o*–quinoline hydroxide (M_R) pair. Note the difference in the equilibrium potential of M_O/M_R and NADH/NAD^+ half–cells.

spontaneous in thermodynamic point of view. The thermodynamic spontaneity can be decided by comparing the equilibrium potentials of two half-cells; here, the formal potentials were used for comparison after assuming a standard state; $E_{eq} \approx E^0 \approx E^{0'}$. The comparison of the two formal potentials reveals that the NADH/NAD$^+$ half-cell has a higher oxidation tendency with more negative $E^{0'}$, whereas the *o*-quinoline/*o*-quinoline hydroxide half-cell has a higher reduction tendency with more positive $E^{0'}$, illustrating that the paired reaction of NADH oxidation and M_O reduction is spontaneous.

$$NAD^+ + H^+ + 2e = NADH \quad E^{0'} = -0.5 \text{ V } (vs. \text{ SCE})$$

$$o\text{-quinoline} + 2H^+ + 2e = o\text{-quinoline hydroxide} \quad E^{0'} = 0.2 \text{ V } (vs. \text{ SCE})$$

Secondly, the electron transfer between NADH and M_O should be kinetically fast enough. In fact, the charge transfer between NADH and the glassy carbon electrode is irreversible as described above (Fig. 8-20-a) because of the large molecular size of NADH. However, there is no reason for the electron transfer between NADH and M_O to be slow. The small-sized *o*-quinoline moiety can easily approaches the redox center of NADH molecule. The third condition for the M_O/M_R redox pair to serve as an electron mediator is that the charge transfer rate between *o*-quinoline hydroxide (M_R) and the glassy carbon electrode should be high. One can readily recognize that this condition is met by considering the molecular size of *o*-quinoline hydroxide. Tunnelling rate seems to be high enough because this molecule can easily approaches the electrode surface.

The reaction involved in electron mediators is called ***EC'-type*** one, where *C'* means a catalytic reaction. In the above example, the electron mediator serves to decrease the overpotential. As shown in Fig. 8-21-b, the M_O/M_R redox pair acts as a catalyst for the NADH oxidation. At the given current, the overpotential for the NADH oxidation was about 1.0 V without the mediator. With the mediator, however, the overpotential decreases to 0.7 V.

If electron mediators are immobilized on electrode surface as in the above example, rather than dissolved in solution, it is not necessary to separate the mediator after the reaction. The above example illustrates that the electrode performance (overpotential and reaction kinetics) can be improved by immobilizing electroactive species (redox pairs) on electrode surface. The electrode exhibiting such improved properties is called a ***modified electrode***.

Self learning 8-4

Electrochemical oxidation of NADH is slower on glassy carbon electrode as compared to the oxidation of *o*-quinoline hydroxide. Explain the difference considering the molecular size.

Self learning 8-5

If the formal potential (that is, equilibrium potential) of M_O/M_R half-cell is more negative than that of $NADH/NAD^+$ half-cell in Fig. 8-21-b, is it possible for the M_O/M_R redox pair to mediate the electron transfer for NADH oxidation?

Self learning 8-6

If the electron transfer rate between *o*-quinoline and NADH is slower than that assumed in Fig. 8-20, how does the cyclic voltammogram change in Fig. 8-20-c?

*8-6 Potential sweep experiment using planar ultra microelectrode (UME)

In Section 7-3, the potential-step experiment using a ***planar ultra microelectrode*** (UME) is discussed. Diffusion in UME is initially dominated by linear diffusion but is converted into spherical diffusion over time. Assume that an electrochemical reaction $O + ne = R$ is a nernstian and that mass transfer limited at all potential range. Using the boundary conditions in Section 7-3, the current-potential relation in linear sweep voltammetry (LSV) is derived as follows.

$$i_d(\text{UME}) = i_d\ (\text{linear}) + \frac{nFAD_O C_O^*}{r_0}\Phi(\sigma t)$$ <8-39>

Here, r_0 is the radius of a disk-shaped UME, and i_d (linear) and σ have been defined as <8-8> and <8-9>, respectively.

Fig. 8-3 shows the potential-dependent $\pi^{1/2}\chi(\sigma t)$ and $\Phi(\sigma t)$. The inspection of <8-8> and <8-9> reveals that the i_d (linear) in <8-39> is proportional to $v^{1/2}$, but the second term in <8-39> is independent on the scan rate but proportional to r_0^{-1}. From this, it is possible to predict the LSV current shape according to the radius (r_0) of UME and scan rate (v). If both r_0 and v are large, i_d (UME) $\approx$ i_d (linear), such that the current has the shape of Fig. 8-4. Conversely, if both r_0 and v are small, <8-40> is derived. Therefore, the current has the shape of $\Phi(\sigma t)$ shown in Fig. 8-3; the shape without a peak current. In addition, the current is independent on the scan rate as expected from <8-40>.

$$i_d(\text{UME}) = \frac{nFAD_O C_O^*}{r_0}\Phi(\sigma t)$$ <8-40>

In cyclic voltammametry, if both r_0 and v are large, i_d (UME) $\approx$ i_d (linear), such that the current-potential has the shape shown in Fig. 8-10. On the other hand, if both r_0 and v are small, the peak current does not exist in the same way as $\Phi(\sigma t)$ in Fig. 8-3 in the forward scan. In the reverse scan, the current overlaps with the forward current. This can be accounted for by two features. Firstly, the mass transfer rate is high because the spherical diffusion is dominant over the linear diffusion, which is more prominent with a smaller r_0. Secondly, it takes longer to reach the switching potential and to return back to the initial potential because v is small, so that the R generated at the electrode surface (assuming that there is no R at the beginning) diffuses out toward the bulk solution. In short, the oxidation current is not detected because R diffuses out for a longer period of time due to slower san rate and at a higher rate due to the spherical diffusion. In summary, when cyclic voltammetry performs with an UME of a fixed radius (r_0), if the scan rate is high, a typical CV (Fig. 8-10) is observed. Here, $i_p \propto v^{1/2}$. In opposite, if the scan rate is low, the voltammograms show no peak current and the reverse current overlaps with the forward one. Also, the current size is constant irrespective of the scan rate.

Problems

01 Answer the following questions about *ECE*-type reactions.

$$A + e = B \qquad E^0 = -0.5 \text{ V } (vs. \text{ SCE})$$

$$B \rightarrow C$$

$$C + e = D \qquad E^0 = -1.0 \text{ V } (vs. \text{ SCE})$$

(a) Sketch two cyclic voltammograms for two different scan rates in the range of 0.0 ~ −1.5 V, and predict how the size of four currents (reduction and oxidation of *A* and *B*, reduction and oxidation of *C* and *D*) change according to scan rate. Assume that compounds *A*, *C*, and *D* are chemically stable and all electrochemical reactions are nernstian.

(b) An RDE experiment was performed in the range of 0.0 ~ −1.5 V with the rotation speed being varied, from which n value was obtained. Predict how the apparent n value (n_{apparent}) changes according to rotation speed.

(c) An RRDE experiment was performed with the disk potential being varied in the range of 0.0 ~ −1.5 V and the ring potential being fixed at 0.0 V. Predict how $N = (-i_R/i_D)$ changes according to rotation speed.

02 Fig. 8-22 shows the cyclic voltammograms taken with a Fe^{3+}-porphyrin complex. The voltammogram taken in argon atmosphere shows a comparable reduction and oxidation current. In air, however, the reduction current increases greatly at the expense of oxidation current. Explain this.

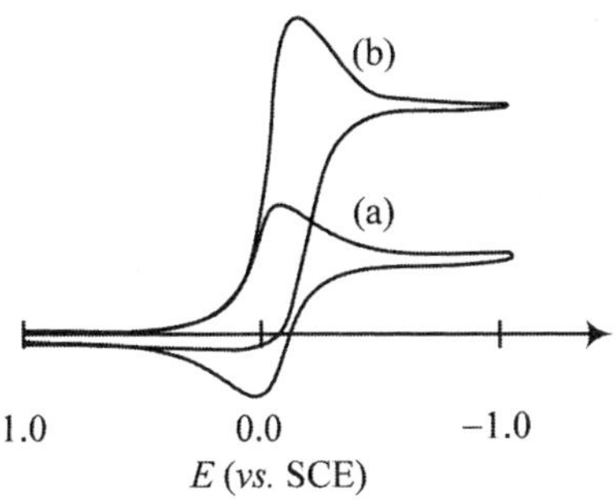

Fig. 8-22 Cyclic voltammogram of a Fe^{3+}–porphyrin complex on an inert electrode (a); in argon atmosphere and (b); in air.

03 The cyclic voltammogram in Fig. 8-23 was obtained by: the potential sweep starts at 200 mV to the negative direction until −300 mV, the positive sweep performs until 800 mV, and finally reaches 200 mV. Th one-electron reduction reaction ($A + e \rightarrow B$) is a nernstian.

(a) Identify the compounds that are accumulated near the electrode surface at the following potential: 200 mV (starting point), −300 mV, 200 mV, 800 mV, and 200 mV (final point).

(b) Draw the expected cyclic voltammogram when the scan rate is faster than that for in Fig. 8-23. Describe how the four current peaks change.

(c) When the scan rate is faster than that for Fig. 8-23, how does the peak potential for $A + e \rightarrow B$ (about -30 mV in Fig. 8-23) move? Rationalize your answer.

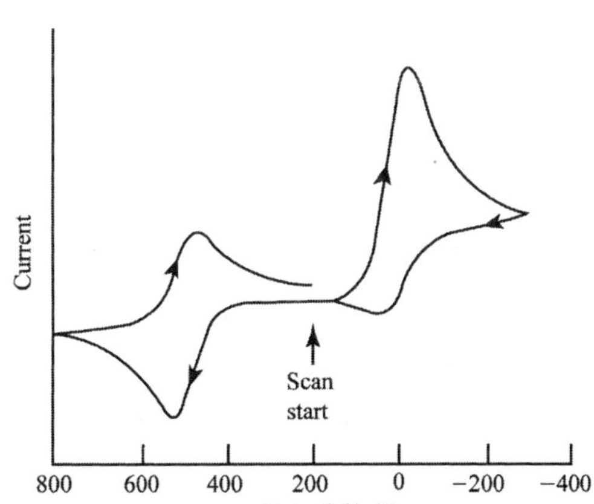

Fig. 8-23 Cyclic voltammogram obtained with compound *A*.

04 R. S. Nicholson and his coworkers predicted how the peak current ratio (i_{pa}/i_{pc}) change according to scan rate for electrochemical reactions (Fig. 8-24). i_{pa}/i_{pc} = 1.0 for some electrochemical reaction (*a*) or i_{pa}/i_{pc} either increases or decreases with an increase in scan rate (*b*, *c*, and *d*). Determine what type of change is observed for the following electrochemical reactions.

(1) $O + e \rightleftarrows R$ (nernstian)

(2) $O + e \rightleftarrows R$, $R \rightarrow Z$ (*EC* reaction)

(3) $O + e \rightleftarrows R$, $R \rightleftarrows Z$ (*EC* reaction)

(4) $O + e \rightleftarrows R$, $R + Z \rightarrow O$ (*EC'* reaction, catalytic regeneration)

(5) $Z \rightleftarrows O$, $O + e \rightleftarrows R$ (Preceding chemical reaction)

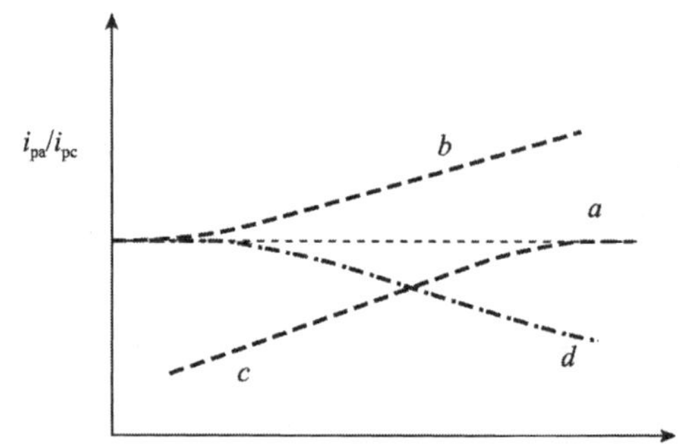

Fig. 8–24 Ratio of two peak currents (i_{pa}/i_{pc}) according to scan rate in cyclic voltammetry. All electrochemical reactions are assumed to be nernstian. The electrode potential was scanned to the negative direction and returned back to the initial value.

05 J. Heinze and his coworkers (*Ber. Bunsenges. Phys. Chem.*, **90**, 1043, 1986) reported the scan-rate-dependent cyclic voltammograms obtained with a planar ultra microelectrode (UME) for $FeCp_2^+ + e = FeCp_2$ (Fig. 8-25). Initially, only $FeCp_2$ was present in the solution.

(a) At the fast scan (iii), the normal cyclic voltammogram is observed, but the reduction current disappears as the scan rate decreases. Explain this.

(b) If the scan rate is further decreased to 0.01 V/sec, how does the cyclic voltammogram differ from that obtained at 0.1 V/sec (i)?

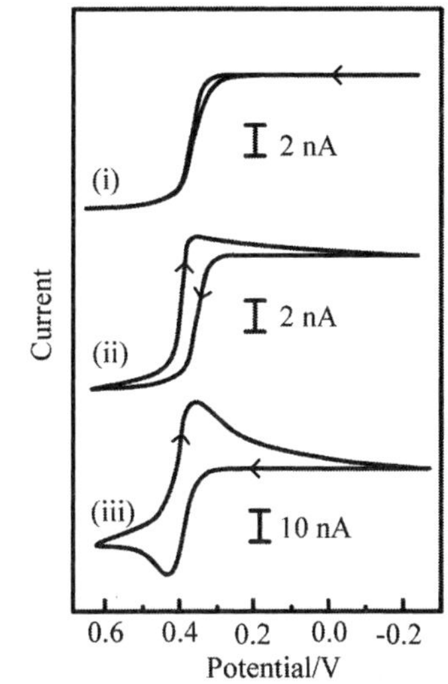

Fig. 8–25 Cyclic voltammograms of ferrocene ($FeCp_2$) obtained with a planar ultra–microelectrode (UME) with a diameter of 6 μm. The san rate was changed: (i); 0.1 V/s, (ii); 1.0 V/s, and (iii); 10 V/s. The arrows indicate the potential sweep direction.

CHAPTER 09

Bulk electrolysis

In the electrochemical techniques introduced so far, it is assumed that the concentration of electroactive species (C_O^* and C_R^*) does not change because the experiments are performed within a short period of time (generally, less than 1 min) with electrodes of small surface area. In bulk electrolysis, however, all of the electroactive species in the cell are oxidized/reduced to measure the electric charges flowing during a given time, from which n value or concentration (C_O^* or C_R^*) are obtained. In order to perform a bulk electrolysis within a reasonably short period of time, the cell volume is minimized but the working electrode of a large surface area is used; e.g. a metal gauze, metal foil, and mercury pool. In addition, the solution is stirred to facilitate mass transfer. It is common to use a separator to prevent the solution mixing; if a reduced species on cathode moves to anode to be oxidized, there evolves an error in the measurement of electric charges. Such a mixing should be prevented by a separator.

There are two methods for bulk electrolysis; a controlled-potential and a controlled-current one. In some cases, the potential or current is controlled to be constant, which are called a constant-potential bulk electrolysis or a constant-current one. Note that the electrode potential and current can not be controlled simultaneously. Hence, in a constant-potential bulk electrolysis, the current change is measured, while the electrode potential is monitored in the constant-current bulk electrolysis. In one constant-current bulk electrolysis, the electrode potential is traced by changing the direction of current flow; e.g. the reduction current first and then oxidation current. The same principle is applied to the galvanostatic charge-discharge cycling of secondary batteries.

Electrochemistry and spectroscopy can be combined for studying redox reactions in solution, which is thus called a spectroelectrochemistry. The electrochemical cell must transmit electromagnetic radiations for spectroscopic measurements. Hence, the cell should be thin and electrode should be transparent for the electromagnetic radiations, so it is called thin-layer cells or optically transparent thin-layer cells.

Meanwhile, industrial electrolysis is a kind of bulk electrolysis because its purpose is to electrochemically oxidize or reduce the reactants present in the electrolysis cells. The purpose of bulk electrolysis in laboratories is, however, the measurement of n value or concentration of electroactive species.

9-1 Controlled-potential bulk electrolysis

Let us assume that a bulk electrolysis is performed on Fe^{2+} dissolved in 1.0 *M* H_2SO_4 to measure the electric charges for $Fe^{2+} \rightarrow Fe^{3+} + e$. The purpose is to determine the concentration of Fe^{2+} from the measured electric charges.

Fig. 9-1 shows the linear sweep voltammogram of 1.0 *M* H_2SO_4 that is used as the supporting electrolyte. The reduction current for H_2 evolution appears from −0.2 V in the negative sweep, and the oxidation current for O_2 evolution from 1.5 V in the positive sweep. This voltammogram illustrates that the usable potential range for controlled-potential bulk electrolysis is about 1.5 V to −0.2 V with 1.0 *M* H_2SO_4 supporting electrolyte. That is, by controlling the electrode potential within this range, the faradaic current (e.g. $Fe^{2+} \rightarrow Fe^{3+} + e$) can be measured without interferences of reduction or oxidation current for H_2 or O_2 evolution. Any compounds that give an oxidation or reduction current in this potential range (1.5 V ~ −0.2 V) can be analyzed by bulk electrolysis. This potential range is called an ***electrochemical stability window*** because there is no oxidation/reduction reaction of supporting electrolytes.

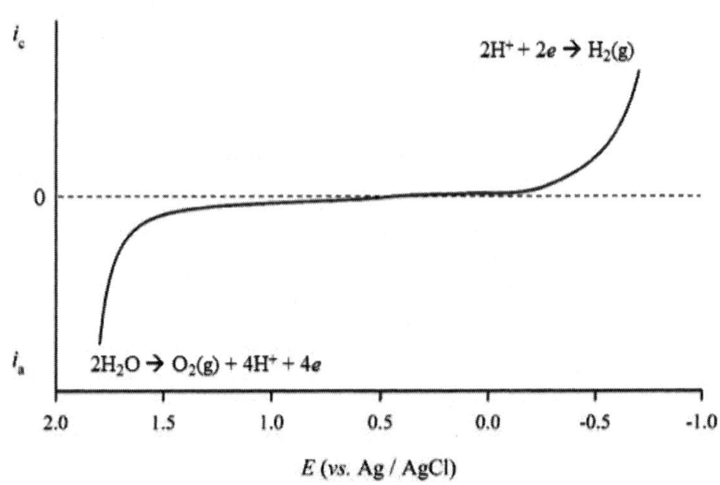

Fig. 9-1 The current-potential curve obtained from 1.0 *M* H_2SO_4 solution using a Pt working electrode.

Self learning 9-1

Estimate the electrochemical stability window of 1.0 M H_2SO_4 from the equilibrium potential (E_{eq}) of hydrogen and oxygen half-cell shown in Fig. 2-12, and compare your result with the data shown in Fig. 9-1. Give an explanation for the discrepancy.

Fig. 9-2 shows the current-potential curves that are expected during a ***constant-potential bulk electrolysis***. The solution is stirred at a constant rate, and thus the current shows a limiting behavior as in the RDE experiment. Before the bulk electrolysis (t_1), Fe^{2+} is present in the solution, but Fe^{3+} is absent. Therefore, the oxidation limiting current ($i_{l,a}$ for $Fe^{2+} \rightarrow Fe^{3+} + e$) is considerable, but the reduction limiting current ($i_{l,c}$ for $Fe^{3+} + e \rightarrow Fe^{2+}$) is zero. Note that the equilibrium potential for $Fe^{3+} + e = Fe^{2+}$ half-cell is 0.53 V (*vs.* Ag/AgCl) by assuming a standard state, such that the oxidation/reduction is supposed to occur within the electrochemical stability window of the supporting electrolyte (1.5 V ~ −0.2 V for 1.0 M H_2SO_4). Also note that the limiting current is proportional to the concentration of electroactive species (refer to <6-7>). As expected, the anodic limiting current ($i_{l,a}$) is appreciable but the cathodic limiting current ($i_{l,c}$) is zero at t_1. The concentration of H^+ and H_2O is much higher than that of Fe^{2+}, such that the limiting

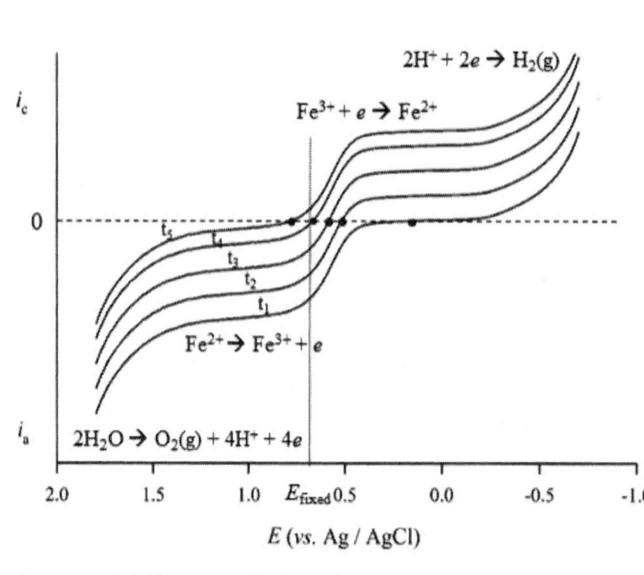

Fig. 9-2 Change of i–V curve with time during a constant-potential bulk electrolysis of Fe^{2+} dissolved in 1.0 M H_2SO_4 solution.

currents for H_2 and O_2 evolution are also very large, thus omitted in Fig. 9-2. As the electrolysis progresses, Fe^{3+} concentration steadily increases at the expense of Fe^{2+}, such that $i_{l,a}$ for $Fe^{2+} \rightarrow Fe^{3+} + e$ gradually decreases while $i_{l,c}$ for $Fe^{3+} + e \rightarrow Fe^{2+}$ keeps increasing. At the end of bulk electrolysis, the concentration of Fe^{2+} is zero.

The total electric charge $Q = nFC_R^*V$ (V is the volume of electrolysis cell) can be obtained by performing bulk electrolysis ($R \rightarrow O + ne$). As the electrolysis goes on, O is produced while R being depleted. As a result, the equilibrium potential (E_{eq}) of the half-cell ($O + ne = R$) moves to the positive direction. That is, the concentration of O and R according to the electrolysis time can be expressed by <9-2>, in which $C_R^*(0)$ is the initial concentration of R (Fe^{2+} in the above example) and x is the fraction of electrolysis. <9-3> illustrates that E_{eq} moves to the positive value with an increase in x (with bulk electrolysis). In Fig. 9-2, the shift of E_{eq} with time ($t_1 \rightarrow t_5$) is indicated by the dots on the x-axis; these are equilibrium potentials because the net current (i_{net}) is zero.

$$E_{eq} = E^{0'} + \frac{RT}{nF} \ln \frac{C_O^*(t)}{C_R^*(t)}$$

$$C_R^*(0) = C_R^*(t) + C_O^*(t) \qquad \text{<9-1>}$$

$$C_O^*(t) = x\, C_R^*(0), \quad C_R^*(t) = (1-x)\, C_R^*(0) \qquad \text{<9-2>}$$

$$E_{eq} = E^{0'} + \frac{RT}{nF} \ln \frac{x}{1-x} \qquad \text{<9-3>}$$

From the current-potential profiles (Fig. 9-2), the evolution of current can be traced. The electrode potential is fixed at E_{fixed} for the constant-potential bulk electrolysis. As is seen, the current value at E_{fixed} gradually decreases with electrolysis ($t_1 \rightarrow t_5$). From t_1 to t_4, $E_{fixed} > E_{eq}$, such that the oxidation reaction ($Fe^{2+} \rightarrow Fe^{3+} + e$) continues at E_{fixed}. At t_5, however, the applied electrode potential (E_{fixed}) is more negative than E_{eq}, under which the reduction current ($Fe^{3+} + e \rightarrow Fe^{2+}$) is supposed to be observed. The purpose of this experiment is to measure the electric charges for Fe^{2+} oxidation. From t_1 to t_4, the targeted oxidation current flows, but the unwanted reduction current from t_5. Hence, the bulk electrolysis should be stopped before reaching t_5. This means that the bulk electrolysis can not be completed until $i_{l,a} = 0$. In Fig. 9-2, $i_{l,a} > 0$ at t_5, indicating that there is still remnant Fe^{2+} in the solution, which in turn indicates that the electrolysis is not performed 100%. One way to solve this problem is to set E_{fixed} to be more positive, but should be careful not to be interfered by the currents of any side reactions.

Self learning 9-2

Using <9-3>, calculate how positive E_{fixed} is as compared to $E^{0'}$ for 99 % electrolysis.

As discussed above, the anodic limiting current decreases with electrolysis. This is explained as follows. The measured current is a limiting current (i_l), which is expressed by <9-6> after considering <9-4> and <9-5>.

$$i_l(t) = \frac{nFAD_R C_R^*(t)}{\delta} = -nF\left[\frac{dN_R(t)}{dt}\right] \quad \text{<9-4>}$$

$$C_R^*(t) = \frac{N_R(t)}{V} \quad \text{<9-5>}$$

$$i_l(t) = -nFV\frac{dC_R^*(t)}{dt} = \frac{nFAD_R C_R^*(t)}{\delta} \quad \text{<9-6>}$$

Further, <9-6> is rearranged to <9-7>. Substituting $C_R^*(t)$ (<9-8>) into <9-6> leads to <9-9> (Fig. 9-3). Here, the larger the p value is, the faster the decreasing rate of current

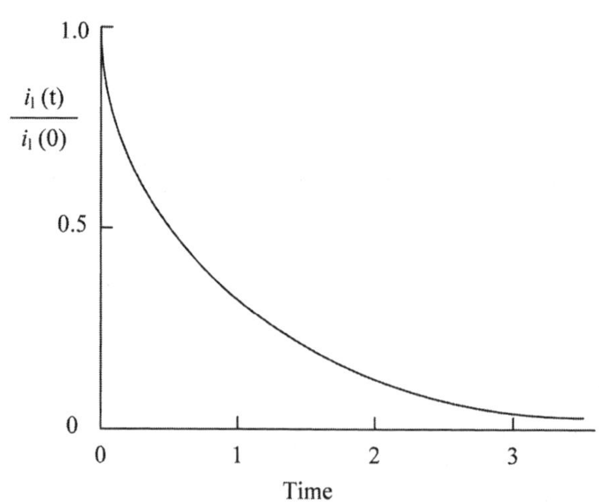

Fig. 9-3 A current–time curve during a constant–potential bulk electrolysis.

is, so the p should be maximized to minimize the electrolysis time in practical bulk electrolysis. Towards this end, the electrode having large surface area (A) is used, the cell volume (V) is minimized, and the stirring speed is increased to reduce the thickness of Nernst diffusion layer (δ) (<9-7>).

$$\frac{dC_R^*(t)}{dt} = -\left(\frac{AD_R}{V\delta}\right)C_R^*(t) = -pC_R^*(t) \qquad \text{<9-7>}$$

$$C_R^*(t) = C_R^*(0)\exp[-pt] \qquad \text{<9-8>}$$

$$i_1(t) = \frac{nFA\ D_R C_R^*(0)}{\delta}\ \exp[-pt] = i_1(0)\exp[-pt] \qquad \text{<9-9>}$$

The electric charges can be obtained by integrating the current obtained by <9-9> with respect to time. With an assumption of 100% electrolysis, one can calculate n value with the known values of $C_R^*(0)$ and cell volume (V). In opposite, if the n value is known, $C_R^*(0)$ can be calculated by using <9-10>.

$$Q = \int_0^t i_1(t)dt = nF\,C_R^*(0)\ V \qquad \text{<9-10>}$$

Self learning 9-3

Using <9-8>, calculate the time required for 99.9% electrolysis with $p = 10^{-2}\ \text{sec}^{-1}$.

9-2 Controlled-current bulk electrolysis

Fig. 9-4 shows the evolution of electrode potential when a ***constant-current bulk electrolysis*** is performed with the current being held at i_{fixed}. At t_1, $i_{l,a} > i_{fixed}$ since Fe^{2+} concentration is sufficiently large (see <6-6>). The electrode potential pins at E_1, and all of current (i_{fixed}) is consumed for Fe^{2+} oxidation; the ***current efficiency*** is 100 %. The anodic

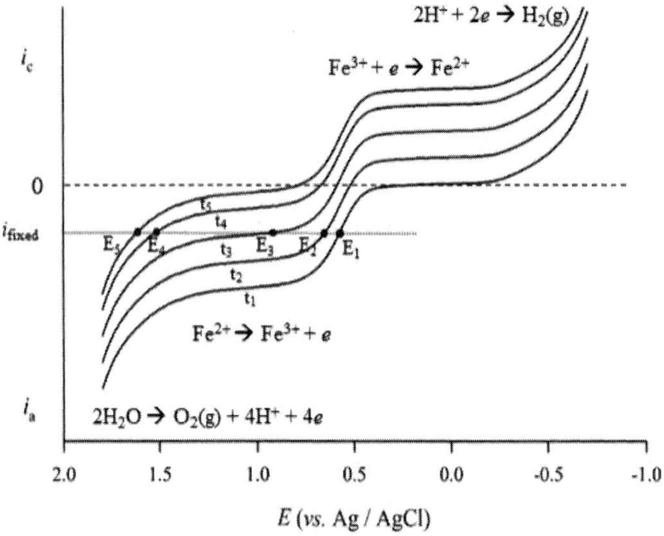

Fig. 9–4 Change of the electrode potential with time during a constant–current bulk electrolysis of Fe^{2+} dissolved in 1.0 *M* H_2SO_4 solution.

limiting current ($i_{l,a}$) decreases with electrolysis as Fe^{2+} concentration decreases together along with a positive shift of electrode potential ($E_1 \rightarrow E_5$). The condition for $i_{l,a} > i_{fixed}$ and 100 % current efficiency is met until t_3. From t_4, however, $i_{l,a} < i_{fixed}$. Hence, only a part of imposed current (i_{fixed}) is used for Fe^{2+} oxidation; the other should be used for other oxidation. The possible oxidation is O_2 evolution by water electrolysis ($H_2O \rightarrow O_2 + 4H^+ + 4e$), which can take place at E_4. Note that the purpose of this experiment is to measure the electric charges for Fe^{2+} oxidation. The current efficiency is less than 100 % from t_4. The change of electrode potential and current efficiency with time are shown in Fig. 9-5-a and 9-5-b, respectively. The less than 100 % current efficiency from t_4 means that there must be an error in the measurement of electric charges for Fe^{2+} oxidation ($Q = nFC_R^*V$, wherein C_R^* is the Fe^{2+} concentration). As can be seen in Fig. 9-4, however, the condition for $i_{l,a} < i_{fixed}$ appears in the later period of electrolysis if the i_{fixed} is lowered; current efficiency can be increased. The problem with smaller i_{fixed} is, however, the prolonged electrolysis time.

The above problem can be solved by adding a redox pair that can serve as an ***electron mediator***. In the constant-current bulk electrolysis of Fe^{2+}, Ce^{3+} can play as the electron mediator. As shown in Fig. 9-6, when $i_{l,a} < i_{fixed}$ from t_4, the electrode potential moves

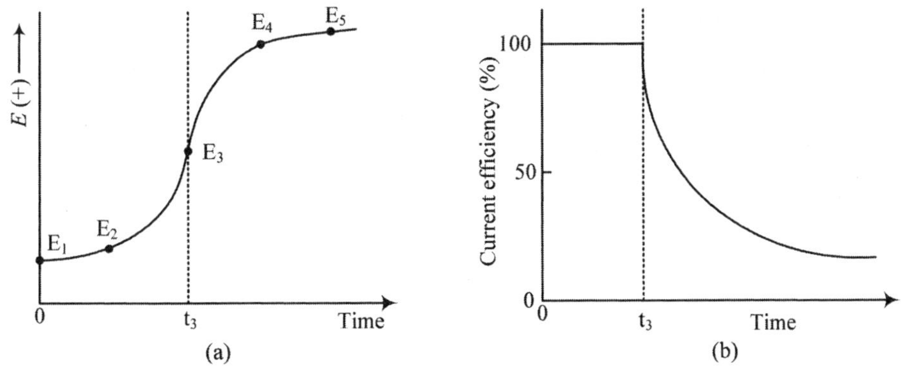

Fig. 9–5 Change of the electrode potential (a) and current efficiency (b) with time during a constant–current bulk electrolysis illustrated in Fig. 9–4.

to E_4, where the oxidation ($Ce^{3+} \rightarrow Ce^{4+} + e$, $E^0 = 1.3$ V *vs.* Ag/AgCl) can take place. Note that the oxidation of Ce^{3+} takes place prior to water oxidation ($H_2O \rightarrow O_2 + 4H^+ + 4e$, $E^0 = 1.5$ V *vs.* Ag/AgCl). Also note that the targeted reaction ($Fe^{2+} \rightarrow Fe^{3+} + e$, $E^0 = 0.53$ V *vs.* Ag/AgCl) is also thermodynamically allowed at E_4, so that the oxidation of Fe^{2+} and Ce^{3+} proceeds simultaneously. One may worry about <100 % current efficiency because Ce^{3+} oxidation takes place from t_4. This is not the case since Ce^{4+} just produced by Ce^{3+} oxidation is utilized to oxidize Fe^{2+} as shown in <9-11>. The re-

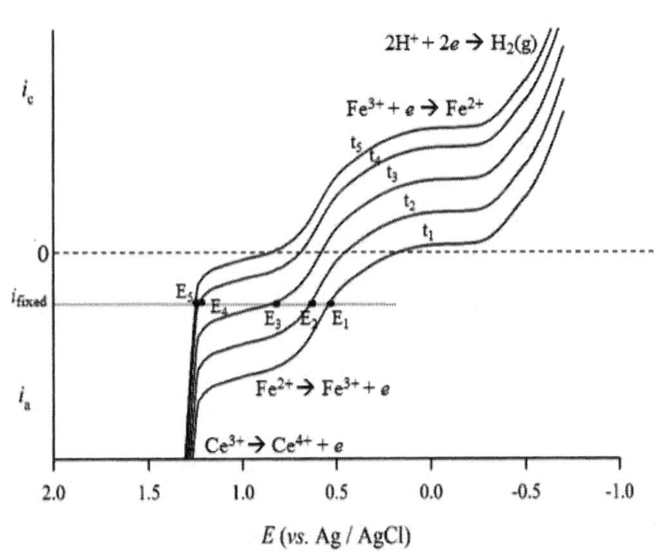

Fig. 9–6 Change of *i–V* curve with time during a constant–current bulk electrolysis of Fe^{2+} dissolved in 1.0 *M* H_2SO_4 solution, into which excess Ce^{3+} is added.

action <9-11> proceeds spontaneously when Ce^{4+} meets with Fe^{2+} in solution. The current consumed for the side reaction (Ce^{3+} oxidation) eventually contributes for Fe^{2+} oxidation that is the targeted one. Therefore, the current efficiency is maintained at 100 %. From the measured Q value, Fe^{2+} concentration can be calculated without error. For the measurement of Q from a constant-current bulk electrolysis, the time required for Fe^{2+} concentration to become zero should be measured separately.

$$Ce^{4+} + Fe^{2+} \rightarrow Ce^{3+} + Fe^{3+} \quad (\Delta G^0 < 0) \qquad \text{<9-11>}$$

9-3 Chronopotentiometry

Chronopotentiometry refers to a method of measuring the electrode potential over time by changing the current. As shown in Fig. 9-7-a, when a constant current is applied, the electrode potential changes as a response. This is equivalent to the potential change from E_1 to E_3 when the current is fixed in Fig. 9-4. Another electrochemical reaction occurs after τ. Fig. 9-7-b shows the potential change with time when an oxidation current

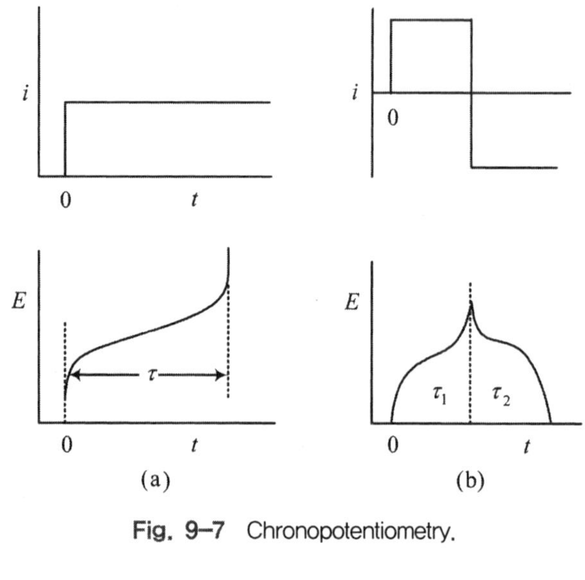

Fig. 9-7 Chronopotentiometry.

is applied for a certain time and switched to a reduction current. The same principle is applied for ***galvanostatic charge-discharge cycling*** of secondary batteries. For example, when a cell is charged for τ_1 and discharged for τ_2, the cell voltage changes as shown in Fig. 9-7-b. The charge and discharge capacity can be calculated from τ_1 and τ_2 because the current is fixed.

Example 9-1

The half-cell reaction of $LiCoO_2$, which is used as a positive electrode material of lithium-ion battery, is as follows.

$$Li_{1-x}CoO_2 + xLi^+ + xe = LiCoO_2$$

Derive E_{cell} in <1-8> ($E_{wk} = E_{cell} - (\eta_a + \eta_c + iR_{total})$) when discharged under a constant current, which represents the change in the equilibrium potential of this half-cell. Confirm that E_{cell} gradually decreases as the discharge progresses.

Solution

The actual half-cell reaction can be written as follows, where < > represents the Li^+ storage sites in $LiCoO_2$ and <Li^+> represents the sites being occupied by Li^+.

$$< > + Li^+_{electrolyte} + e = <Li^+>$$

From the half-cell reaction, Nernst equation can be derived as follows, where y represents the fraction of Li^+ occupied sites.

$$E_{eq} = E^0 + \frac{RT}{F}\ln\frac{a_{Li^+}a_{< >}}{a_{<Li^+>}} = E^0 + \frac{RT}{F}\ln a_{Li^+} + \frac{RT}{F}\ln\frac{a_{< >}}{a_{<Li^+>}}$$

$$= E^0 + \frac{RT}{F}\ln a_{Li^+} + \frac{RT}{F}\ln\frac{1-y}{y}$$

As y increases (as it discharges), the equilibrium potential decreases.

*9-4 Spectroelectrochemistry

When spectroscopy is used to identify the nature of electroactive species and their concentration while electrochemical reaction progresses on electrodes, available information can be enlarged.

Fig. 9-8 shows the structure of an electrochemical cell that is used for ***spectroelectrochemistry*** utilizing ultraviolet-visible (UV-VIS) or infrared (IR) absorption spectroscopy. Since the working electrode must transmit the electromagnetic radiations, a metal grid is used. This is because the electromagnetic radiations can transmit through the gap of metal grid. A counter electrode and a reference electrode, if necessary, are placed inside the cell in the region where electromagnetic radiations do not pass through. Meanwhile, the cells must be fabricated from the materials that do not absorb electromagnetic radiations. A glass, quartz, and germanium can be used when the spectroscopy utilizes the visible light, ultraviolet, and infrared radiation, respectively. The glass, quartz or germanium plates are used as the working electrode after depositing a thin film of platinum, gold, or SnO_2-In_2O_3 (indium tin oxide, ITO). Note that when a metal is deposited to a

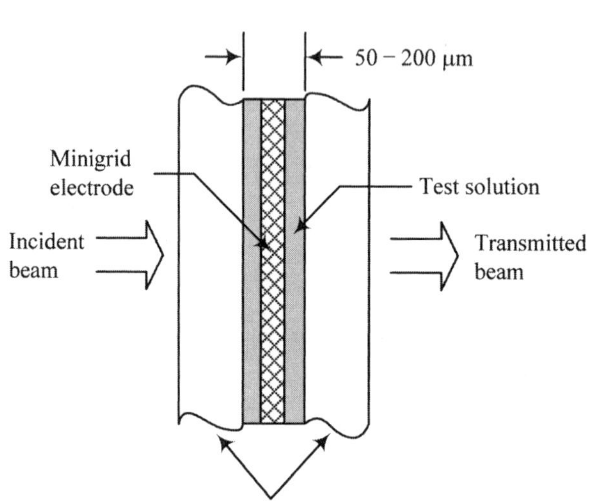

Fig. 9-8 Structure of a thin-layer spectroelectrochemical cell.

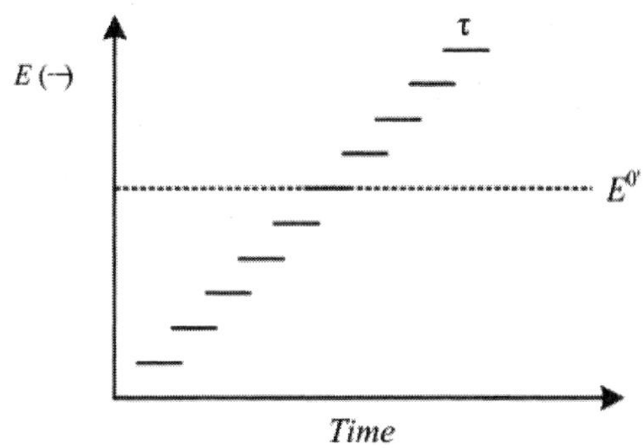

Fig. 9–9 Variation of electrode potential for thin–layer spectroelectrochemical experiments.

thickness of 5000 Å or less, a large portion of electromagnetic radiations are transmitted. Also, since the ITO thin film has high electric conductivity, it can be used as an electrode. The ITO does not absorb the visible light (transparent). As shown in Fig. 9-8, the solution to be analyzed is injected into a very thin space, hence it is called a ***thin-layer cell***.

For the thin-layer spectroelectrochemical experiment, the potential steps are continuously given to the working electrode as shown in Fig. 9-9. After a certain time (τ) elapse in each potential step, the electroactive species in the cell, of which the concentration changes according to the step potential, are analyzed by using the spectroscopy. Fig. 9-10 shows the change in the concentration profiles of O and R in a thin-layer cell after a step potential being applied. Let us assume that the electrochemical reaction, $O + ne = R$, is a nernstian and the concentration profiles of O and R are those indicated by the dotted lines before

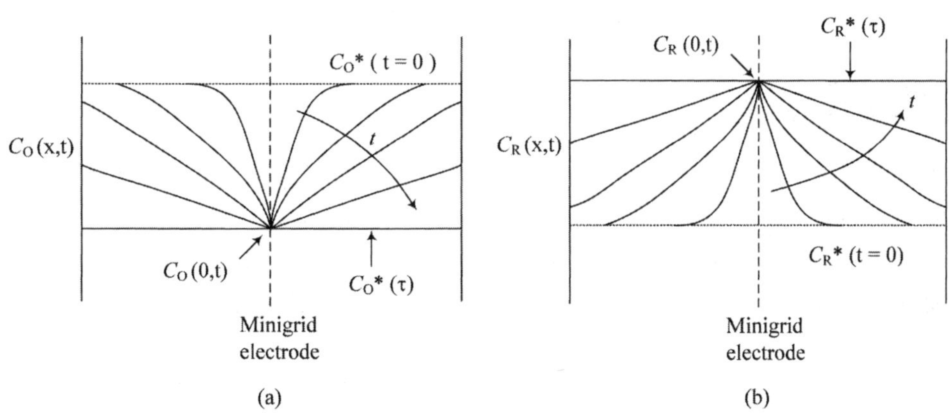

Fig. 9–10 Concentration profile of O and R during reduction of O in a thin–layer spectroelectrochemical cell.

the step. When the potential step is given, the surface concentrations being determined by the Nernst equation are maintained during the step (τ). In Fig. 9-10-a, $C_O(0,t) < C_O^*$ right after the potential step, such that O diffuses to the electrode surface and is reduced to R. The diffusion layer of O expands at a rate of $l = (2D_O t)^{1/2}$ with the step time. This feature progresses simultaneously in both solutions divided by the metal grid electrode. Since the cell thickness is very thin, the condition for $C_O(0,t) = C_O^*$ reaches within a short period of time (τ) as in <9-12>. In the case of R, $C_R(0,t) > C_R^*$ at the beginning, but $C_R(0,t) = C_R^*$ within a short time (τ) since R is continuously generated by the reduction of O.

Since the concentration of O and R in the cell is uniform across the solution after τ, their concentrations can be analyzed using the spectroscopy. In the successive potential steps, the surface concentrations are determined by the Nernst equation, and the bulk concentrations approach them within a short time (τ). Substitution of <9-12> into Nernst equation gives <9-13>, which shows the relation between the step potential (E) and C_O^*/C_R^*. After measuring C_O^*/C_R^* using spectroscopy with a variation of step potential, $\log(C_O^*/C_R^*)$ can be plotted against the step potential (E) (this is called a ***nernstian plot***), from which n value and $E^{0'}$ can be obtained.

$$E = E^{0'} + \frac{RT}{nF} \ln \frac{C_O(0,t)}{C_R(0,t)}$$

$$C_O(0,t) = C_O^*(\tau), \quad C_R(0,t) = C_R^*(\tau) \qquad \text{<9-12>}$$

$$E = E^{0'} + \frac{RT}{nF} \ln \frac{C_O^*(\tau)}{C_R^*(\tau)} \qquad \text{<9-13>}$$

Self learning 9-4

As shown in Fig. 9-10, the concentration of O and R in the cell is the same across the solution after τ. If τ is not long enough, the concentration profile of O and R shows a slope as shown in Fig. 9-10. What is the problem with the sloped concentration profile when measuring the concentration using a UV-VIS spectrophotometer?

Self learning 9-5

Assuming that the gap between the metal grid electrode and the glass plate 0.1 mm in Fig. 9-8, calculate the time required for the bulk concentration (C_O^* and C_R^*) to be equalized with the surface concentration after a potential step; $D_O = D_R = 5 \times 10^{-6}$ cm^2/sec.

Fig. 9-11-a shows an example of measuring n value and $E^{0'}$ of a half-cell ($O + ne = R$) using a UV-VIS spectroscopy. O exhibits a strong absorption at 500 nm and R at 420 nm. The spectra were obtained by varying the step potential from 0 mV (*vs.* Ag/AgCl) to −400 mV. At the very positive step potential of 100 mV, the reduction reaction does not proceed, such that the concentration of O is maximal. Therefore, it shows a maximum absorption (A_3) at 500 nm. As the step potential moves to a more negative value, the concentration of R increases at the expense of O, so the absorption

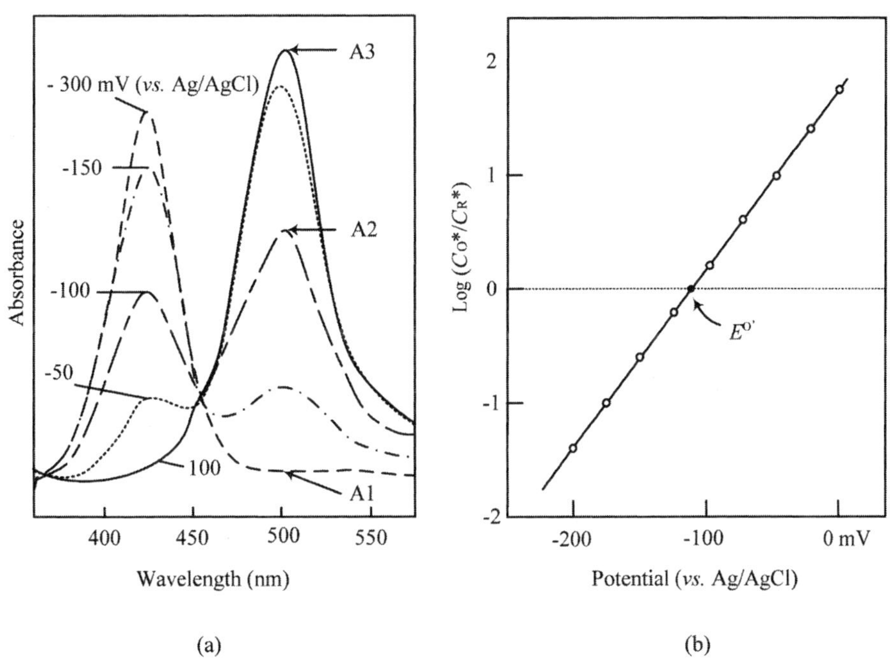

Fig. 9-11 (a); UV-VIS spectra during a thin-layer spectroelectrochemical experiment on an electroactive species and (b); The nernstian plot derived from the 500 nm absorption band.

at 420 nm gradually increases but absorption intensity becomes weaker at 500 nm. There is no spectral change in −300 ~ −400 mV because all of *O* are reduced to *R*. Here, the absorption at 500 nm (A_1) is not associated with *O* as there is no remnant *O*; it is a background absorption. (A_2-A_1) represents the concentration of *O* at a given step potential and (A_3-A_2) corresponds to the concentration of *R*.

$$\frac{C_O^*(E)}{C_R^*(E)}=\frac{A_2 - A_1}{A_3 - A_2} \qquad \text{<9-14>}$$

After measuring the absorption intensity at 500 nm (A_1, A_3, and A_2) according to the step potential, the potential-dependent $C_O^*(E)/C_R^*(E)$ is obtained by using <9-14>. The nernstian plot is drawn, from which $E^{0'}$ = −108 mV and n = 0.98 (= 1.0) are obtained.

Self learning 9-6

Derive the intercept and slope in nernstian plot (Fig. 9-11-b) from Nernst equation.

Problems

01 To which of the following electrochemical experiments is the Cottrell equation applied?

(a) cyclic voltammetry, (b) chronocoulometry, (c) bulk electrolysis, and (d) rotating disk electrode experiment

02 Let us assume that $n = 1$ and $E^{0'} = 0.2$ V (*vs.* SCE) for a half-cell ($O + ne = R$). A constant-potential bulk electrolysis is performed with 250 cm^3 solution containing 0.1 M of O. The surface area of working electrode is 250 cm^2.

(a) What is the applied potential for 99 % electrolysis?

(b) If the initial current is 50 A when the electrolysis is performed at the potential calculated in (a), what is the time required for 99 % electrolysis?

03 An overpotential of 1.0 V is required for the oxidation of ascorbic acid (AH_2) on a Pt electrode ($A + 2H^+ + 2e = AH_2$, $E^{0'} = -0.4$ V *vs.* SCE). Determine which of the following two redox pairs can be used as the electron mediator to reduce the overpotential.

$Fe(bpy)_3^{2+/3+}$	$E^{0'} = 0.8$ V (*vs.* SCE)
$Co(CN)_6^{3-/4-}$	$E^{0'} = -0.8$ V (*vs.* SCE)

04 Biomaterials such as cytochrome *c* and ferredoxins show slow or negligible charge transfer rates at solid electrodes. In such cases, a redox pair (electron mediator) can be added to convey electrons between biomaterials and solid electrode. $E^{0'}$ and n value for cytochrome *c* half-cell were reported (*Anal. Chem.*,

47, 79, 1975), which were obtained by using optically transparent thin-layer spectroelectrochemical cell. A gold mini-grid sandwiched between two slides (distance = 0.1 mm) was used as a working electrode. 2,6-dichloroindophenol (Fig. 9-12-a) was used as the electron mediator. The concentration ratio C_O^*/C_R^* was calculated with the absorbance data at 550 nm for the reduced cytochrome *c*.

(a) Give an explanation why large-sized biological redox molecules show slow charge transfer rates at solid electrodes.

(b) The mediated electron transfer reaction ($M_{red} + B_{ox} \rightarrow M_{ox} + B_{red}$, M and B are the electron mediator and biomaterial, respectively) should be spontaneous. That is, thermodynamically feasible. Explain the meaning of "spontaneous".

(c) Determine $E^{0'}$ and n value of cytochrome *c* half-cell from Fig. 9-12-b.

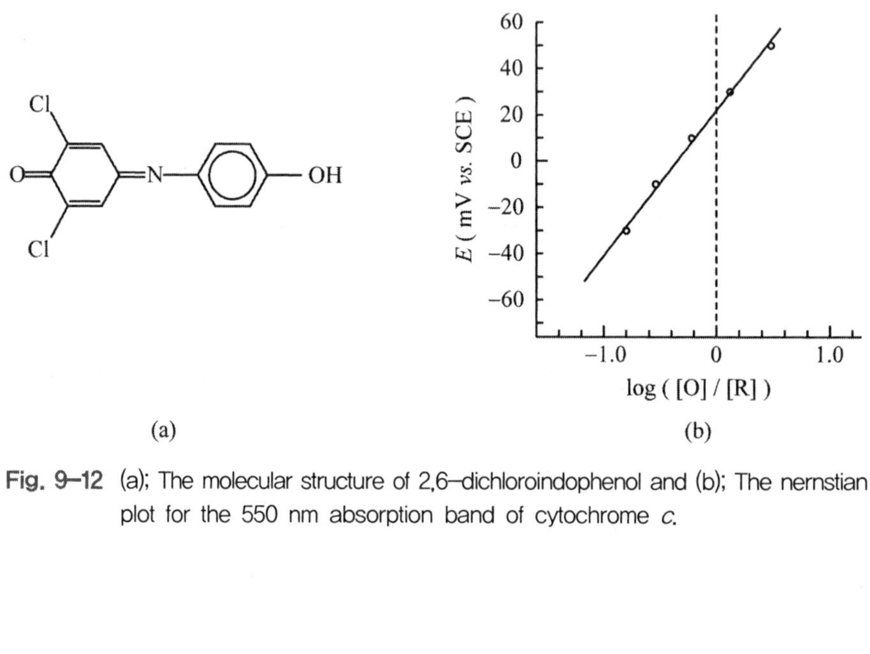

Fig. 9-12 (a); The molecular structure of 2,6-dichloroindophenol and (b); The nernstian plot for the 550 nm absorption band of cytochrome *c*.

CHAPTER 10

Impedance method

All electrochemical systems can be represented by equivalent circuits by combining resistors, capacitors and inductors; e.g. Fig. 1-14. These three components have a function of impeding current flow. The degree of current impediment for a resistor is represented by resistance, and those for capacitor and inductor are referred to as capacitive reactance and inductive reactance, respectively. The sum of resistance, capacitive reactance, and inductive reactance is called an ***impedance***. Resistor gives a resistance that is independent on the frequency of applied voltage; direct or alternating voltage. However, the capacitive reactance and inductive reactance are dependent on the frequency of applied voltage.

When impedance method is applied to an electrochemical system, the values of resistance (R), capacitance (C), and inductance (L) that comprises the equivalent circuit can be obtained. That is, the alternating voltage having a wide range of frequencies is sequentially applied to the working electrode, and the electrochemical parameters can be obtained after determining the impedance from the measured alternating current. The advantageous feature for impedance analysis over the other electrochemical methods is that, in addition to the values of resistance and capacitance in given equivalent circuits, the parameters related to charge transfer (for example, R_{ct}) and mass transfer (for example, D_O) can be obtained by a single measurement.

10-1 Impedance

Resistor gives a resistance (R) regardless of the frequency of applied voltage. When an alternating voltage is applied to a capacitor, whose capacitance is C, the alternating current is determined by $i = e/X_C$. Here, i and e are the instantaneous current and voltage that vary with time. The capacitive reactance (X_C) acts as a resistance and its frequency-dependency is expressed by <10-1>, wherein ω (= $2\pi f$) is the **angular frequency** that is 2π times the conventional frequency (f) in Hz.

$$X_C = \frac{1}{\omega C}$$ <10-1>

From <10-1>, it is seen that X_C increases with decreasing frequency, and $X_C = \infty$ for direct voltage ($\omega = 0$). That is, when the direct voltage is applied across a capacitor, the current can not flow after being completely charged. Meanwhile, when an alternating voltage is applied across an inductor, inductive reactance ($X_L = \omega L$) is generated. For a circuit comprising a serially connected resistor (R), capacitor (C) and inductor (L) shown in Fig. 10-1-a, the total impedance is given as <10-2>. However, if the frequency range used for the impedance analysis is not very large, the contribution of X_L is ignored. The total impedance is then simplified as <10-3>.

$$Z = \sqrt{R^2 + (X_L - X_C)^2}$$ <10-2>

$$Z = \sqrt{R^2 + {X_C}^2}$$ <10-3>

When an alternating voltage $e = E\sin\omega t$ is applied across a resistor, the current changes as shown in <10-4>, where E and I are the amplitude of the voltage and current, respectively.

$$i = \frac{e}{R} = \frac{E}{R}\sin\omega t = I\sin\omega t$$ <10-4>

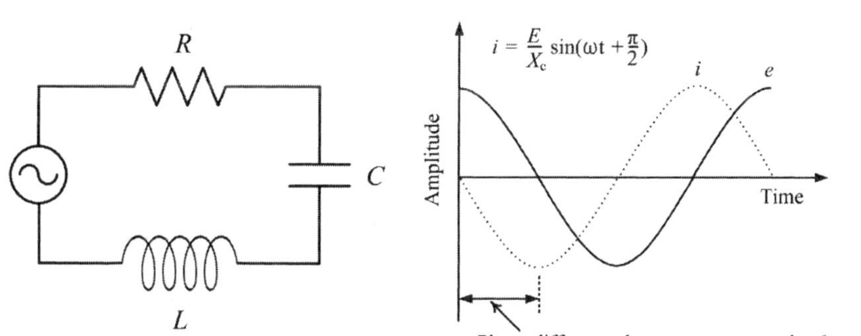

(a) (b)

Fig. 10–1 (a); Resistor (R), capacitor (C), and inductor (L) that impede current flow when an alternating voltage is applied, and (b); the relationship between an alternating voltage across a capacitor and the alternating current. Note the phase difference between the voltage and current.

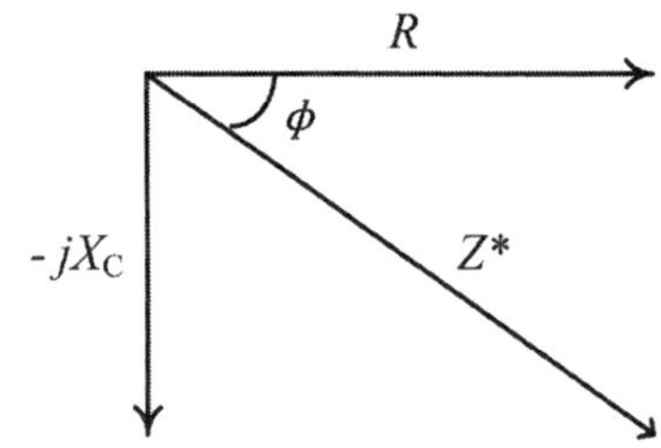

Fig. 10–2 An impedance vector diagram. The total impedance (Z^*) is the vector sum of the real part (R) and the imaginary part (X_C).

Upon an alternating voltage being applied across a capacitor, the current changes as shown in <10-5>.

$$\begin{aligned} i &= \frac{dq}{dt} = C\frac{de}{dt} = \omega CE\cos\omega t \\ &= \omega CE\sin\left(\omega t + \frac{\pi}{2}\right) = \frac{E}{X_C}\sin\left(\omega t + \frac{\pi}{2}\right) \end{aligned} \qquad \text{<10-5>}$$

Comparison of <10-4> and <10-5> reveals that X_C plays the same role as R; retarding current flow. Also, it is seen that the alternating current has a phase difference of $\pi/2$ with the alternating voltage. Therefore, when an alternating voltage is applied to a serially connected RC circuit, a phase difference of $\pi/2$ occurs between two impedance components (X_C and R). It is expressed by a complex notation as shown in <10-6> and Fig. 10-2. Here, the component along the ordinate is assigned as imaginary and is multiplied by $j = \sqrt{-1}$, whereas the component along the abscissa is real. The total impedance (Z^*) is the sum of two vectors.

$$Z^* = R - jX_C \qquad \text{<10-6>}$$

Total impedance is given by <10-7>, and the phase difference (Φ) has the relationship shown in <10-8>. Both total impedance and phase difference change with ω.

$$Z = \sqrt{R^2 + X_C^{\,2}} = \sqrt{R^2 + \left(\frac{1}{\omega C}\right)^2} \qquad \text{<10-7>}$$

$$\tan\phi = \frac{X_C}{R} = \frac{1}{\omega RC} \qquad \text{<10-8>}$$

10-2 Impedance plot

Impedance is expressed as a complex notation; it has a real part and imaginary part as shown in <10-6>. The real part (Z_{Re}) and imaginary part (Z_{Im}) are plotted along the abscissa (x-axis) and the ordinate (y-axis) in a two-dimensional plane, respectively. This is called a **Nyquist plot**.

Fig. 10-3-a shows the Nyquist plot for a resistor. The impedance of a resistor is constant irrespective of ω, and since there is no imaginary part, it appears as a point on the Z_{Re} axis. When an alternating voltage is applied to a capacitor and current is measured, the impedance value decreases as the frequency increases by <10-1>. Since the impedance has only the imaginary part, it gradually decreases along the Z_{Im} axis (Fig. 10-3-b).

For a serially connected RC circuit, there are both the real and imaginary part. The impedance due to the resistor is plotted on the Z_{Re} axis and the impedance due to the capacitor is plotted on the Z_{Im} axis. The total impedance is then represented by the sum of two vectors as shown in Fig. 10-4. Here, at ω'', the impedance of the resistor is Z'_{Re} and the impedance of the capacitor is Z''_{Im}, so the total impedance is the sum of two vectors. At ω' ($> \omega''$), Z'_{Re} does not change, but the impedance of the capacitor decreases to Z'_{Im}. The total impedance, which is the sum of two vectors, also decreases at higher ω. In short, the total impedance falls down along y-axis at the given Z'_{Re} as shown in Fig. 10-4. The phase angle (Φ) also decreases with increasing ω.

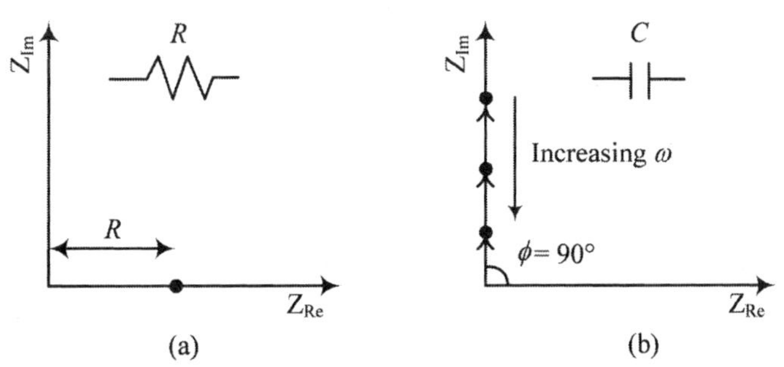

Fig. 10-3 Nyquist plot for a resistor (R) and capacitor (C).

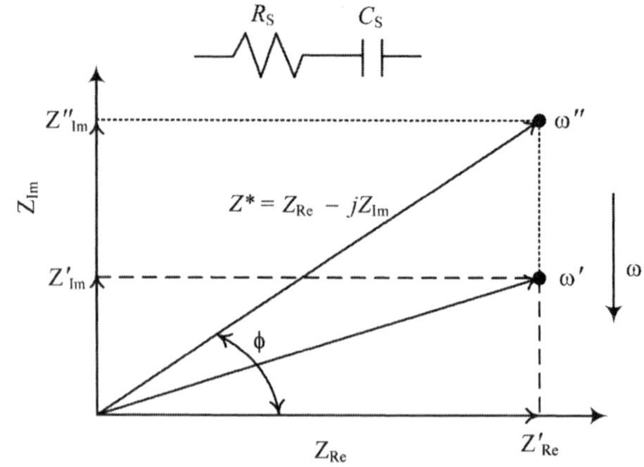

Fig. 10-4 Change of total impedance according to ω in a series *RC* circuit.

Fig. 10-5-a shows the impedance change for a parallel connected *RC* circuit. Since the admittance of a parallel *RC* circuit is expressed as <10-9>, the impedance is obtained as <10-10>.

$$A^* = \frac{1}{R_P} + j\omega C_P \qquad \text{<10-9>}$$

$$Z^* = \frac{1}{A^*} = \frac{1}{\dfrac{1}{R_P} + j\omega C_P} = \frac{R_P}{1 + j\omega C_P R_P} = \frac{R_P(1 - j\omega C_P R_P)}{1 + (\omega R_P C_P)^2}$$

$$= \frac{R_P}{1 + (\omega R_P C_P)^2} - j\,\frac{\omega C_P R_P^2}{1 + (\omega R_P C_P)^2} \qquad \text{<10-10>}$$

Total impedance is $Z^* = Z_{Re} - jZ_{Im}$, such that Z_{Re} and Z_{Im} are expressed as <10-11> and <10-12> from <10-10>, respectively.

$$Z_{Re} = \frac{R_P}{1 + (\omega R_P C_P)^2} \qquad \text{<10-11>}$$

$$Z_{Im} = \frac{\omega C_P R_P^2}{1 + (\omega R_P C_P)^2} \qquad \text{<10-12>}$$

Now, if one finds out the relationship between Z_{Re} and Z_{Im}, the total impedance can be plotted in two-dimensional plane with Z_{Re} as the *x*-axis and Z_{Im} as the *y*-axis. From <10-11> and <10-12>, <10-13> is derived, which can be rewritten as <10-14>. <10-14>

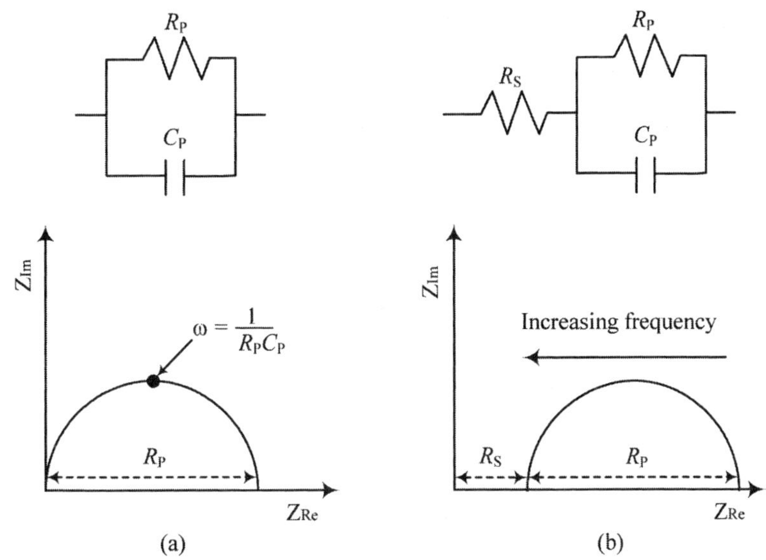

Fig. 10–5 Nyquist plot for a parallel *RC* circuit.

is a formula for a circle: $(x-a)^2 + (y-b)^2 = c^2$. Hence, the relationship between Z_{Re} and Z_{Im} appears as a circle with an origin of $(R_P/2, 0)$ and a radius of $R_P/2$. Total impedance according to ω is presented as a semi-circle as shown in Fig. 10-5-a.

$$Z_{Im} = Z_{Re}\omega R_P C_P \quad \text{<10-13>}$$

$$Z_{Im}^2 = Z_{Re}^2 \omega^2 R_P^2 C_P^2 = Z_{Re}^2\left(\frac{R_P}{Z_{Re}} - 1\right)$$

$$Z_{Im}^2 = Z_{Re}R_P - Z_{Re}^2$$

$$Z_{Im}^2 - Z_{Re}R_P + Z_{Re}^2 = 0$$

$$\left(Z_{Re} - \frac{R_P}{2}\right)^2 + Z_{Im}^2 = \left(\frac{R_P}{2}\right)^2 \quad \text{<10-14>}$$

Since the maximum value of Z_{Im} in the semicircle has the relation of <10-15>, substituting this into <10-13> yields <10-16>.

$$Z_{Im} = Z_{Re} = \frac{R_P}{2} \quad \text{<10-15>}$$

$$\omega C_P R_P = 1, \quad \omega = \frac{1}{R_P C_P} \quad \text{<10-16>}$$

From the Nyquist plot for a parallel *RC* circuit in Fig. 10-5-a, R_P can be obtained from the diameter of the semicircle. Also since one knows the ω value at which Z_{Im} has the maximum value (that is, ω at the top of the semicircle), C_P can be obtained by substituting the as-obtained R_P into <10-16>. Fig. 10-5-b shows the Nyquist plot for a circuit comprising a resistor and a parallel *RC*. It appears as the addition of R_S to a semicircle.

Self learning 10-1

Sketch the Nyquist plot with an assumption of $R_1C_1 > R_2C_2$ in Fig. 10-6-a. How do you obtain R_1, C_1, R_2, and C_2 from the plot?

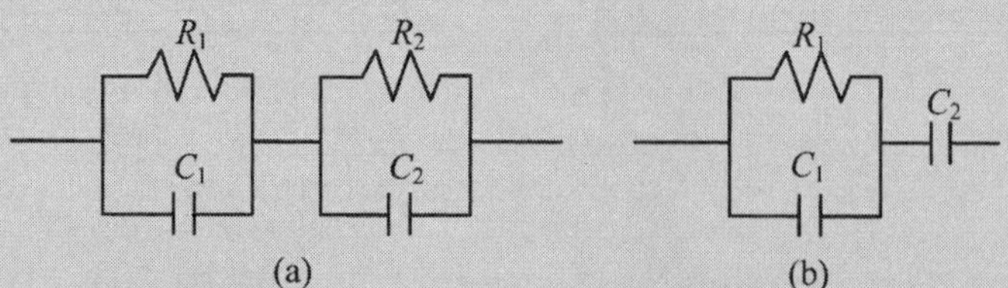

Fig. 10-6 (a); A serially connected two parallel *RC* circuits and (b); a serial connection of a parallel *RC* with a capacitor.

Self learning 10-2

Sketch the Nyquist plot for the circuit shown in Fig. 10-6-b.

Example 10-1

The C_{dl} in electrochemical cells is approximately 10^{-4} F, whereas the capacitance value of ferroelectric materials is very small, approximately 10^{-9} to 10^{-10} F. Explain the following. (a) Due to this difference, the frequency range of impedance analyzers for the study of electrochemical reactions (to obtain R_{ct} and C_{dl}) is relatively lower (< 10 MHz) than that for ferroelectric studies (> 10 MHz). (b) As shown in <10-3>, the contribution of inductive reactance ($X_L = \omega L$) to total impedance is ignored when impedance measurement is made on electrochemical systems, but its contribution cannot be ignored for the measurement of ferroelectric materials.

Solution

(a) The R and C values can be obtained from the impedance spectrum as shown in Fig. 10-5. As is seen in <10-16>, when the capacitance (C) is small, the angular frequency ($\omega = 2\pi f$) corresponding to the vertex of semicircle is large. Therefore, in order to obtain a semicircle like Fig. 10-5 in ferroelectric studies, impedance must be measured in a very large frequency range (f > 10 MHz). However, the semicircle can be obtained even in relatively low frequency ranges (f < 10 MHz) in the study of electrochemical reactions.

(b) For ferroelectric studies, impedance must be measured in a very large angular frequencies (ω). As shown in <10-1>, as ω increases, X_C decreases, but X_L (= ωL) increases. The contribution of inductive reactance (L) cannot be ignored in ferroelectric studies that should be performed at high frequency range.

10-3 Impedance for electrochemical reactions

Electrochemical parameters such as R_{ct}, C_{dl}, and D_O can be obtained from impedance measurements. The current flow and equivalent circuit for a three-electrode cell is displayed in Fig. 10-7. Here, R_Ω is the solution resistance between working and reference electrode, and Z_f is the impedance for faradaic reaction. The total current is the sum of non-faradaic (electric double-layer charging) current (i_C) and faradaic one (i_f). In im-

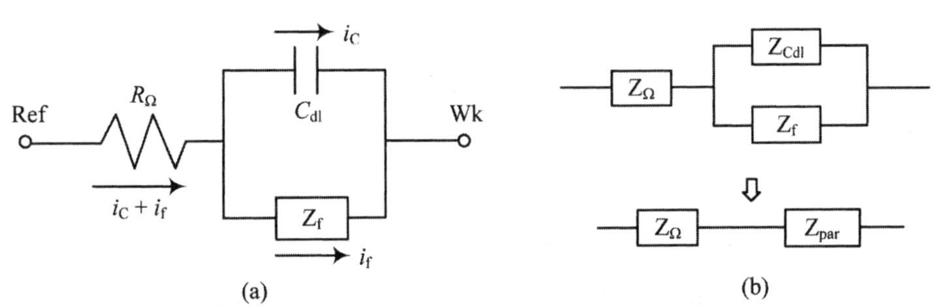

Fig. 10-7 (a); An equivalent circuit for a three-electrode electrochemical cell. (b); Impedance equivalent circuit.

pedance measurements, an alternating voltage as shown by <10-17> is applied between working and reference electrode, and the alternating current expressed by <10-18> is measured with a sweep of frequency. The total impedance is derived using the relation of <10-19>. Generally, ω = 10^{-3} Hz ~ 10 MHz for the study of electrochemical reactions. E_{dc} can be E_{eq} or any arbitrary value. The amplitude of alternating voltage ΔE = 5 ~ 50 mV.

$$e = E_{dc} + \Delta E \sin \omega t \qquad \text{<10-17>}$$

$$i = I_{dc} + \Delta I \sin(\omega t + \phi) \qquad \text{<10-18>}$$

$$Z = \frac{\Delta E}{\Delta I} \qquad \text{<10-19>}$$

The faradaic impedance (Z_f) shown in Fig. 10-7-a can be subdivided into a circuit, in which R_S and C_S are connected in series as shown in Fig. 10-8. R_S has the meaning of <10-20>, wherein σ is related with mass transfer as expressed by <10-21>.

$$R_S = R_{ct} + \frac{\sigma}{\omega^{1/2}} = R_{ct} + \frac{1}{\omega C_S} \qquad \text{<10-20>}$$

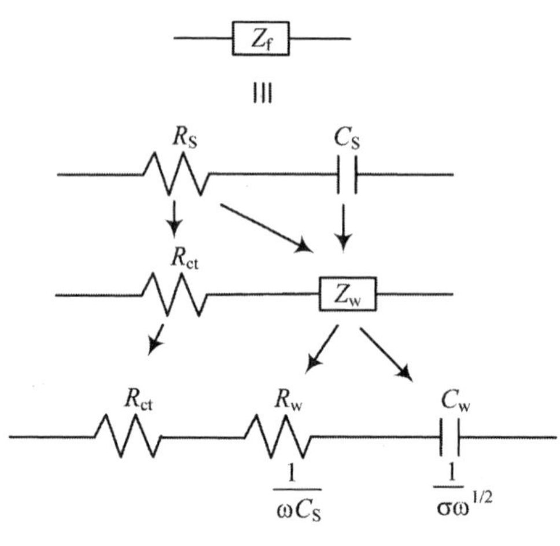

$\omega \to 0;\ Z_f \to Z_w$

$\omega \to \infty;\ Z_f \to R_{ct}$

Fig. 10-8 Subdivision of faradaic impedance (Z_f) into R_S and C_S, and further into R_{ct} and Z_W.

$$\sigma = \frac{RT}{n^2 F^2 A\sqrt{2}} \left(\frac{1}{D_O^{1/2} C_O^*} + \frac{1}{D_R^{1/2} C_R^*} \right)$$ <10-21>

Meanwhile, C_S is expressed by <10-22>, and can be converted into <10-24> using the relation of <10-23>.

$$C_S = \frac{1}{\sigma \omega^{1/2}}$$ <10-22>

$$\sigma = \frac{1}{\omega^{1/2} C_S}$$ <10-23>

$$\frac{\sigma}{\omega^{1/2}} = \frac{1}{\omega C_S}$$ <10-24>

In Fig. 10-8, Z_W is expressed by the following equation, which is a function of σ that is associated with mass transfer parameter (<10-21>). Z_W is called ***Warburg impedance***, from which mass transfer parameters can be obtained.

$$Z_W = R_W + C_W = \frac{1}{\omega\, C_S} + \frac{1}{\sigma \omega^{1/2}} = \frac{\sigma}{\omega^{1/2}} + \frac{1}{\sigma \omega^{1/2}}$$

The three impedance parameters that are derived by subdividing Z_f vary with ω (Fig. 10-8). In the case of $\omega \rightarrow 0$, $Z_W >> R_{ct}$, therefore $Z_f = Z_W$. In opposite, in case of $\omega \rightarrow \infty$, $Z_W << R_{ct}$, and thus $Z_f = R_{ct}$. Thereby, Z_W can be obtained at lower ω region, whereas R_{ct} at higher ω region.

The equivalent circuit of Fig. 10-7-a can be redrawn as shown in Fig. 10-7-b. The total impedance $Z = Z_\Omega + Z_{par}$, where Z_{par} is given by <10-25>. As Z_f and Z_{Cdl} are given as <10-26> and <10-27> respectively, Z_{par} is derived as <10-28>, where $B = \omega R_S C_S$.

$$\frac{1}{Z_{par}} = \frac{1}{Z_{C_{dl}}} + \frac{1}{Z_f}$$ <10-25>

$$Z_f = R_S - \frac{j}{\omega C_S}$$ <10-26>

$$Z_{C_{dl}} = \frac{-j}{\omega C_{dl}}$$ <10-27>

$$\frac{1}{Z_{par}} = \frac{(1+B)^2 R_S[B^2 - j\omega R_S\{C_S + (1+B^2)C_{dl}\}]}{B^4 + \omega^2 R_S^2 C_S^2 \left\{1 + (1+B^2)\frac{C_{dl}}{C_S}\right\}^2} \quad \text{<10-28>}$$

The total impedance ($Z = Z_\Omega + Z_{par}$) is obtained by adding Z_Ω to Z_{par} given as <10-28>. Since the total impedance is $Z^* = Z_{Re} - jZ_{Im}$, Z_{Re} and Z_{Im} can be separated and expressed as <10-29> and <10-30>, respectively.

$$Z_{Re} = Z_\Omega + \frac{R_{ct} + \frac{\sigma}{\omega^{1/2}}}{(C_{dl}\sigma\omega^{1/2} + 1)^2 + \omega^2 C_{dl}^2\left(R_{ct} + \frac{\sigma}{\omega^{1/2}}\right)^2} \quad \text{<10-29>}$$

$$Z_{Im} = \frac{\omega C_{dl}\left(R_{ct} + \frac{\sigma}{\omega^{1/2}}\right)^2 + \frac{\sigma}{\omega^{1/2}}(\omega^{1/2}C_{dl}\sigma + 1)}{(C_{dl}\sigma\omega^{1/2} + 1)^2 + \omega^2 C_{dl}^2\left(R_{ct} + \frac{\sigma}{\omega^{1/2}}\right)^2} \quad \text{<10-30>}$$

Taking the relationship between Z_{Re} and Z_{Im} given by <10-29> and <10-30>, one can obtain a Nyquist plot. The Nyquist plot is obtained by dividing the frequency range. If $\omega \to 0$, <10-29> and <10-30> are simplified into <10-31>, from which the relationship between Z_{Re} and Z_{Im} is obtained as <10-32>. This is shown in Fig. 10-9. In the low frequency region, a linear relationship is observed between Z_{Re} and Z_{Im} (45°) and the intercept at Z_{Re} axis is $R_\Omega + R_{ct} - 2\sigma^2 C_{dl}$.

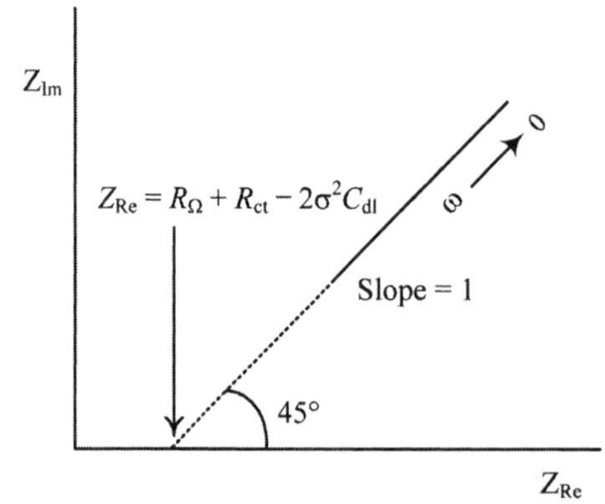

Fig. 10-9 Impedance plot at low frequencies for the electrochemical cell shown in Fig. 10-7.

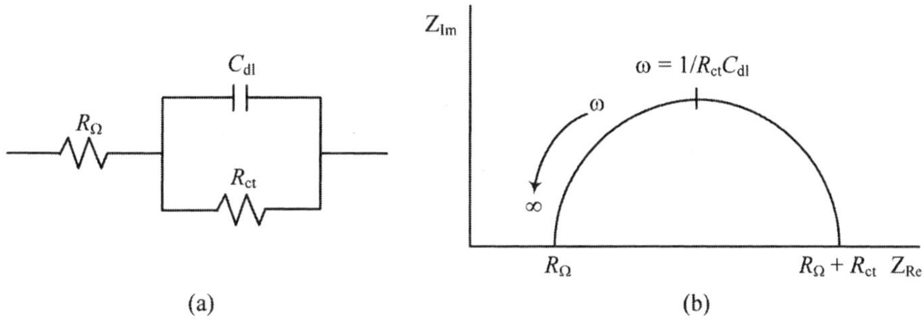

Fig. 10-10 (a); Simplified equivalent circuit at high frequencies for the electrochemical cell shown in Fig. 10-7. (b); Nyquist plot.

$$Z_{Re} = R_\Omega + R_{ct} + \frac{\sigma}{\omega^{1/2}}, \quad Z_{Im} = 2\sigma^2 C_{dl} + \frac{\sigma}{\omega^{1/2}} \tag{10-31}$$

$$Z_{Im} = Z_{Re} + 2\sigma^2 C_{dl} - R_\Omega - R_{ct} \tag{10-32}$$

Meanwhile, if $\omega \rightarrow \infty$, $R_S = R_{ct}$ from <10-20> and $C_S = 0$ from <10-22>. Hence, $Z_f = R_S + C_S = R_{ct}$; the faradaic impedance is the charge transfer resistance at high frequency region. Hence, the equivalent circuit in Fig. 10-7-a can be simplified as shown in Fig. 10-10-a at higher ω range. In this circuit, Z_{Re} and Z_{Im} are derived as <10-33>, and <10-34> holds between Z_{Re} and Z_{Im}. Since this is a formula for a circle, $(x-a)^2 + (y-b)^2 = c^2$, with the origin of $(R_\Omega + R_{ct}/2, 0)$ and the radius of $R_{ct}/2$, the total impedance appears as a semicircle as shown in Fig. 10-10-b. The ω at which Z_{Im} has the maximum value in the semicircle is shown in <10-35>.

$$Z_{Re} = Z_\Omega + \frac{R_{ct}}{1 + \omega^2 R_{ct}^2 C_{dl}^2}, \quad Z_{Im} = \frac{\omega R_{ct}^2 C_{dl}}{1 + \omega^2 R_{ct}^2 C_{dl}^2} \tag{10-33}$$

$$\left(Z_{Re} - R_\Omega - \frac{R_{ct}}{2}\right)^2 + Z_{Im}^2 = \left(\frac{R_{ct}}{2}\right)^2 \tag{10-34}$$

$$\omega = \frac{1}{R_{ct} C_{dl}} \tag{10-35}$$

Fig. 10-11 combines the impedance spectrum at low frequencies (Fig. 10-9) and high frequencies (Fig. 10-10-b). As is seen, the total impedance decreases linearly as ω increases with 45° in the low frequency range, and a semicircle appears in the high frequency range. R_Ω and R_{ct} can be obtained from the values on the Z_{Re} axis, and C_{dl} by

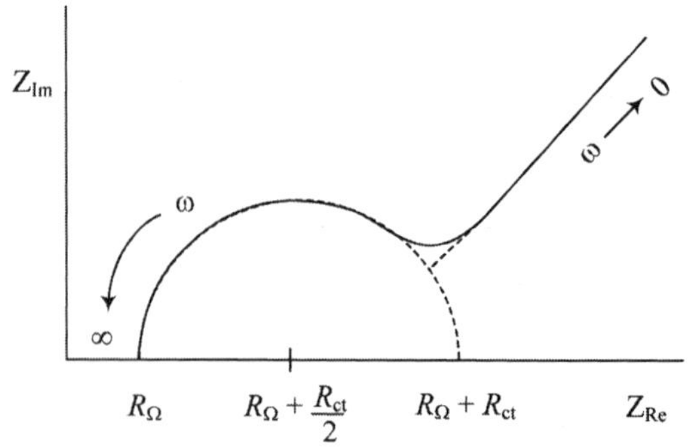

Fig. 10–11 Impedance plot combined Fig. 10–9 and Fig. 10–10–b.

using <10-35>. From the Z_{Re} axis intercept of the straight line, σ and diffusion coefficient can be calculated (<10-21>).

Impedance method has several advantages over other electrochemical methods (hydrodynamic method, cyclic voltammetry, etc.). Firstly, the parameters related to charge transfer and mass transfer in electrochemical reactions can be measured simultaneously. In addition, the values of resistance and capacitance in a given equivalent circuit can be obtained. Secondly, electrochemical parameters can be obtained by minimizing electrode deformation compared to other electrochemical methods. For example, in cyclic voltammetry, the working electrode is driven to a condition far from equilibrium, so that the electrode itself can be oxidized or reduced. However, in impedance method, electrode deformation is relatively small because the voltage perturbation is very small; ±(5~50) mV near equilibrium potential. However, in a very low frequency range, the measurement time is so long that the electrode or solution may change during the measurement; the reliability of impedance data is poor.

Self learning 10-3

If $\omega = 10^{-2}$ Hz, what is the time required to apply one alternating voltage cycle?

Self learning 10-4

Confirm the difference between the angular frequency of a sinusoidal oscillation ($\omega = 2\pi f$) and the angular frequency of rotation ($\omega = 2\pi$ x rotation rate). The latter has been explained in Chap. 6.

Problems

01 From the Nyquist plot in Fig. 10-12, derive the equivalent circuit and calculate the electrochemical parameters comprising the equivalent circuit.

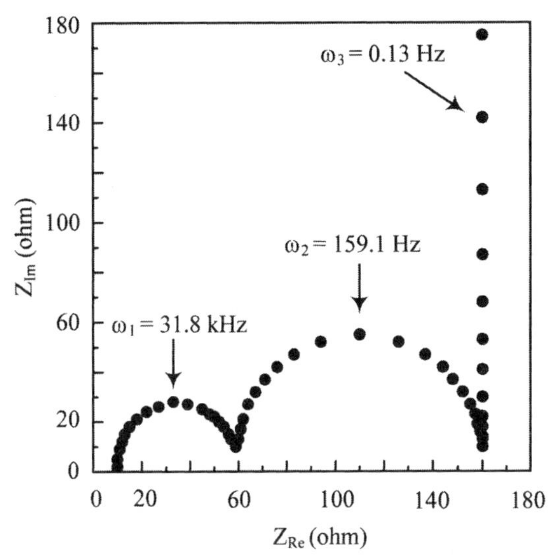

Fig. 10-12 A Nyquist plot of an electrochemical cell.

02 Draw the equivalent circuit and expected impedance spectrum (Nyquist plot) for the following electrochemical system.

(a) Hg/aqueous KCl solution/reference electrode (refer to Fig. 1-16)

(b) Li metal/$LiClO_4$ in acetonitrile/reference electrode

03 Consider that R_{ct} is larger than that for Fig. 10-11. How does the Nyquist plot differ when impedance is measured in the same frequency range?

04 If R_{ct} is smaller than that for Fig. 10-11, the semicircle may not appear. How can you obtain the semicircle in Nyquist plot?

05 J. H. Sluyters and coworkers (*Rec. Trav. Chim. Pays-Bas*, **79**, 1101, 1960) obtained the Nyquist plots for two electrochemical reactions (Fig. 10-13). The frequency range was; $f = \omega/2\pi$ = 20 Hz ~ 20 kHz. The numbers in the figure indicate the frequency (f in kHz).

(a) Assuming that both electrochemical reactions undergo a mass transfer and charge transfer process, determine which step is rate-limiting for the reaction (a) and (b), respectively. Rationalize your answer.

(b) Calculate C_{dl} from Fig. 10-13-a.

(c) The maximum frequency was 20 kHz for the measurement. Draw the expected Nyquist plots for the two cases when the frequency is infinitely extended.

(d) In Fig. 10-13-a, $C^*_{Zn^{2+}} = C^*_{Zn(Hg)} = 8 \times 10^{-3}$ M. Calculate the standard rate constant (k^0) for $Zn^{2+} + 2e = Zn(Hg)$. The electrode area (A) was 1 cm^2.

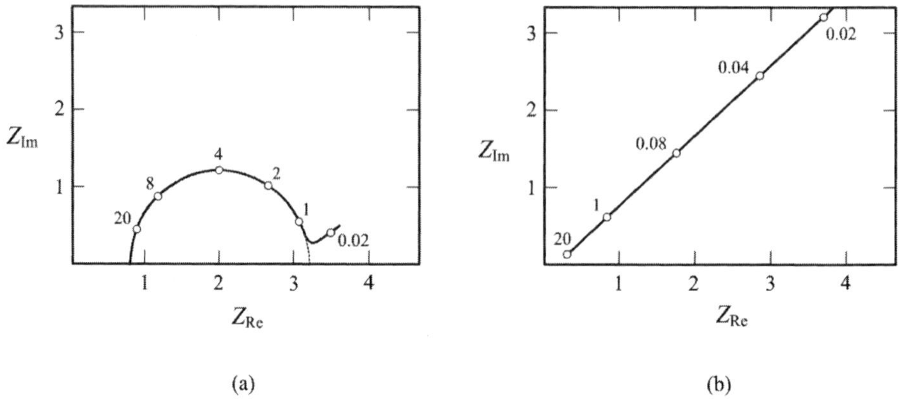

Fig. 10-13 Nyquist plots for two electrochemical reactions: (a) $Zn^{2+} + 2e = Zn(Hg)$ and (b) $Hg_2^{2+} + 2e = Hg$.

06 Fig. 10-14 shows the equivalent circuit that includes the capacitance (C_Ω) of the solution between a reference and working electrode.

(a) In general, $C_\Omega << C_{dl}$. Explain why (refer to problem 3-8).

(b) Draw the expected Nyquist plot.

(C) Normally, C_Ω is ignored for constructing equivalent circuits (for instance, Fig. 10-7-a). Give an explanation.

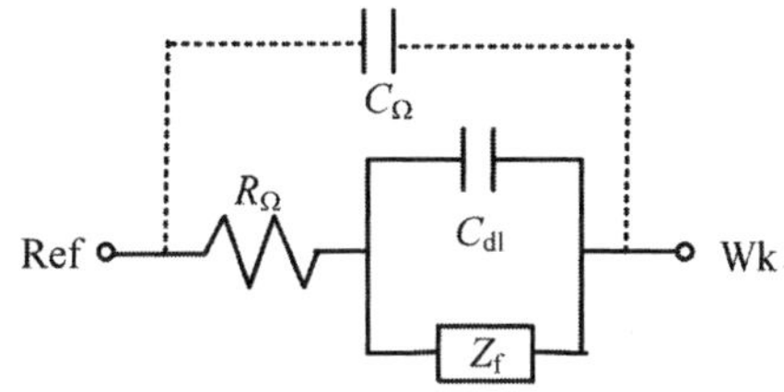

Fig. 10–14 The equivalent circuit including the capacitance (C_Ω) of the solution between a reference and working electrode.

CHAPTER 11

Electrochemical industries

In electrolysis processes, electrical energy is supplied to oxidize or reduce reactants to synthesize desired products. The typical examples are the brine electrolysis, in which Cl_2 and NaOH are obtained, and extraction of aluminum (Al). Globally, enormous amounts of Cl_2, NaOH, and aluminum are produced through electrolysis. In metal-related industries, the electrowinning converts ores into the corresponding metals through electrolysis, electrorefining purifies impure metals. Electroplating, anodizing of metal surfaces, electrophoretic coating, and electrochemical machining are the other examples of electrolysis. Meanwhile, electrolytic syntheses are applied for the production of inorganic materials ($KMnO_4$, MnO_2, H_2O_2, and so on) and organic compounds (adiponitrile, an intermediate of vitamin C, and fluorine compounds). In electrodialysis, ions are removed by using electrochemical principles.

The electric energy required for electrolysis is $Q \times E_{appl}$. In order to reduce the energy consumption, two measures are needed. Firstly, the consumption of electric charges for parasitic reactions should be minimized; the selectivity (***Coulombic efficiency***) for the targeted electrochemical reaction should be maximized. Secondly, E_{appl} should be minimized. Fig. 11-1 shows how the applied voltage (E_{appl}) between two half-cells is determined when

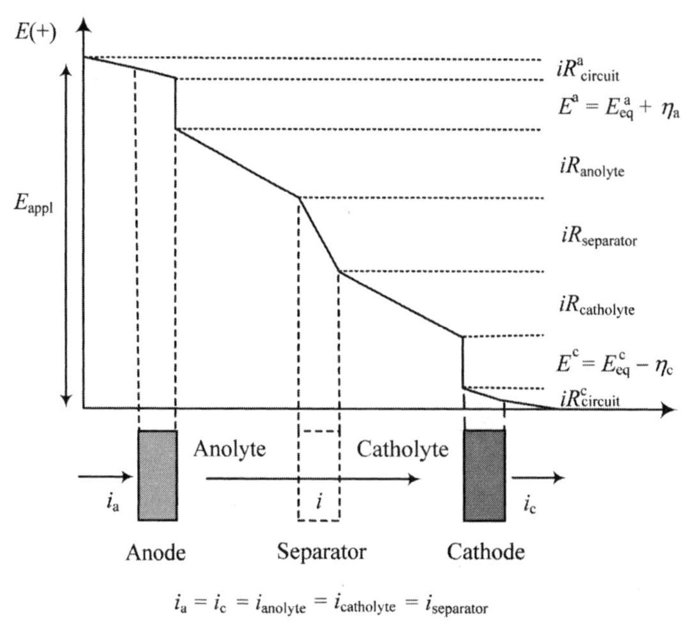

Fig. 11–1 *iR* drops and overpotentials for determining applied voltage (E_{appl}) in an electrolytic cell.

electrolysis is performed under a constant-current condition. As is seen in Fig. 11-1, an iR drop ($iR^a_{circuit}$) evolves due to the resistance of external circuit and anode itself. The potential of anode is $E^a = E^a_{eq} + \eta_a$ because of anodic overpotential (η_a) at a given current condition. In addition, there evolves an iR drop due to the solution resistance of anolyte (solution in the anode chamber that is separated from catholyte by a separator) and an iR drop due to separator resistance. For the same reason, $iR^c_{circuit}$ and $iR_{catholyte}$ evolve at the given current condition. Similarly, the potential of cathode is $E^c = E^c_{eq} - \eta_c$ because of cathodic overpotential (η_c). Summation of all iR drops and overpotentials gives E_{appl} expressed as <11-1> (refer to <1-6>).

$$E_{appl} = E_{cell} + \eta_a + \eta_c + iR_{total} \qquad \text{<11-1>}$$

In <11-1>, E_{cell} is called a ***thermodynamic decomposition voltage***. It is the difference between the equilibrium potentials of two half-cells ($E_{cell} = E^a_{eq} - E^c_{eq}$); it is a thermodynamic value because it is obtained under equilibrium condition ($i_c = |i_a| = 0$). Further, it is expressed as E^0_{cell} if standard states are assumed. E_{cell} may change with a progress of electrolysis or be fixed depending on the electrode reactions. In practical electrolysis, however, $|i_a| = i_c \neq 0$. Hence, E_{appl} is larger than E_{cell} by the amount of ($\eta_a + \eta_c + iR_{total}$), which is called a ***cell polarization***. The three components (η_a, η_c and iR_{total}) increase with an increase in current, such that a larger voltage (E_{appl}) must be applied to speed up the electrolysis. Note that the cell polarization terms (η_a, η_c and iR_{total}) are kinetic parameters since they are current (rate)-dependent.

To minimize E_{appl}, the overpotential (η_a and η_c) and iR_{total} should be reduced. The overpotential is the sum of activation overpotential, concentration overpotential and others if there are any. Since the activation overpotential is related with the charge transfer process, the methods to increase the charge transfer rate should be sought. Since R_{ct} and i_0 shown in <4-19> and <4-12>, respectively, determine the activation overpotential, in order to reduce this, the electrodes with a large standard rate constant (k^0) should be selected. Also, the electrode performance can be improved (k^0 is increased) by any modifications. Other way for reducing activation overpotential is the expansion of electrode surface area (A) and increase of the electroactive species concentration (C_O^* and C_R^*). To reduce concentration overpotential, as is seen in <6-31> and <6-7>, the ways to increase mass transfer rate should be sought; expanding the electrode area (A), increasing the electroactive species concentration (C_O^* and C_R^*), and increasing the stirring rate.

The resistances that impede the current flow through a closed loop cause an iR drop, which is one of the contributors for cell polarization; $R_{total} = R_{separator} + R_{solution} + R_{circuit} + R_{others}$. In general, $R_{separator}$ that is related to ion conduction in separator is the largest. Hence, the ***undivided cells*** are advantageous, in which the electrolyte solution is not divided into anolyte and catholyte by a separator. Since $R_{separator}$ is absent, the cell polarization and thus E_{appl} is smaller as compared to the ***divided cells*** that are fabricated with a separator. In some electrochemical cells, however, the use of a separator (divided cells) is inevitable to prevent the intermixing of electrochemically generated products. Namely, let us assume that an electrochemical reduction ($O + ne \rightarrow R$) takes place on a cathode and the as-produced R can move to an anode by any ways of mass transfer. Then, R can be oxidized back to O. The net result is the waste of electric charges consumed for the reduction; Coulombic efficiency must be less than 100 %.

In general, the resistance of electrolyte solution ($R_{solution}$), which is related to ion conduction in electrolytes, is larger than $R_{circuit}$ comprising the resistance of external circuit and electrode. $R_{solution}$ is determined by three factors as <11-2>; the distance between two electrodes (l), the ionic conductivity of electrolyte solution ($\kappa_{solution}$) and the electrode area (A).

$$R_{solution} = \frac{l}{\kappa_{solution} A} \qquad \text{<11-2>}$$

As is seen in <11-1>, the factor deciding cell polarization is not $\kappa_{solution}$, but $R_{solution}$. Apparently, a larger $\kappa_{solution}$ gives a smaller $R_{solution}$, which gives a smaller cell polarization and ultimately a smaller E_{appl}. However, even if $\kappa_{solution}$ is large, $R_{solution}$ can be controlled to be small enough by increasing A and decreasing l as expected from <11-2>.

Self learning 11-1

If only the electrode area is enlarged in an electrochemical cell while the other variables (the type of electrodes and electrode reactions) being fixed, how do the four terms on the right side of <11-1> change?

11-1 Design of electrolysis cells

The design of electrolysis cell (or electrolytic cell) can be divided into flow cell and batch cell (Fig. 11-2). In the former, the reactants are continuously fed into the cell, but not in the latter. Flow cells have the advantage of dissipating the heat generated by electrolysis but require facilities for pumping. If the inter-electrode gap is too narrow, the pumping of liquids is difficult; there is a limit in minimizing the inter-electrode gap.

Electrodes are fabricated in the form of a parallel plate or a cylinder. In either cases, the ***inter-electrode gap*** should be uniform. If this is not the case, under a constant-current electrolysis, the voltage difference between two electrodes (E_{appl}) differs on the two-dimensional electrode surface. Let us say a constant-current electrolysis is performed with a cell in which the inter-electrode gap is not uniform (refer to Fig. 11-10). Since $iR_{solution}$

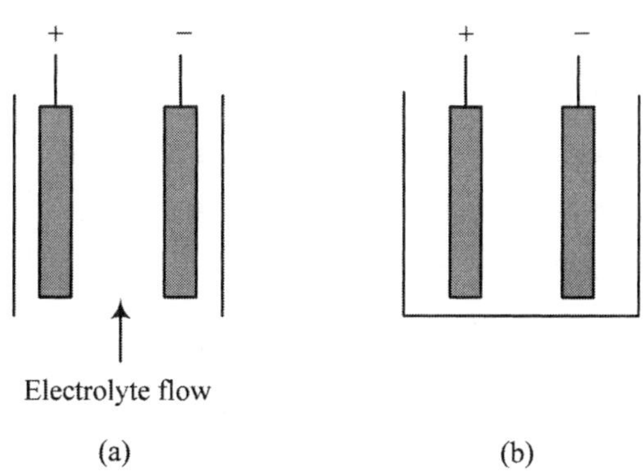

Fig. 11-2 Types of electrolytic cell: (a); flow cell (b); batch cell.

is larger between two points facing each other at a longer distance than between two points facing closer, the cell polarization and thus E_{appl} is larger in the former. Such a non-uniform E_{appl} can cause a non-uniform electrochemical reaction on two-dimensional electrode surface. For example, a desired electrochemical reaction on one point but side reaction on the other.

Self learning 11-2

The internal temperature of an electrolytic cell may rise due to heat evolution during electrolysis. Identify the origins for heat evolution in both thermodynamic and kinetic point of view. Is it possible that the cell temperature decreases during electrolysis? (refer to self learning 12-6).

Methods of aligning electrodes include monopolar and bipolar types. In the ***monopolar design*** shown in Fig. 11-3-a, three electrodes are positive and the other three are negative. If E_{appl} = 3.5 V is needed for electrolysis, 3.5 V is applied between the two groups of electrodes. This alignment has the effect of increasing the current (reaction rate) by enlarging electrode area. In the ***bipolar design*** shown in Fig. 11-3-b, voltage is applied only to the electrodes located at both ends. Since it is composed of five unit cells, the voltage to be applied is 3.5 V × 5 = 17.5 V. This is called bipolar because the electrodes have

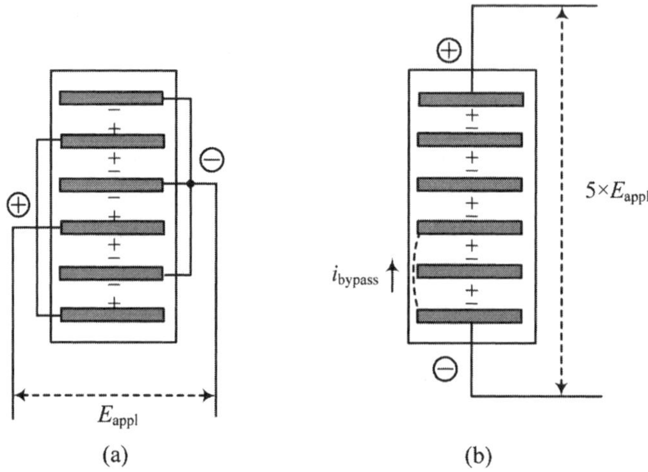

Fig. 11-3 (a); Monopolar cell and (b); bipolar cell.

two polarities; one side of an electrode serves as a positive but the other side as a negative. This is contrasted by the monopolar design, in which the electrodes have a single polarity; the electrodes are either positive or negative. In the monopolar design, the electrode area is large, such that electrolysis can be performed at a large current (speed). A larger voltage is applied in the bipolar design, but the electrolysis rate is relatively slow because the electrode area is not increased as in the case of monopolar one. The advantageous feature for the bipolar design is that the contact resistance is smaller and the structure is simpler because the contact points between the external lead and electrodes is smaller. However, problems due to bypass current may occur. That is, when the current is bypassed as shown in Fig. 11-3-b, 7 V (3.5 V × 2) is applied between the two electrodes. Side reactions are possible on the two electrodes.

Self learning 11-3

Consider the principle of "a closed-loop must be formed for current flow in electrolytic cells" and propose a way to suppress the bypass current in the bipolar design.

11-2 Brine electrolysis (chlor-alkali industry)

Chlorine (Cl_2) obtained from ***brine electrolysis*** is used for the production of PVC (poly vinylchloride), and NaOH is used as a neutralizing agent. Currently commercially available processes include diaphragm cells, mercury cells, and membrane cells.

11-2-1 Diaphragm cell

Fig. 11-4 shows the structure of a ***diaphragm cell***. When brine is fed into the cell, chlorine

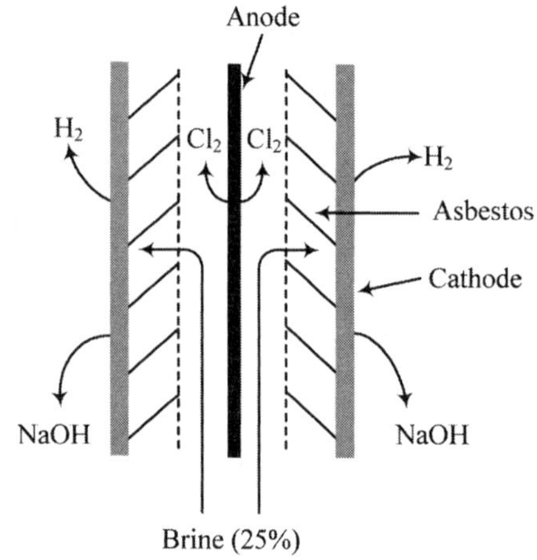

Fig. 11–4 Structure of a diaphragm cell.

is generated on the anode. The brine can pass through the asbestos layer to the cathode, on which OH^- is generated by reduction of water. Table 11-1 lists the equilibrium potential of several half-cells that are involved in brine electrolysis. Note that pH = 4.0 for the brine solution (NaCl concentration; 25%) and pH = 14 for the catholyte as OH^- is generated there. Gaseous chlorine (Cl_2) is generated by the reaction (*A*) in Table 11-1 (pH = 4.0), whereas the reaction (*C-ii*) prevails in the cathode to produce OH^- and H_2 (pH = 14.0). Therefore, E_{cell} = 2.15 V at room temperature. Meanwhile, the reaction (*B-i*) with more negative E_{eq} is thermodynamically favored compared to the reaction (*A*) at pH = 4.0; oxygen generation is more probable than Cl_2 generation in thermodynamic point of view. Furthermore, there is a plentiful amount of H_2O that is the reactant for (*B-i*). However, kinetics select Cl_2 generation; since the anodes (typically metal oxides) exhibit

Table 11–1 Equilibrium potentials for some half-cells*

Reaction	pH	E_{eq} (*vs.* NHE)
(A): $2Cl^- = Cl_2 + 2e$	4	+1.31
(B): (i) $2H_2O = O_2 + 4H^+ + 4e$	4	+0.99
(ii) $O_2 + 2H_2O + 4e = 4OH^-$	14	+0.39
(C): (i) $2H^+ + 2e = H_2$	4	−0.24
(ii) $2H_2O + 2e = H_2 + 2OH^-$	14	−0.84
(D): $Na^+ + Hg + e = NaHg$	4	−1.85

*Concentration of NaCl = 25 %, T = 298 K

a higher standard rate constant (k^0) for reaction (*A*) over (*B-i*), Cl_2 generation rate is much higher than O_2 evolution rate. These metal oxide electrodes are also referred to as ***dimensionally-stable anodes*** (DSA). These new electrodes are very stable to corrosion (oxidation), which is contrasted by the previously used carbon electrodes that are oxidized to be worn under the positive working potential of the chlorine half-cell. The DSA is dimensionally stable due to absence of wear. Their service life can be many years.

The asbestos used in this process allows the transfer of water and ions. As a result, the following problems may occur. Firstly, Cl^- can move to the catholyte and thus the aqueous NaOH solution is contaminated by NaCl impurity. Secondly, OH^- generated on the cathode can move toward the anode, causing the following problems. Firstly, the chlorine produced by electrolysis is wasted by $Cl_{2(g)} + 2OH^- \rightarrow H_2O + OCl^- + Cl^-$. Secondly, the E_{eq} for (*B-i*) becomes more negative with pH increase (inflow of OH^-). This causes the oxygen generation, which was ignored at pH = 4.0, to be more feasible; oxygen (O_2) can be introduced into Cl_2 as an impurity. To minimize all these problems, the NaOH concentration in the catholyte is adjusted to be less than 12 % to suppress the diffusion of OH^- toward the anolyte; an additional cost is necessary to evaporate water to meet the normal commercial concentration of 50 %. Another drawback for this process is an appreciable separator (asbestos) resistance to cause a relatively high applied voltage (E_{appl}). Despite these problems, this process has become commercially viable because the impure NaOH has a diverse application; e.g. as a neutralizing agent for acids, in which the purity of NaOH is not critical.

11-2-2 Mercury cell

In this process, Cl_2 is generated by the reaction (*A*) in Table 11-1, and NaHg amalgam is generated by the reaction (*D*), which gives E_{cell} = 3.16 V. Addition of water to the NaHg amalgam produces aqueous NaOH and regenerates Hg; $NaHg + H_2O \rightarrow 1/2H_2 + Na^+ + OH^- + Hg$. It is not necessary to control the NaOH concentration like diaphragm cell, so that it produces 50 % NaOH directly. Here, H_2 generation by (*C-i*) at pH = 4 is thermodynamically more favorable than the NaHg formation. However, H_2 generation

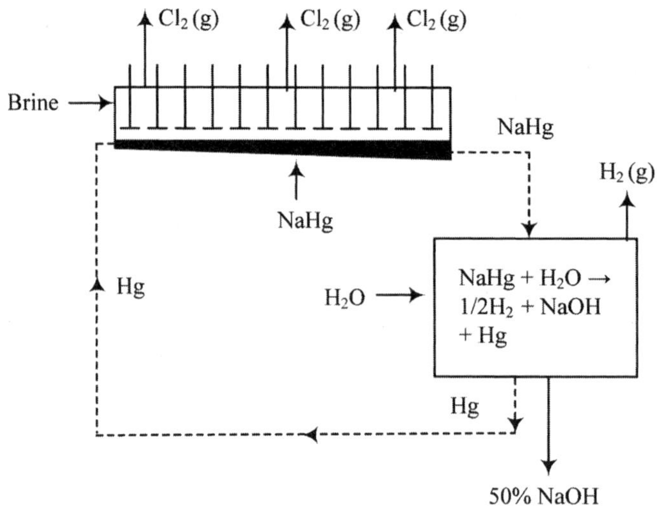

Fig. 11-5 Mercury cell and regeneration of mercury.

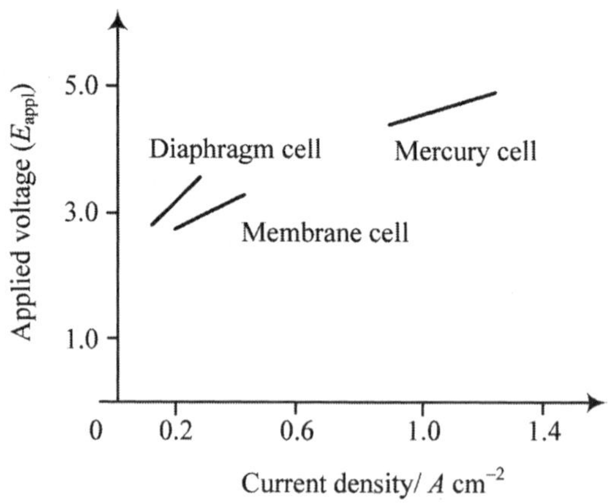

Fig. 11-6 Applied voltage (E_{appl}) according to current density in three brine electrolysis cells.

is not serious since k^0 for H_2 generation is very small on Hg electrode (see Fig. 1-15 and Fig. 1-16).

In the ***mercury cell***, as shown in Fig. 11-5, the anode and cathode place at the top and bottom of the cell, respectively, without a separator. Such an undivided cell is possible in that chlorine gas leaves the cell at the top while the NaHg amalgam leaves at the bottom; the chance for the anodic (cathodic) product to move to the cathode (anode) is low. The absence of separator allows a high speed (current) electrolysis. A comparison of the current density-applied voltage relation for three brine electrolysis cells (Fig. 11-6) reveals that the applied voltage (E_{appl}) is not substantially larger than that of the other cells even if this cell operates under the much higher current density. This is because the separator resistance does not contribute to the cell polarization. Due to the pollution problem of mercury, however, this process is gradually being discarded.

Self learning 11-4

As shown in Fig. 11-6, E_{appl} of mercury cell is only about 1.5 times larger than that of the other cells even if the working current density is four times larger. Explain this using <11-1>.

Self learning 11-5

Fig. 11-6 shows a linear relationship between the current density and applied voltage for three cells. In electrochemical cells, current flows through a closed-loop and the current flow is impeded by charge transfer, mass transfer, and ion-conduction in electrolyte and separator. Among the current-impeding factors, which is most prevailing to give such a linear relationship? Explain the meaning of the slopes in Fig. 11-6. Explain why the slopes differ for three cells. Refer to Example 6-1.

11-2-3 Membrane cell

The electrode reactions in a ***membrane cell*** is identical with those of diaphragm cell, so E_{cell} = 2.15 V. The asbestos in diaphragm cell is replaced by cation-exchange resins in membrane cells. There are two types of cation-exchange resins: strong-acid form ($-SO_3H$, trade-name ***Nafion***) and weak-acid form ($-COOH$). The chemical composition of the strong-acid cation-exchange resin is shown in Fig. 11-7. The acid functional groups present in the cation-exchange resins are hydrated and dissociated into anion and cation. The anion is immobilized in the polymer membrane, but the cation is mobile to migrate and thus ion-conduction is possible by the mobile cations. Furthermore, the cations can

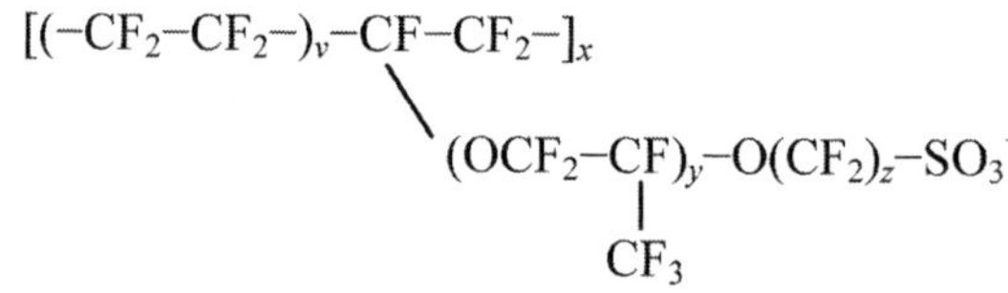

$v = 5 \sim 15 \quad x = 1000, \quad y = 1 \sim 3 \quad z = 1 \sim 4$

Fig. 11–7 Chemical composition of strong–acid type polymer membrane used in membrane cell.

be exchanged by other cations, such that they are called as cation-exchange resins. Since the ionic conductivity is larger for the strong-acid type, it is used in most of the processes at present.

The use of cation-exchange resins, which allows the permeation of cations but prevents anions, solves many of the problems that are caused by the movement of anions (Cl^- and OH^-) through the separator in the diaphragm cells. There is no problem of NaCl impurities, which is caused by the movement of Cl^- to the catholyte in the diaphragm cells. Also, the problems associated with OH^- diffusion from the catholyte to anolyte can be eliminated. However, if the NaOH concentration is higher than 35 %, OH^- diffusion is appreciable; hence, the NaOH concentration in the catholyte should be adjusted to less than 35 %. This process is currently the most widely used in the world.

Fig. 11-8 shows the schematic diagram of the membrane cell fabricated using the ***zero-gap principle*** and the enlarged view near the membrane/electrode interface. As shown in Fig. 11-8-b, the gap between the membrane and electrodes is minimized; thus minimizes the solution resistance. This arrangement is particularly advantageous in that the formation of gas bubbles (hydrogen and chlorine gas) is avoided at the spaces between the membrane and electrodes. When gas bubbles form, the solution resistance increases significantly. By minimizing the gap between the membrane and electrode (zero-gap), the gas bubble formation can be prevented.

Self learning 11-6

Explain the adverse effects of gas bubbles on cell polarization, which are placed at the membrane/electrode gap.

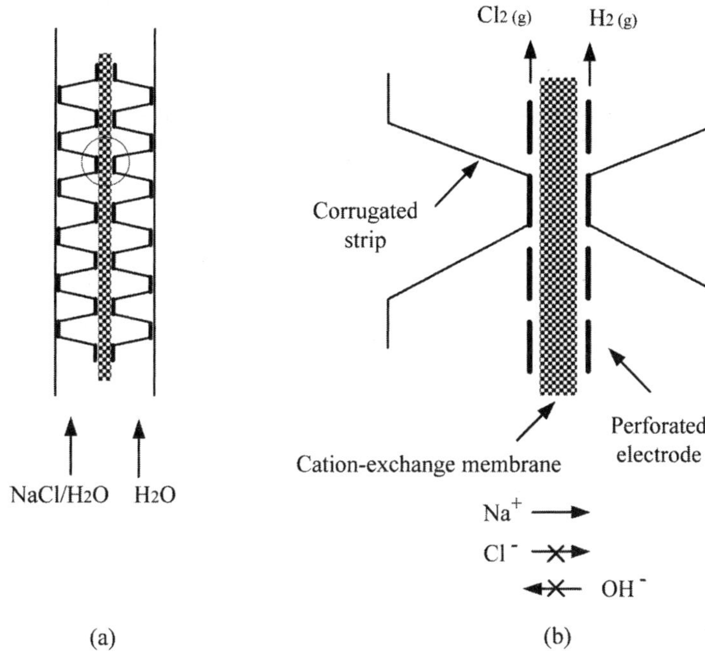

Fig. 11-8 (a); Structure of membrane cell applying zero-gap principle and (b); an enlarged view between membrane and electrode.

Another membrane cell has been developed, in which the cathodic reaction (*B-ii*) in Table 11-1 is utilized along with the anodic reaction (*A*). Aqueous NaOH is produced by using this combination without the by-product of hydrogen. E_{cell} = 0.92 V, which is remarkably smaller compared to that for the conventional membrane cells (E_{cell} = 2.15 V), wherein the cathodic reaction (*C-ii*) is adopted. One can thus expect a lower applied voltage (E_{appl}). However, the actual E_{appl} is not so small for this combination because the reaction (*B-ii*) needs a larger activation overpotential; in <11-1>, E_{cell} is smaller but η_c is larger as compared to those for the conventional one. Until now, a platinum/carbon electrode shows the best performance as the cathode. To further reduce η_c, the electrodes that exhibit a superior activity for four-electron oxygen reduction are desperately needed.

11-3 Electrowinning

Electrowinning is the process of obtaining metals by electrochemical reduction of metal ions that are generated from metal-containing ores. When this process performs in aqueous solutions, hydrogen evolution may be accompanied as a side reaction. Since the potential-determining equilibrium depends on the solution pH, the equilibrium potential of hydrogen half-cell at standard state is also different according to the pH as follows.

$$\text{pH} = 0;\quad 2H^+ + 2e = H_2,\quad E^0 = 0.0\ \text{V}\ (vs.\ \text{NHE})$$

$$\text{pH} = 14;\quad 2H_2O + 2e = H_2 + 2OH^-,\quad E^0 = -0.83\ \text{V}\ (vs.\ \text{NHE})$$

Self learning 11-7

Confirm that $E^0 = -0.83$ V (*vs.* NHE) for hydrogen half-cell at pH = 14 using <2-20> and Fig. 2-12.

If the equilibrium potential of a metal/metal ion half-cell is more negative than that of the hydrogen half-cell, hydrogen evolution is thermodynamically more favored than the reduction of metal ion; the electrowinning of that metal is impossible in aqueous solution because of H_2 evolution. In this case, a molten salt that is free from proton (H^+) and water can be used as an electrolyte. In practice, Li, Na, Mg, Al, etc. are obtained through electrowinning in molten salt electrolytes. A representative example is ***aluminum extraction***. In this process, cryolite (Na_3AlF_6) is mixed with CaF_2 and AlF_3, and heated at 970 °C to obtain a liquid molten salt, then used as an electrolyte. The following reactions are involved in Al extraction; dissolution of bauxite (the main component is Al_2O_3) and electrolysis in the molten salt.

Dissolution: $2Al_2O_3 + 2AlF_6^{3-} \rightarrow 3Al_2O_2F_4^{2-}$

Anode: $Al_2O_2F_4^{2-} + 4F^- + C \rightarrow CO_2 + 2AlF_4^- + 4e$

Cathode: $AlF_6^{3-} + 3e \rightarrow Al + 6F^-$

Overall: $2Al_2O_3 + 3C \rightarrow 4Al + 3CO_2$

In the anode, the carbon electrode itself is oxidized and converted into CO_2; the carbon electrode is consumed during electrolysis. In this electrolytic cell, the inter-electrode gap is designed to be relatively wide. Hence, the electrolyte resistance is appreciable to cause a substantial heat generation during the electrolysis due to a substantial cell polarization (refer to Q_{Joule} discussed in 12-1-7). The heat is used for cell heating up to 970 °C.

If the equilibrium potential of a metal/metal ion half-cell is more positive than that of the hydrogen half-cell, the electrochemical reduction of metal ion takes a thermodynamic precedence over hydrogen evolution in aqueous solutions; the electrowinning is possible. This is called a ***hydrometallurgy***. Representative metals include Cu and Zn. In the case of copper, E^0 = 0.34 V (*vs.* NHE) for $Cu^{2+} + 2e = Cu$, which is more positive than E^0 = 0.0 ~ −0.83 V (*vs.* NHE) (pH = 0 ~ 14) for hydrogen half-cell. Note that the comparison should be made between two equilibrium potentials but the two E^0 were used assuming standard state. The comparison reveals that Cu electrowinning (Cu^{2+} reduction) is allowed without hydrogen evolution irrespective of solution pH. In the case of Zn, however, since E^0 = −0.76 V (*vs.* NHE), the Zn electrowinning is impossible at lower pH; e.g. E^0 = 0.0 V (*vs.* NHE) at pH = 0.0. In real, Zn is produced through hydrometallurgy, which is allowed for the following reasons. Firstly, the solution pH is adjusted to a high value to inhibit hydrogen evolution; E^0 = −0.83 V (*vs.* NHE) at pH = 14, at which the Zn electrowinning precedes hydrogen evolution. As shown in Fig. 2-13, at higher pH, both hydrogen evolution and Zn dissolution decrease, so that the re-dissolution of Zn obtained through hydrometallurgy is also reduced. Secondly, the k^0 for hydrogen generation is small on Zn electrode; hydrogen evolution rate is quite low on Zn electrode. Table 11-2 compares the exchange current density of hydrogen evolution reaction on various inert metal electrodes. The larger the number in the list, the smaller the k^0 for hydrogen evolution (see <4-12>). In the case of Zn, the k^0 value is relatively small like Hg, Pb and Cd. Thirdly, some additives are added into the electrolyte solution to control the k^0 value to be further smaller.

Table 11-2 Exchange current density of hydrogen evolution reaction (1.0 *M* H_2SO_4)

Metal	$-\log I_0(A/cm^2)$	Metal	$-\log I_0(A/cm^2)$
Ru	2.1	Fe	6.0
Pd	2.3	Cu	6.7
Rh	2.8	W	7.0
Pt	3.6	Cr	7.4
Co	5.2	Zn	10.5
Ni	5.2	Cd	11.0
Ag	5.4	Pb	12.2
Au	5.5	Hg	12.5

11-4 Electrorefining

The process of increasing the purity of metal by removing impurities is called an ***electrorefining***. A representative example is the electrolytic copper manufacturing process. The impurities in copper are classified into four groups: (*a*); Ag, Au, Pt, (*b*); Sn, Bi, Sb, (*c*); Pb, and (*d*); Fe, Ni, Co, Zn. The impure copper is used as an anode, from which Cu^{2+} is dissolved out, and the dissolved Cu^{2+} is reduced on the cathode to be a high-purity copper. A mixture of aqueous solution of $CuSO_4$ and H_2SO_4 is used as the electrolyte. The impurity metals belonging to (*a*) are not dissolved because they have a lower oxidation tendency than copper; their equilibrium potentials at standard state (E^0) are more positive than that of Cu half-cell. The metals belonging to (*b*) are dissolved because their E^0 is more negative than E^0 of Cu half-cell, but the dissolved ions are converted into oxides or hydroxides to precipitate on the bottom of electrolytic cell; the concentration of these metal ions is very small in the solution. The Pb impurity in group (*c*) dissolves, but converted into insoluble $PbSO_4$ to be deposited on the bottom of the cell. The metals in group (*d*) also dissolves due to more negative E^0 compared to Cu half-cell. However, since the reducing tendency is poorer than that of copper, the metal ions remain in the solution when copper is electrodeposited. In summary, only copper is electroplated on the cathode to produce high-purity copper. Meanwhile, after the electrorefining of copper, the metals in group (*a*) are left on the anode, from which Ag and Au are purified in order through a similar process.

Self learning 11-8

After electrorefining of copper, the remaining silver and gold can be obtained in order. Explain why electrorefining is possible in this order.

11-5 Electroplating and electroless plating

Electroplating is a technique in which metals (e.g. Sn, Cu, Ni, Cr, and Au), alloys (e.g. Cu-Zn, Cu-Sn, and Ni-Cr), or composites (e.g. metal-teflon and metal-Al_2O_3) are deposited on an electrically conductive surface through an electrolytic process. The purpose of electroplating is diverse; e.g. prevention of corrosion, reduction of electrical contact resistance, decoration, and suppression of abrasion.

In general, the metal to be plated is used as the anode. The anode metal dissolves into an electrolyte solution, and the as-generated metal ion is electrodeposited on the surface of a target material. However, some metals (e.g. Cr) are passivated when used as an anode (see Section 2-5 and Fig. 8-18), thus can not be used as the anode. In this case, the salts of the metal to be electrodeposited are dissolved into an electrolyte solution. On an inert anode, oxygen evolution by water oxidation proceeds.

The quality of plating layers is affected by the plating rate (current). As shown in Fig. 11-9, when metals are electrodeposited at a lower current, they have excellent crystallinity. In opposite, if the plating current is large, the metals are plated in the form of dendrite or powder; further, the plated layers are not strongly bonded to the substrate.

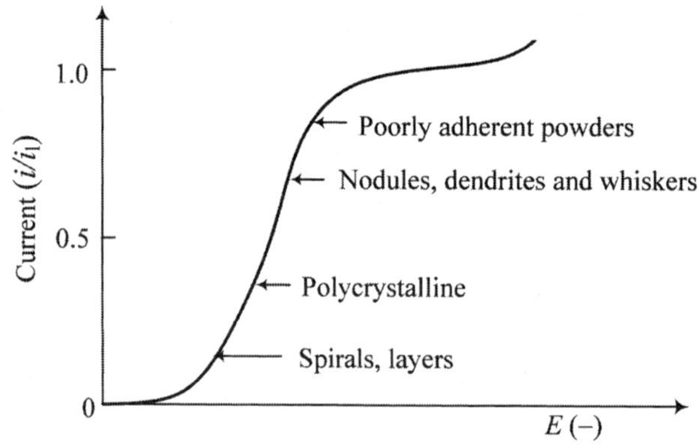

Fig. 11–9 Characteristics of plated layers according to current density.

In the electroplating process, not only the quality of plated layer described above, but also the uniformity of plating thickness (also referred to as ***throwing power***) is important. Namely, if an electroplating is performed for corrosion prevention, the plated layer on thinly plated area can not exert its role. This is important especially when expensive noble metal is plated; the plating should be carried out to secure uniform thickness with a minimum amount of precious metal.

When an electroplating is performed on the surface of a substrate, an ideal uniform plating thickness can be obtained when the same reduction rate (current) is applied to anywhere of the two-dimensional surface. The throwing power is decided by the types of current distribution, which is estimated using the ***Wagner number*** (Wa).

$$Wa = \frac{d\eta/di}{L/\kappa}$$

Here, $d\eta/di$ represents the activation overpotential according to the current for the plating reaction, while κ the ionic conductivity of plating solution. If the activation overpotential is smaller (R_{ct} for the plating reaction is smaller and k^0 is larger) and ionic conductivity is smaller ($R_{solution}$ is larger), the Wagner number (Wa) is smaller. A ***primary current distribution*** prevails with a smaller Wa but a ***secondary current distribution*** with a larger Wa.

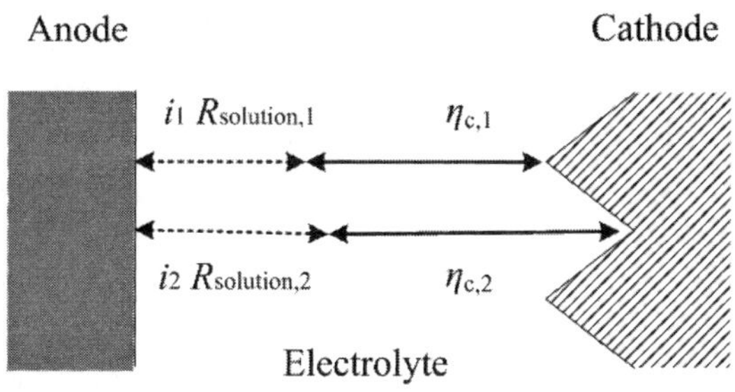

Fig. 11-10 Primary- and secondary-current distributions.

In order to achieve a high uniformity in plating, the secondary current distribution is desired. This is explained as follows. As shown in Fig. 11-10, the two-dimensional surface of the to-be-plated substrate is not smooth (see the side view in Fig. 11-10); the inter-electrode gap differs from one place to the other. When plating is performed under a constant-voltage condition, $E_{appl} = E_{cell} + \eta_c + \eta_a + iR_{solution}$ is applied between the two electrodes. Since E_{cell} and η_a are constant (it can be assumed to be the case), the variables that determine the current are η_c and $R_{solution}$. Note that it is expressed as $R_{solution}$ instead of R_{total} because undivided cells are commonly used for electroplating. Under the condition of a primary current distribution, $iR_{solution}$ has a larger value than the activation overpotential (η_c) because the solution resistance is large; $iR_{solution} > \eta_c$. Thus, when current flows along a closed-loop, the ion conduction in solution ($R_{solution}$) determines the overall rate. Here, a larger current flows if inter-electrode gap is narrower because $R_{solution}$ is smaller. Then, a larger amount of metal is plated at this point because the plating rate is higher there; the uniformity is poor under a primary current distribution.

On the contrary, when η_c is larger and κ is larger ($R_{solution}$ is smaller), a secondary current distribution prevails. Under this condition, $\eta_{c,1} + i_1R_{solution,1} = \eta_{c,2} + i_2R_{solution,2}$ when a constant-voltage plating is performed because E_{cell} and η_a are constant irrespective of the inter-electrode gap. Here, if $R_{solution}$ is small enough to ignore the iR drop, $\eta_{c,1} \approx \eta_{c,2}$. In this case, since the activation overpotential is comparable regardless of the inter-electrode gap, the current is also comparable. Note that current is decided by the overpotential. Since the plating rates are comparable, the thickness of plating layer is also comparable. In summary, in order to increase the uniformity in the plating thickness, the plating conditions should be adjusted to have a secondary current distribution; both $R_{solution}$

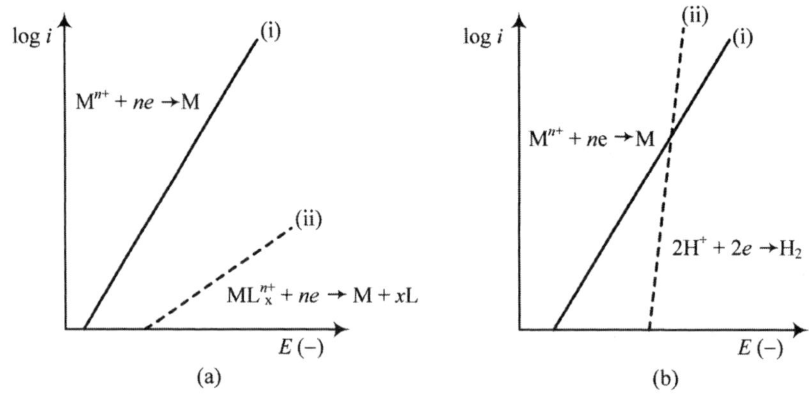

Fig. 11–11 (a); Change of current–potential relationship when a metal ion forms a complex (ML_x^{n+}) and (b) combination of plating and hydrogen evolution.

and k^0 of the plating reaction should be adjusted to be small.

As a method for decreasing k^0 (a larger activation overpotential), a ligand capable of forming a strong complex with the to-be-plated metal ion is added into the plating solution. For example, the electrolyte solution used in copper plating is composed of $Cu(CN)_2$ (40~50%), KCN (20~30%) and K_2CO_3 (10%). Cu^{2+} forms a strong complex with CN^-. As explained in Example 4-1, when the complex forms, the activation energy for plating reaction increases. The activation overpotential is larger at a given current when a metal-ligand complex forms as shown in Fig. 11-11. Another point in Fig. 11-11 is that the complex formation leads to an improvement of the quality of plated layers by decreasing the current (Fig. 11-9) in addition to the improvement of throwing power by leading to a secondary current distribution. Another way for improving the quality of plated layers is the co-generation of hydrogen together along with plating. As shown in Fig. 11-11-b, the current for plating is reduced because some of the current is consumed for hydrogen evolution at the potential where two reactions occur simultaneously; since the plating proceeds at a slower rate, the quality of plating layer is improved.

Several types of additives are used to improve the quality of plated layers. Examples are the accelerators for increasing the plating rate, inhibitors for decreasing the plating rate, and brighteners capable of suppressing irregular reflection by making the surface of plating layer uniform.

When metal ions are reduced by a chemical reducing agent, the metal layer can be deposited on a substrate without electrolysis. This is called an ***electroless plating***. For example, Ag metal can be deposited on a glass surface by a chemical reaction between Ag^+ and formaldehyde (a chemical reducing agent). While the electroplating is only possible on electrically conductive substrates, electroless plating is possible on non-conductive substrates; e.g. glass and plastic. For example, in the printed circuit board (PCB) manufacturing process, a seed layer of copper is prepared by electroless plating on an epoxy resin that is electrically non-conductive, and then copper is further electroplated on the copper seed layer to complete the wiring.

Metals may be precipitated on another metal surface by a replacement plating; e.g. gold can be deposited on copper surface. When a piece of copper is immersed into a solution containing Au^{3+}, copper is oxidized to Cu^{2+} and Au^{3+} is reduced to Au; a galvanic cell forms; this paired electrochemical reactions are thermodynamically allowed. However, since the plating layer is uneven and the bonding force between two metals is poor, this plating is not practically useful.

A CoFe magnetic alloy used for a read/write head of a hard disk is manufactured by electroplating. In addition, copper is wired through electroless plating and electroplating in the integrated circuit manufacturing process. In details, after a trench is prepared on a non-conductive silicon substrate, a copper seed layer is firstly formed by electroless plating, and the trench is filled by electroplating. As the final step, the copper-wired silicon surface is flattened by chemical mechanical polishing (CMP). This process is called a damascene process, which is also called an inlaid technique, similar to the process of manufacturing an inlaid celadon.

Self learning 11-9

From the standard electrode potential values of the half-cells composed of metal/metal ion listed in Table 2-1, suggest the metals that can be deposited by replacement plating on Zn substrate. Assume the standard state for all half-cells.

11-6 Anodizing

Anodizing refers to the formation of an oxide layer on Al, Ti, Cu, or other metal surface by electrochemical oxidation. When Al is anodized (oxidized), a porous Al_2O_3 layer forms on the surface. Since it is porous and has a large surface area, it is used as an electrode of an electrolytic capacitor. In addition, a dye may be injected into the pores to produce a colored Al.

11-7 Electrophoretic coating

Electrophoretic coating is a process of coating polymeric resins on metal surface, and is currently used for corrosion prevention in car manufacturing process. Two coating methods exist; anodic and cathodic coating. The polymer resins, electrode reactions, and the materials to be coated are compared in Fig. 11-12. In the anodic coating, a polymer resin containing a carboxylate ($RCOO^-$) functional group is used after dissolving in water. The electrode reaction on the anode is the formation of H^+ by water oxidation near the electrode

	Anodic	Cathodic
Resin	$RCO_2^- + K^+$	$R_3NH^+ + RCO_2^-$
Electrode reaction	$2H_2O \rightarrow 4H^+ + O_2 + 4e$	$2H_2O + 2e \rightarrow H_2 + 2OH^-$
Deposited film	RCO_2H	R_3N

Fig. 11-12 Polymer resins used for anodic and cathodic electrophoretic coating, the electrode reaction and deposited films.

surface. The enrichment of H^+ shifts the equilibrium between carboxylic acid (RCOOH) and carboxylate ($RCOO^-$) toward the acid. The acid has a poorer solubility in water, so that the polymer resin precipitates as a form of acid (RCOOH) on the anode surface. In the cathodic coating, a polymer resin containing quaternary ammonium ions is used. Here, OH^- is enriched near the cathode surface by the water reduction. Since the pH increases, the equilibrium between the quaternary ammonium ion and amine shifts to the amine that has a poorer solubility in water, such that the resin precipitates as a form of amine on the cathode surface.

11-8 Electrochemical machining

A metal substrate can be processed into a desired shape by dissolving out some part of the substrate, which is called an ***electrochemical machining***. As shown in Fig. 11-13, the metal substrate to be processed is used as the anode along with an inert cathode. A process for producing a metal plate having fine holes used in a razor can be exemplified. The mechanical machining for making fine holes is impractical. However, when a metal plate

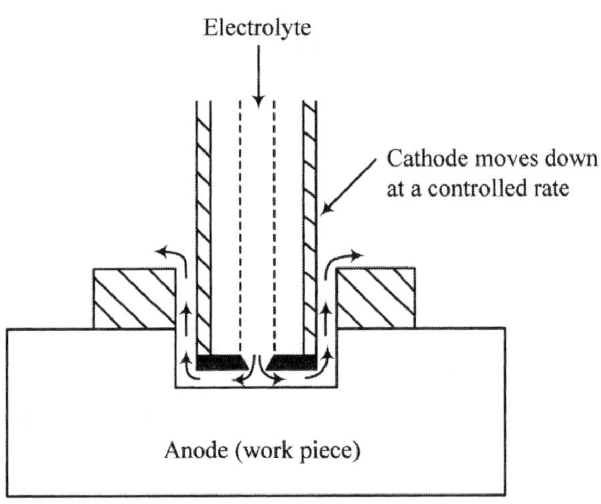

Fig. 11-13 Cell structure used for an electrochemical machining.

is used as an anode and a fine needle-aligned electrode is used as a cathode, metal dissolution occurs to produce holes at the position facing the needles because the inter-electrode gap is narrower. Electrochemical machining is used for the fabrication of metal objects that are very complicated in structure and difficult to process by mechanical machining, and for the processing of materials that are too hard to be machined. It is also used for precise machining of micro-objects (e.g. MEMS, micro-electromechanical systems) that are too small to be processed mechanically.

In the electrochemical mechanical polishing (ECMP) process used in the semiconductor manufacturing process, the rough surface is mechanically flattened by grinding and simultaneously dissolving the metal by electrochemical oxidation. Here, a copper coated with a hard material (e.g. diamond) is used as the cathode. The diamond wears the object mechanically and the copper plays as a cathode to electrochemically dissolve the anode to be flattened.

11-9 Electrodialysis

Fig. 11-14 shows the cell stack for ***electrodialysis***. A cation-exchange membrane (C) and an anion-exchange membrane (A) are alternately arranged between two electrodes at both ends. The electrode reactions are hydrogen and oxygen evolution by water decomposition on the cathode and anode, respectively. When a NaCl solution flows into the cell, the cation (Na^+) moves by migration toward the negative electrode under the given electric filed and the anion (Cl^-) toward the positive electrode. However, the ion migration is restricted to the adjacent compartment since the two different membranes are alternately positioned; the Na^+ that crosses the cation-exchange membrane can not cross the anion-exchange membrane. In Fig. 11-14, the ions are concentrated in one compartment, but depleted in the adjacent rooms. Using this principle, demineralization of beverage products (e.g. milk), purification of salts, separation of amino acids, and recovery of metal ions are possible.

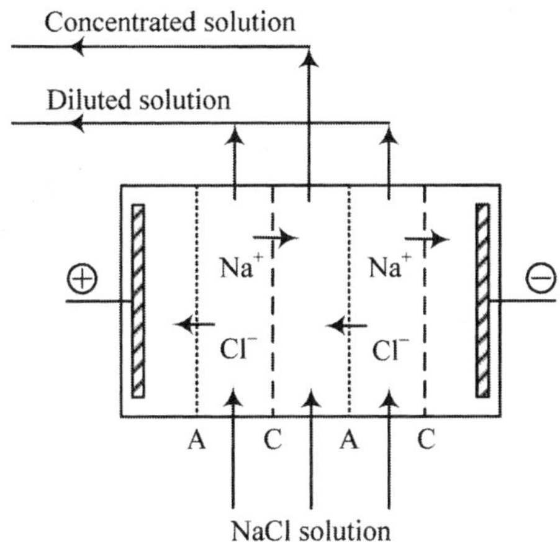

Fig. 11-14 Electrolytic cell used for electrodialysis. *A* and *C* denote an anion-exchange and cation-exchange membrane, respectively.

11-10 Electrosynthesis

Inorganic or organic materials can be synthesized using electrochemical reactions. Electrons act as a reducing agent in ***electrosynthesis***, while chemical reducing agents are used in chemical reactions.

Hydrogen and oxygen can be obtained by water electrolysis. Although it is not cost-competitive due to the mass production of hydrogen from fossil fuels, it is used in food or semiconductor manufacturing processes requiring high-purity hydrogen, small-scale on-site production, and heavy water (D_2O) production which is used as a moderator in nuclear reactor. A trace amount of D_2O is present in water. Since the electrolysis (decomposition) rate of H_2O is faster than that of D_2O, D_2O can be concentrated with a prolonged water electrolysis. In addition, O_3, H_2O_2, $NaClO_3$, $NaClO_4$, $KMnO_4$, $K_2Cr_2O_7$, MnO_2 (EMD, electrolytic manganese dioxide), Cu_2O are electrochemically synthesized.

Fig. 11-15 shows the electrosynthetic process of adiponitrile, which can be used as a starting material for hexamethylenediamine and adipic acid, both of which are the raw

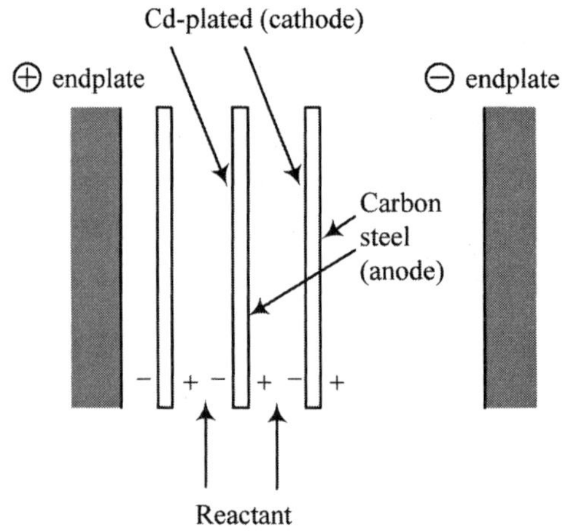

Fig. 11–15 Electrolytic cell for hydrodimerization of acrylonitrile (new Monsanto process).

material for Nylon 66. Acrylonitrile is hydrodimerized to obtain adiponitrile.

$$2CH_2 = CHCN + 2H_2O + 2e \rightarrow (CH_2CH_2CN)_2 + 2OH^-$$

Here, the following two parasitic reactions are involved.

$$CH_2 = CHCN + e \rightarrow [CH_2{=}CHCN]^- \xrightarrow{H^+} CH_3CH_2CN$$

$$CH_2 = CHCN + OH^- \rightarrow HOCH_2HCN \xrightarrow{H_2O} HOCH_2CH_2CN + OH^-$$

Since H^+ involves in the first side reaction, this reaction can be suppressed by blocking the approach of H^+ to the electrode surface. The addition of quaternary ammonium ions containing long-chain aliphatic hydrocarbons to the solution can inhibit the H^+ approach because the solution near the electrode surface is hydrophobic. The second side reaction is the reaction by OH^-, such that this side reaction can be suppressed by lowering the solution pH.

Since the quaternary ammonium ion is a surfactant, it is difficult to be separated from the aqueous solution containing the product. Therefore, the addition of quaternary ammonium ion should be minimized. Fig. 11-15 shows the ***new Monsanto process*** that solves the problems associated with the precedent. The electrodes are designed to be a bipolar, and a voltage is applied to the two end electrodes. To control the pH, Na_2HPO_4 is added at 15 % and a quaternary ammonium salt is added to provide hydrophobicity but only a minimum amount (0.4 %) because of separation problem after the reaction. Carbon steel is used as the anode, on which the main reaction is the oxidation of water. Cd is electrodeposited on the carbon steel to use as a cathode. This is because the standard rate constant (k^0) for hydrogen evolution, which is the major side reaction on the cathode, is very small on Cd electrode (Table 11-2). As a side reaction on the anode, Fe and Ni contaminated in carbon steel can be oxidized to be dissolved into the electrolyte solution, which can then be electrodeposited on the cathode (Cd) surface. If Fe and Ni are deposited on Cd surface, H_2 evolution is greatly increased because the k^0 for this reaction is much larger on Fe or Ni (Table 11-2). Therefore, both the dissolution of Fe and Ni from the anode, and the electrodeposition on the cathode should be prevented. Borax is added to inhibit the metal dissolution and sodium ethylenediaminetetraacetic acid (Na_4EDTA), which can form a strong complex with Fe^{2+} and Ni^{2+}, is added to the solution to inhibit the electrodeposition. If metal ions form complexes, their reduction (plating) becomes difficult (see Section 2-2, Section 11-5 and Fig. 11-11).

Other examples of organic electrosynthesis include hydrogenation of heterocyclic compounds and nitrile compounds, epoxidation of olefins, fluorination of organic compounds, methoxylation and the oxidation of polynuclear aromatic hydrocarbons.

11-11 Direct and indirect electrolysis

Electrolysis can be performed with the aid of an ***electron mediator*** if the selectivity is poor, side reactions are severe or k^0 is small. This is called indirect electrosynthesis. For example, a Ti^{4+}/Ti^{3+} redox pair is used as an electron mediator in the quinone reduction to hydroquinone. $Ti^{4+} + e \rightarrow Ti^{3+}$ proceeds on an inert electrode, and the as-generated Ti^{3+} transfers electrons to quinone to be Ti^{4+}. Hydroquinone is produced as a result of quinone reduction. The as-generated Ti^{4+} is reduced again to Ti^{3+}. In order for any redox pairs to serve as an electron mediator, two conditions must be satisfied. Firstly, the mediation reaction should be thermodynamically feasible; since the equilibrium potential at standard state (E^0) of Ti^{4+}/Ti^{3+} half-cell is more negative than that of quinone/hydroquinone half-cell, electron transfer from Ti^{3+} to quinone is thermodynamically possible when they make a close contact in solution. Secondly, the electron transfer from Ti^{3+} to quinone should be fast, and the charge transfer rate of the mediator (Ti^{4+}) on the electrode surface must be fast; that is, the k^0 of $Ti^{4+} + e \rightarrow Ti^{3+}$ must be large. In real, two conditions are met for the indirect electrosynthesis of hydroquinone.

In indirect electrolysis, electron mediators are commonly dissolved in the solution together with the reactants. If the electron mediator is immobilized on an inert electrode surface, however, the process is simple since it is not necessary to separate the product and the mediator after completion of the reaction. One of the goals of ***modified electrodes*** is to fix the mediator on electrode surface to increase the selectivity or to facilitate the reaction rate.

Problems

01 The electrode reaction of one electrolytic cell differs from that of the others. Which one is it?

(a) Electrophoretic coating, (b) Electrodialysis, (c) Water electrolysis, and (d) Anodizing of metals

02 Of the following electrolysis processes, which one can be performed by using an undivided cell? If a membrane is required, give an explanation why.

(a) Mercury cell for brine electrolysis, (b) Hydrometallurgy of zinc, (c) Electrorefining of copper, (d) Electroplating of chromium, (e) New Monsanto process for adiponitrile, (g) Anodizing of Al, and (h) Conversion of Fe^{2+} to Fe^{3+} by bulk electrolysis.

03 The dichromate ($Cr_2O_7^{2-}$) is prepared by oxidizing Cr^{3+} in a sulfuric acid electrolyte at a current density of 1500 Am^{-2}. The electrode reactions are as follows.

$$\text{Anode: } 2Cr^{3+} + 7H_2O \rightarrow Cr_2O_7^{2-} + 14H^+ + 6e$$

$$\text{Cathode: } 6H^+ + 6e \rightarrow 3H_2$$

$$\text{Overall: } 2Cr^{3+} + 7H_2O \rightarrow Cr_2O_7^{2-} + 3H_2 + 8H^+$$

The characteristics of the electrolytic cell are as follows.

(i) The distance between the anode and separator, cathode and separator is 4 mm, and the solution conductivity is 42 $\Omega^{-1}\ m^{-1}$.

(ii) The separator resistance is $3.0\times10^{-4}\ \Omega m^2$.

(iii) The overpotential for the anode and cathode are 340 mV and 210 mV at a current density of 1500 A m^{-2}, respectively.

Answer the following questions.

(a) The free energy change (ΔG) of the overall reaction is 852 kJ mol^{-1} ($Cr_2O_7^{2-}$). Calculate the thermodynamic decomposition voltage (E_{cell}).

(b) Calculate the applied voltage (E_{appl}) when electrolyzing under the above conditions.

(c) Electrolytic cell is designed as a monopolar-type with 6 square electrodes (1 m × 1 m) as shown in Fig. 11-3-a. How many kilograms of $K_2Cr_2O_7$ can be produced per year using this electrolytic cell?

04 In the electrosynthesis of ethylene glycol in an aqueous electrolyte, the cathodic reaction is the conversion of formaldehyde into ethylene glycol, but there are three possible anodic reactions: (i) oxygen evolution, (ii) hydrogen oxidation, and (iii) conversion of methanol to formaldehyde. The free energy of formation of methanol, formaldehyde, ethylene glycol, and water is −165, −130, −320, and −237 kJ mol^{-1}, respectively. Assuming that the solution pH is all 7.0, present the reduction reaction, oxidation reaction, and overall reaction for three electrolytic cells, and calculate the E_{cell} values.

05 Hydrochloric acid (HCl) is produced as a by-product in many organic syntheses. One wants to design a process to produce chlorine (Cl_2) and water (H_2O) by electrolysis of HCl in a membrane cell. The following oxidation or reduction reactions are possible.

$$O_2 + 4H^+ + 4e = 2H_2O \qquad E^0 = 1.23\ \text{V}\ (vs.\ \text{NHE})$$
$$2H^+ + 2e = H_2 \qquad E^0 = 0.0\ \text{V}\ (\text{vs. NHE})$$
$$Cl_2 + 2e = 2Cl^- \qquad E^0 = 1.36\ \text{V}\ (\text{vs. NHE})$$
$$2H_2O + 2e = H_2 + 2OH^- \qquad E^0 = -0.83\ \text{V}\ (vs.\ \text{NHE})$$

(a) What reaction is desirable on the cathode?

(b) What would be the side reactions on the cathode?

(c) Explain how electric energy consumption differs from the conventional chlor-alkali membrane cell.

06 In the electrorefining of copper, E_{appl} is 0.23 ~ 0.27 V, and the cathodic overpotential is larger than the anodic one. The Tafel slope of the cathodic and anodic reaction is $(120\ mV)^{-1}$ and $(50\ mV)^{-1}$, respectively. There appears a limiting current in the cathodic current-potential curve.

(a) Explain why the cathodic overpotential is larger.

(b) Explain why the Tafel slope for the cathodic reaction is larger.

(c) Give an explanation why there appears a limiting current only in the cathodic branch.

07 When electroplating a conductive object with a rough surface, the electrolysis conditions should be adjusted to increase the throwing power and quality of the plated layer. Answer the following questions.

(a) How is the salt concentration in the electrolyte solution adjusted?

(b) What kind of additives can be used?

(c) How can we control the Tafel slope of the plating reaction?

(d) How can we utilize the side reactions?

08 Zinc is to be electroplated on Pt and Zn electrode in an aqueous solution; pH = 7 and $[Zn^{2+}] = 0.01\ M$. Hydrogen is generated as a side reaction. The exchange current density for hydrogen evolution is 8×10^{-4} A/cm^2 on Pt and 5×10^{-11} A/cm^2 on Zn electrode.

(a) Calculate the equilibrium potential for the 0.01 M Zn^{2+} solution, and calculate the hydrogen evolution current at this potential on the Pt and Zn electrode, respectively.

(b) Determine if Zn plating is possible on Pt and Zn electrode under the above conditions. T = 298 K, $\alpha_c = 0.5$ in <4-23>, and the activity coefficient for all ions is unity.

09 The $Ni(OH)_2/NiOOH$ redox pair (the equilibrium potential at the standard state, $E^0 = 0.52$ V *vs.* NHE) can be used as an electron mediator (or electrocatalyst) for alcohol (RCH_2OH) oxidation. What should be the equilibrium potential at the standard state (E^0) of alcohols for this catalytic reaction to proceed? Answer by comparing the equilibrium potentials of $Ni(OH)_2/NiOOH$ half-cell and $RCH_2OH/RCOO^-$ half-cell.

$$Ni(OH)_2 + OH^- \rightarrow NiOOH + H_2O + e$$

$$4NiOOH + RCH_2OH \rightarrow RCOO^- + 4Ni(OH)_2$$

CHAPTER 12

Galvanic cells

Galvanic cells are the devices that obtain electrical energy through spontaneous electrochemical reactions; e.g. batteries and fuel cells. The corrosion is also included in galvanic cells in that it proceeds spontaneously. In contrast to batteries, which obtain electrical energy through electrochemical reactions of loaded electroactive materials, fuel cells obtain electrical energy by supplying fuel (e.g. hydrogen or carbon monoxide) and air continuously into fuel cell stack. Batteries are classified into primary (non-rechargeable) and secondary (rechargeable) ones. The secondary batteries have a function of electric energy storage because they can be moved (or carried) after charging and used. Somewhat differently, fuel cells have a function of electricity generation; electrical charging is not needed. Supercapacitors are the device for electric energy storage.

12-1 Batteries

The reaction <12-1> proceeds spontaneously to the right because $\Delta G^0 = -212.3$ kJ/mol < 0.

$$Zn + Cu^{2+} \rightarrow Zn^{2+} + Cu \qquad \text{<12-1>}$$

Since ΔG^0 is a thermodynamic value, it is determined only by the initial and final state; independent on the reaction pathway. Therefore, even if the reaction proceeds through an electrochemical pathway, which is one of possible pathways, like the ***Daniell cell*** shown in Fig. 12-1-a, and if the initial and final state are the same, ΔG^0 also has the same value and must proceed spontaneously. The electrochemical pathway forms as a full-cell consisting of two half-cells as follows. As shown below, the overall reaction of the electrochemical pathway is the same as in <12-1>; the paired electrochemical reactions of Zn oxidation and Cu^{2+} reduction is spontaneous in thermodynamic point of view.

Anode: $Zn \rightarrow Zn^{2+} + 2e$ $\qquad E^0 = -0.76$ V (*vs.* NHE)
Cathode: $Cu^{2+} + 2e \rightarrow Cu$ $\qquad E^0 = 0.34$ V (*vs.* NHE)
Overall: $Zn + Cu^{2+} \rightarrow Zn^{2+} + Cu$ $\qquad E^0_{cell} = 1.1$ V ($\Delta G^0 = -212.3$ kJ/mol)

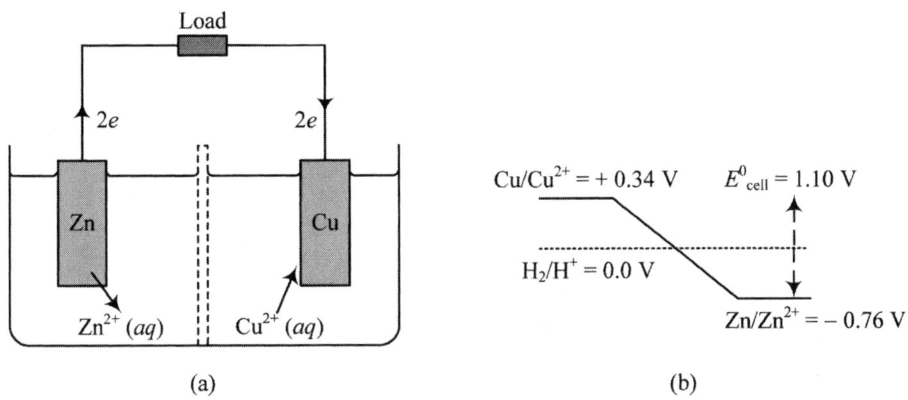

Fig. 12–1 (a); Configuration of Daniell cell, and (b); The standard electromotive force(E^0_{cell}) derived from the difference in the equilibrium potentials for two half–cells under standard state.

The ***standard electromotive force*** (or open-circuit voltage in the standard state) of Daniel cell can be calculated from $\Delta G^0 = -nFE^0_{cell}$. E^0_{cell} = 1.1 V is calculated from n = 2 and F = 96,485 C/mol. This is identical with that calculated from the standard electrode potential values; $E^0_{cell} = E^0_{cathode} - E^0_{anode} = 0.34-(-0.76) = 1.1$ V. Through this electrochemical pathway, the electrons generated from Zn oxidation move to the copper half-cell through the external load to be consumed for Cu^{2+} reduction; it is a galvanic cell because it supplies electric energy to an external load by the paired spontaneous electrochemical reactions.

When the Daniel cell is discharged, the difference of equilibrium potentials for the two half-cells (electromotive force) is expressed as <12-2>. Here, E^c_{eq} denotes the equilibrium potential of the half-cell undergoing the reduction reaction, whereas E^a_{eq} is the equilibrium potential for the oxidation electrode. Obviously, $E^c_{eq} = E^0_{cathode}$ and $E^a_{eq} = E^0_{anode}$ in the standard state because $a_{Cu^{2+}} = 1.0$ and $a_{Zn^{2+}} = 1.0$ in <12-2>. As discharge goes, however, the activity of ions deviates from unity, so that the two half-cells no longer show the standard electrode potential values. The electromotive force is thus $E_{cell} = E^c_{eq} - E^a_{eq}$, which is given by <12-2>. The value in parenthesis on the right-hand side of <12-2> is constant, but $a_{Cu^{2+}}$ decreases and $a_{Zn^{2+}}$ increases with discharge. As a result, the electromotive force (E_{cell}) steadily decreases with discharge.

$$E_{eq}^{c} = E_{Cu/Cu^{2+}}^{0} + \frac{RT}{2F}\ln a_{Cu^{2+}}, \quad E_{eq}^{a} = E_{Zn/Zn^{2+}}^{0} + \frac{RT}{2F}\ln a_{Zn^{2+}}$$

$$E_{cell} = E_{eq}^{c} - E_{eq}^{a} = \left(E_{Cu/Cu^{2+}}^{0} - E_{Zn/Zn^{2+}}^{0}\right) + \frac{RT}{2F}\ln\frac{a_{Cu^{2+}}}{a_{Zn^{2+}}} \qquad \text{<12-2>}$$

As shown above, batteries consist of two half-cells. Here, if there is a difference in the equilibrium potentials of two half-cells; that is, if $E_{cell} = E^{c}_{eq} - E^{a}_{eq} \neq 0$, batteries can form by combining two half-cells; thus, numerous batteries are possible in theory. However, commercialized batteries are limited due to practical reasons; e.g. the rate (current) of electrochemical reactions on two half-cells, the ease of closed-loop formation as a galvanic cell, the availability of appropriate electrolytes, and the magnitude of electromotive force.

The terminology relevant to batteries is defined as follows.

① Working voltage (E_{wk}) or discharge voltage: It is defined by <12-3>. E_{cell} either changes (e.g. Daniel cell expressed by <12-2>) with discharge or remains constant (e.g. Zn-Ag_2O battery in Section 12-1-2). It is alway true that $E_{wk} < E_{cell}$ because of the cell polarization ($\eta_a + \eta_c + iR_{total}$). Since the cell polarization is larger at higher discharge current, E_{wk} decreases with an increase in discharge current.

$$E_{wk}\ (E_{discharge}) = E_{cell} - (\eta_a + \eta_c + iR_{total}) \qquad \text{<12-3>}$$

② Capacity: For full-cells, it denotes the amount of electricity (Q in Ah, amperes × hours) that can be obtained by charge or discharge. The unit of capacity is Ah/kg or Ah/L. For half-cells, however, the capacity indicates the electric charge that can be delivered by an electrode material (also frequently referred to as active material). It is represented by $Q = (w/M)nF$, wherein w is the weight of the given electrode material, and M is its atomic or molecular weight.

③ Specific energy and energy density: The electric energy obtained by discharge is expressed by $Q \times E_{wk}$; Q is the discharged capacity of batteries and E_{wk} is the average discharge voltage. The specific energy (in Wh/kg) and energy density (in Wh/L) refer to the discharge energy divided by battery weight or volume, respectively.

④ Specific power and power density: The power (Watt in unit) of batteries is expressed by $i \times E_{wk}$; i refers to the discharge current and E_{wk} is the average discharge voltage. The specific power (in W/kg) and power density (in W/L) indicate the battery power divided by weight or volume, respectively.

⑤ Cycle performances: It refers to the number of cycles that secondary batteries can be charged/discharged.

⑥ Self-discharge rate: It refers to the rate, with which batteries discharge by themselves when not in use; also called a shelf-life.

⑦ Overdischarge and overcharge characteristics: Secondary batteries are overdischarged (over-charged) after a full discharge (full charge). In these cases, oxidation or reduction of other cell components (typically electrolyte) proceeds, which has a significant influence on cell life and safety.

⑧ Energy efficiency: Energy efficiency = voltage efficiency × Coulombic efficiency; voltage efficiency = discharge voltage/charging voltage and Coulombic efficiency = discharged capacity/charged capacity. Coulombic efficiency is smaller than 1.0 due to parasitic reactions, whereas voltage efficiency due to voltage hysteresis; in general, discharge voltage is smaller than charging voltage. The voltage hysteresis is caused by cell polarization. Energy efficiency is an important factor for the secondary batteries to be used for energy storage.

Fig. 12-2 shows a charge/discharge voltage profile and charged/discharged cell capacity of a secondary battery, which were obtained with a constant-current charge/discharge. As shown, the charged capacity is 10 Ah and the discharged one is 8.5 Ah, which gives a Coulombic efficiency of 0.85. Meanwhile, the charging voltage (E_{charge}) is larger than the discharge one ($E_{discharge}$) due to cell polarization. The electric energy stored upon charging is the product of charging voltage and charged capacity; the charged energy is the area integrated the charging voltage profile with respect to the charged capacity. On the other hand, the discharged energy is the area obtained by integrating the discharge voltage curve with respect to the discharged capacity. From Fig. 12-2, it is seen that the larger the difference between E_{charge} and $E_{discharge}$, the smaller the energy efficiency. Here, since both E_{charge} and $E_{discharge}$ are determined by cell polarization, the cell polarization should

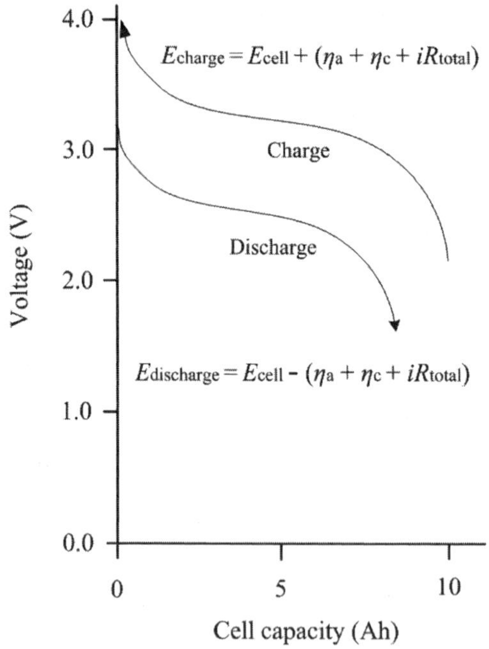

Fig. 12–2 A galvanostatic (constant–current) charge–discharge voltage profile for a secondary cell. Note that the discharge voltage ($E_{discharge}$) is smaller than the charging voltage (E_{charge}) due to cell polarization.

be minimized to have a maximum voltage efficiency. In addition, side reactions during charge/discharge lead to a decrease in Coulombic efficiency and thus eventually energy efficiency. High energy efficiency is a demanding characteristic that secondary batteries should have to be used for energy storage devices.

Self learning 12-1

Calculate the capacity (in mAh/g) of lithium (Li) metal electrode.

Self learning 12-2

Let us assume that the charging voltage of a secondary battery is constant at 2.0 V throughout the charging process and the charged capacity is 10.0 Ah. Also assume that the discharge voltage is constant at 1.6 V during discharge and the discharged capacity is 9.5 Ah. Draw a picture showing the capacity on the x-axis and the voltage on the y-axis as shown in Fig. 12-2. Calculate the energy efficiency after obtaining the voltage efficiency and Coulombic efficiency. Graphically indicate the energy consumed for charging and the energy recovered by discharge. Confirm that the discharged energy is smaller than the charged energy. Where is the energy not recovered consumed (refer to 12-1-7)?

The size and shape of batteries are diverse. Miniature batteries are used as power sources for small electronic devices or medical devices. Medium- or large-sized batteries are used for SLI (starting, lighting and ignition) for automobiles, emergency power supply (e.g. uninterrupted power supply, UPS), and load-leveling for electricity generation.

Batteries have different performance requirements depending on their use. For example, as shown in Table 12-1, SLI batteries for automobiles should have high operating voltage, high energy density, low cost, and operate at < -20 °C. However, the batteries for heart pacemaker are not demanding for energy density and price, but the reliability should be high. The batteries for memory backup are an emergency power source, so that their self-discharge rate should be low; shelf-life should be long.

Batteries are fabricated with many shapes; coin, cylindrical, prismatic, and pouch. A cathode, an anode, a separator, and an electrolyte solution are contained inside of batteries, and a protective circuit for safety may be attached to the outside in some cases.

Table 12-1 The performance requirements according to applications

	SLI	Heart pacemaker	Computer memory backup
Voltage	12 V	1~3 V	1~3 V
Cell capacity	20~200 Ah	0.5 Ah	0.1~0.5 Ah
Specific energy	50 Wh/kg		
Specific power	100 W/kg	1~5 mW/kg	1~4 mW/kg
Usage	Uneven demand	Constant demand	Long shelf life
Temp. (°C)	−20 ~ 40	36	25
Cost	Important	Unimportant	Unimportant
Reliability	Very high	Extremely high	High

12-1-1 Alkaline batteries

Alkaline battery is a primary cell, which is named as like because it uses an aqueous alkaline (KOH) electrolyte solution. Fig. 12-3 shows the structure of this cell. The negative electrode material is Zn powder that is dispersed into a gelling agent. MnO_2 (EMD) is the positive electrode material, which is mixed with carbon powder for conductivity enhancement. Two electrode materials are electrically separated by a porous non-woven fabric, into which aqueous KOH solution is soaked. The steel can is used as the current collector for the positive electrode, whereas a metal pin for the negative electrode. This design is called a ***bobbin-type cell***.

Alkaline battery uses a strong alkaline solution to suppress the self-discharge of Zn negative electrode and H_2 evolution (see Fig. 2-13). In order to further suppress the Zn self-discharge and H_2 evolution, mercury-added batteries have been produced. Since the standard rate constant (k^0) for H_2 evolution is very small on mercury electrode (Table 11-2), the H_2 evolution can be suppressed by adding mercury, which can thus suppress the Zn self-discharge. In addition, mercury has an effect of strengthening the adherence between Zn particles. The Zn particles being dispersed in a gelling agent are vulnerable upon mechanical impact to destruction of electrically conductive networks made between Zn particles. The measure for this was the mercury addition. Now, the mercury-free cells

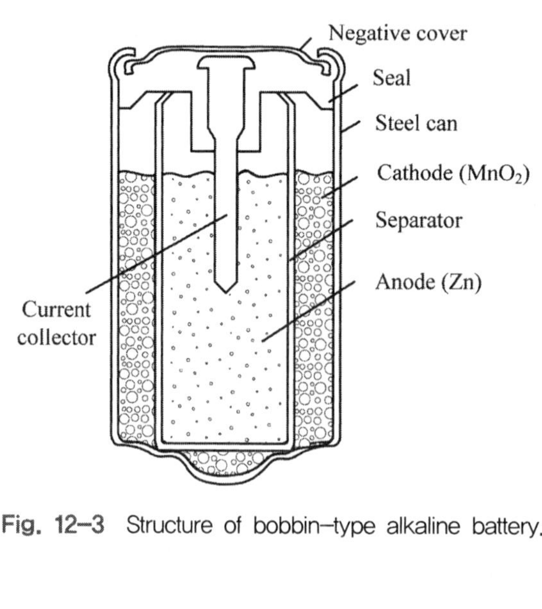

Fig. 12-3 Structure of bobbin-type alkaline battery.

have been commercialized by solving the above-mentioned problems encountered without mercury. The H_2 evolution/Zn self-discharge was solved by developing the alloys based on Zn; some alloys in which In, Pb, or Bi is added into Zn matrix have a very small k^0 for H_2 evolution, so it is used as a negative electrode material for a mercury-free alkaline battery. The other problem caused by not adding mercury, that is, the poor adhesion between Zn particles, was solved by developing a new gelling agent that can localize Zn particles. That is, when a highly cross-linked acrylic polymer with a large grain size is used as a gelling agent, the dispersed Zn particles are squeezed to be localized at the interstitial space made by the gelling agent grains, which enforces a close contact between Zn particles; it is called a localization. Further, since the gelling agent grains play as a strong mechanical support, the localization is not destructed even at a strong mechanical shock; the mutual electrical contact is maintained between Zn particles.

12-1-2 Zinc-silver oxide battery

The following electrochemical reactions proceed in ***zinc-silver oxide primary battery.***

Negative electrode: $Zn(s) + 2OH^- \rightarrow Zn(OH)_2(s) + 2e$ $\quad E^0 = -1.25$ V(*vs.* NHE)

Positive electrode: $Ag_2O(s) + H_2O + 2e \rightarrow 2Ag(s) + 2OH^-$ $\quad E^0 = 0.34$ V(*vs.* NHE)

Overall: $Ag_2O(s) + Zn(s) + H_2O \rightarrow 2Ag(s) + Zn(OH)_2(s)$ $\quad E^0_{cell} = 1.59$ V

The electromotive force ($E_{cell} = E^c_{eq} - E^a_{eq}$) is expressed by <12-4>. The solid electrode materials have a unit activity and the activity of water does not change much; the thermodynamically determined E_{cell} does not change during discharge.

$$E_{cell} = \left(E_c^0 - E_a^0\right) + \frac{RT}{2F}\ln\frac{a_{Ag_2O(s)}\, a_{H_2O}\, a_{Zn(s)}}{a^2_{Ag(s)}\, a_{Zn(OH)_2(s)}} \qquad \text{<12-4>}$$

As shown in <12-3>, if the cell polarization ($\eta_c + \eta_a + iR_{total}$) does not change during discharge, the discharge voltage (E_{wk}) will show a flat profile since E_{cell} does not change with discharge. In fact, this battery meets these conditions to give a flat discharge voltage profile. This coin-type cell is used as a power source for small electronic devices; e.g. watches.

12-1-3 Zinc-air battery

Fig. 12-4 shows the structure of a ***zinc-air primary battery***. The negative electrode material (Zn powder) is dispersed into a gelling agent along with a KOH aqueous solution. The positive electrode material (MnO_2/carbon) is placed on a metal screen, onto which the ***oxygen reduction reaction (ORR)*** proceeds; $O_{2(g)} + 2H_2O + 4e \rightarrow 4OH^-$. There is an air access port and the incoming air is supplied to the positive electrode through an air diffusion paper/hydrophobic porous membrane. In this battery, oxygen is the positive electrode material, but is supplied from outside. Therefore, Zn powder can be loaded into the cell to the maximum value, giving a volumetric cell capacity much larger than that of the other primary cells. This cell is widely used in hearing aids. Also, since the volumetric cell capacity is large, studies are underway to convert it into rechargeable one.

For the oxygen reduction reaction (ORR), $O_{2(g)}$, H_2O, and electrons (e) must meet together on the catalyst (MnO_2) surface. Electrons (e) are supplied from the current collector through the carbon particles supporting the catalyst, and H_2O is provided from the electrolyte solution. $O_{2(g)}$ is supplied from outside. Here, it must be noted that the solubility of $O_{2(g)}$ in aqueous electrolyte solution is very low and the diffusion coefficient of dissolved O_2 in aqueous solution is about $10^{-5} \sim 10^{-6}$ cm^2/s, which is much smaller that the diffusion coefficient in the gaseous state ($10^{-1} \sim 1$ cm^2/s). It is apparent that, if the catalyst surface is fully wetted by the electrolyte solution, the mass transfer (diffusion)

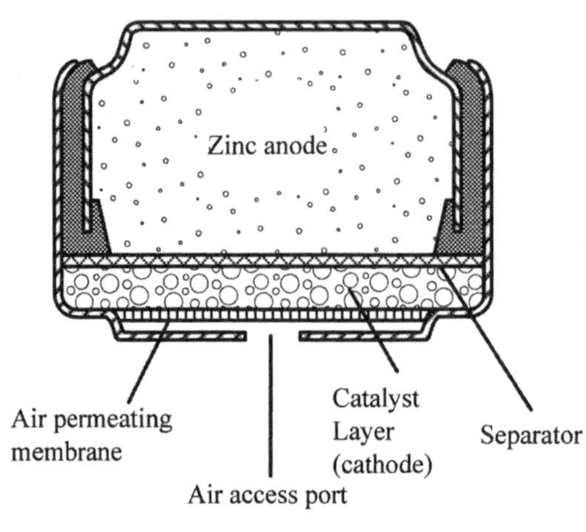

Fig. 12-4 Structure of button-type zinc-air battery.

rate of O_2 is very low because O_2 must diffuse through solution; the concentration overpotential for ORR is expected to be large due to the poor O_2 solubility and slow diffusion rate. To mitigate this, void spaces must be provided for O_2 travel; then the gas-phase diffusion rather than solution-phase is allowed for O_2; the concentration overpotential for ORR can be reduced. This principle is applied in the practical cells, in which ORR takes place on the catalyst surface to which carbon, electrolyte solution, and void space are contacted. This is called ***three-phase boundary site***. In order to maximize the three-phase boundary sites, it is necessary to control the microstructure of air electrode; e.g. minimizing the catalyst particle size, expanding the contact area between the catalyst and electrolyte solution, and securing void spaces for gas-phase O_2 diffusion. Meanwhile, carbon dioxide can enter from outside along with air; it reacts with KOH to form an insoluble $KHCO_3$, which deposits on the separator and catalyst surface to deteriorate the cell performances.

12-1-4 Lithium primary cell

In ***lithium primary cell***, lithium metal is used as the negative electrode while a variety of positive electrodes (MnO_2, $(CF_x)_n$, $Ag_2V_4O_{11}$, SO_2, and $SOCl_2$) can be employed. The lithium salts dissolved in organic solvents are used as an electrolyte, which is soaked within a porous polymer (e.g. polyethylene) film. The lithium salts comprising the large anions such as $LiPF_6$ are used because ion bond strength between Li^+ and large anions is weak to be easily dissociated. As the organic solvent, EC (ethylene carbonate), PC (propylene carbonate), γ-butyrolactone, and ethers are used. In most cases, the solvents of high dielectric constant and low viscosity are used. The capacity of Li metal electrode is as high as 3,861 mAh/g. The discharge voltage is 3 ~ 4 V with a very low self-discharge rate.

Lithium primary batteries exhibit a long shelf-life; low self-discharge rate. This is because a ***solid electrolyte interphase (SEI)*** forms on Li metal surface. Lithium half-cell has a very negative standard electrode potential of −3.04 V (*vs.* NHE), and thus the oxidizing tendency of lithium metal is very high; it is a strong reducing agent. When lithium metal makes a contact with organic solvent, almost organic solvents are not stable under such a strong reducing condition to be decomposed by reduction; lithium metal is oxidized by $Li \rightarrow Li^{+} + e$ and the solvent molecules are reduced by taking the electrons. Here, the reductively decomposed products (*e.g.* lithium alkyl carbonate, oligomers, polymers, Li_2CO_3, and LiF) deposit on the lithium metal surface. When the deposition layer is thicker than a certain level, electron tunneling rate through the deposition layer is greatly reduced; the electrochemical reduction (decomposition) of solvent molecules is now negligible; the dissolution of lithium metal (self-discharge) is also negligible. This is why this cell exhibits such a low self-discharge rate. Here, the Li metal electrode is said to be ***passivated***. The deposited layer is called a solid electrolyte interphase (SEI) in that it is a solid film, it has an ion-conduction property like common electrolytes, and it places at the electrode/electrolyte interface. Due to the ion conduction property, Li^{+} generated by the Li discharge can move to the electrolyte through the SEI layer. Considering the fact that Li metal electrode cannot avoid self-discharge (oxidation, dissolution) no matter what electrolyte solution is used due to the very high oxidation tendency, it is said that the commercialization of lithium primary batteries was not successful without SEI layer formation.

12-1-5 Lead-acid cell

Lead-acid cell is an rechargeable cell and currently shares the largest market for SLI, uninterrupted power supply (UPS), and load levelling. The electrode reactions for discharge are as follows; the charging reactions are the reverse ones.

- electrode: $Pb(s) + HSO_4^{-} \rightarrow PbSO_4(s) + 2e + H^{+}$ $\qquad E^0 = -0.32\ V(vs.\ NHE)$

+ electrode: $PbO_2(s) + 3H^{+} + HSO_4^{-} + 2e \rightarrow PbSO_4(s) + 2H_2O(l)$ $\qquad E^0 = 1.75\ V(vs.\ NHE)$

Overall: $Pb(s) + PbO_2(s) + 2H_2SO_4 \rightarrow 2PbSO_4(s) + 2H_2O(l)$ $\qquad E^0_{cell} = 2.07\ V$

Lead-acid batteries have the following operating characteristics. First, as shown above, sulfuric acid is consumed during discharge, so that electrolyte conductivity decreases and the freezing point increases with discharge. Second, as the discharge proceeds, $PbSO_4(s)$

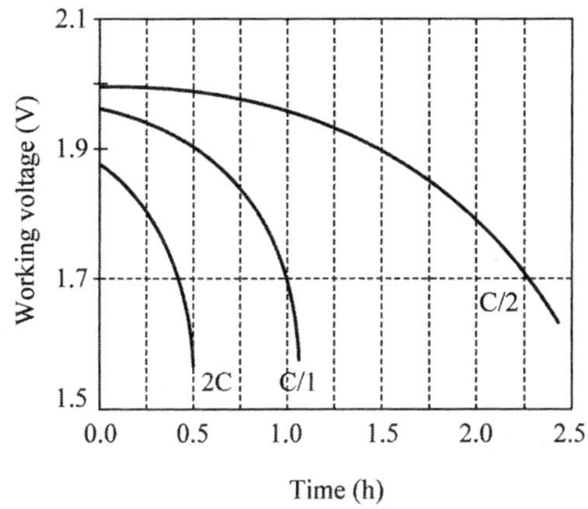

Fig. 12-5 The galvanostatic (constant-current) discharge voltage profiles for a lead-acid battery according to discharge current (rate).

deposits on both electrodes, which shows a poor electrical conductivity to give rise to a large electrode resistance ($R_{electrode}$) for both electrodes. Fig. 12-5 shows the galvanostatic discharge voltage profiles according to the discharge current. In general, the charge/discharge rate is expressed as c/t, where c is the practical cell capacity and t is the charge/discharge time (unit: hr). For example, when c = 10 Ah, if the discharge rate is $c/10$, the cell is discharged for 10 h; the discharge current is 1 A. If it is $2c$ ($c/0.5$), the cell is discharged for 0.5 h; the discharge current is 20 A. The discharge current at $2c$ is 20-times larger than that at $c/10$. Two features are apparent in Fig. 12-5. First, the overall discharge voltage becomes smaller with an increase in the discharge current. This must be due to an increase of cell polarization with an increase of discharge rate: $E_{discharge} = E_{cell} - (\eta_a + \eta_c + iR_{total})$. Second, the drop of discharge voltage becomes more apparent approaching the end of discharge. This can be accounted for by a steady increase of cell polarization. As mentioned above, sulfuric acid is depleted with discharge, such that both electrolyte resistance ($R_{solution}$) and separator resistance ($R_{separator}$) increase toward the end of discharge. Also, the poorly resistive $PbSO_{4(s)}$ deposits thicker with discharge to lead an increase of electrode resistance ($R_{electrode}$). In short, $R_{solution}$, $R_{separator}$, and $R_{electrode}$ steadily increase to give a large cell polarization at the end of discharge.

Self learning 12-3

Let us assume that a lead-acid cell is discharged until 1.7 V as shown in Fig. 12-5, which is called the cut-off voltage. Practically, batteries are discharged down to the cut-off voltage; the minimum discharge voltage is the cut-off voltage. Calculate the discharged capacity at three different discharge currents assuming a cut-off at 1.7 V. The practical cell capacity, which is the maximum (latent) discharge capacity of a given cell, is determined by the loading amount of active materials within the cell. The discharged capacity corresponds to the capacity delivered until the cut-off voltage. Hence, the discharged capacity is always smaller than the practical capacity. Furthermore, the discharged capacity decreases with an increase of discharge current due to cell polarization. Confirm these features from your calculation.

Fig. 12-6 presents the potential ranges where the electrode reactions and side reactions can take place, which are predicted from thermodynamics. The charging reaction can take place at <-0.32 V (*vs.* NHE) on the negative electrode and the discharge reaction at > -0.32 V. On the positive electrode, the charging reaction can take place at > 1.75 V, whereas the discharge reaction at < 1.75 V. Thermodynamics predict that H_2 evolution is possible at < 0.0 V (*vs.* NHE) and the reverse reaction (H_2 oxidation) at > 0.0 V.

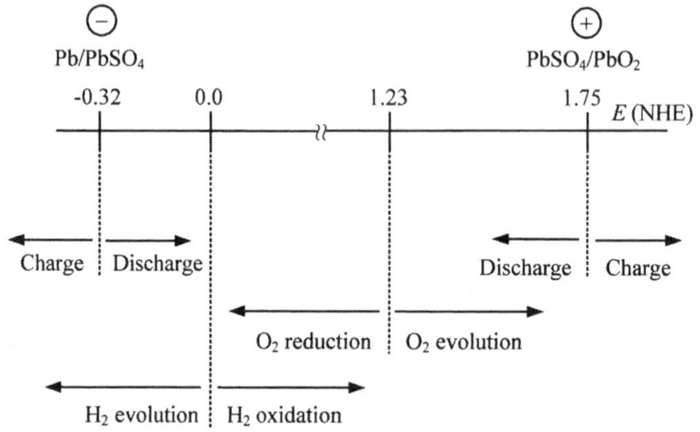

Fig. 12-6 Thermodynamically predicted potential ranges for the electrode reactions and side reactions at standard state. -0.32 V, 0.0 V, 1.23 V and 1.75 V are the equilibrium potentials (E_{eq}) of the four half-cells. The arrows indicate the potential ranges where each reaction is thermodynamically allowed.

O_2 evolution is thermodynamically feasible at > 1.23 V and O_2 reduction at < 1.23 V. From this, it is apparent that H_2 evolution and O_2 evolution by water decomposition are thermodynamically feasible during the charge of lead-acid batteries. For charging, the positive electrode potential should be higher than 1.75 V, under which condition O_2 evolution on the positive electrode is thermodynamically possible. For charging, the negative electrode potential should be lower than −0.32 V, under which condition H_2 evolution on the negative electrode is thermodynamically possible. Note that there is a plentiful amount of reactants; H_2O for O_2 evolution and H^+ for H_2 evolution. Due to H_2O consumption for O_2 evolution and H_2 generation, in the past, distilled water was added periodically. Recently, however, the maintenance-free lead-acid batteries, which do not require the addition of distilled water, have been developed. The key technology is the ***internal oxygen cycle***, in which the oxygen generated by water decomposition ($2H_2O \rightarrow O_{2(g)} + 4H^+ + 4e$) is transferred to the negative electrode through the separator and converted back to water. As shown in Fig. 12-6, O_2 reduction ($O_{2(g)} + 4H^+ + 4e \rightarrow 2H_2O$) is thermodynamically possible at the charging potential of the negative electrode (<−0.32 V). The periodical water-addition is not required as water is regenerated by the internal oxygen cycle.

As discussed above, H_2 evolution is thermodynamically feasible during the charge. The internal hydrogen cycle is also thermodynamically possible; H_2 gas evolved at the negative electrode can cross over to the positive electrode, and oxidized ($H_{2(g)} \rightarrow 2H^+ + 2e$) since the positive electrode potential during charge (>1.75 V) is more positive than the equilibrium potential of hydrogen half-cell (0.0 V *vs.* NHE). In real, however, this thermodynamically allowed H_2 oxidation reaction is kinetically very slow on the positive electrode ($PbSO_4/PbO_2$); k^0 for H_2 oxidation on the $PbSO_4/PbO_2$ electrode is very small. The pressure build-up due to continued H_2 evolution but negligible H_2 oxidation (decomposition) seems to be unavoidable. Practically, however, the problem of pressure build-up has been solved by two approaches. First, H_2 generation rate itself was controlled to be very small by the development of Pb-based alloys that show a very small k^0 for hydrogen evolution. However, since H_2 generation can not be completely prevented, an increase of internal pressure and electrolyte solution leakage can not be avoided. This has been solved by using the second approach, in which the accumulated H_2 is released intermittently through a small valve being attached to the cell. Hence, a completely sealed lead-acid battery with a valve is possible. This is called a ***valve-regulated lead-acid (VRLA) cell.***

The ***theoretical specific capacity***, which is calculated with an assumption that the cell is composed only of the materials participating in the discharge reaction (active materials), is calculated for lead-acid battery. The discharge reaction is $Pb(s) + PbO_2(s) + 2H_2SO_4 \rightarrow 2PbSO_4(s) + 2H_2O(l)$; one mole of Pb(s) (207.2 g), one mole of $PbO_2(s)$ (239.2 g) and two moles of H_2SO_4 (196.2 g) produce two equivalents (moles) of electric charges (2F) through the discharge reaction. The total weight of active materials required to produce 2F (2 mol × 26.8 A h/mol) of electric charge is 642.6 g, from which the theoretical specific capacity is calculated to be 83.4 A h/kg.

In general, the ***practical specific capacity*** is smaller than the theoretical one because normal cells include not only the active materials but also the passive components; e.g. water, separator, current collector and metal can. In order to increase the practical specific capacity, it is necessary to load the active materials as much as possible at the expense of the passive components. The practical specific capacity defined here refers to the maximum (latent) cell capacity available by discharge. It is decided by the cell size and loading amount of active materials; a given cell has a fixed value of practical specific capacity. In real discharge, however, the discharged capacity (actually delivered capacity by discharge) is less than this maximum value. This is because the discharge voltage is affected by cell polarization and cells are discharged only down to the cut-off voltage. Let us assume that a given cell is discharged at two different currents. At the higher discharge current, the cell polarization is larger giving rise to a smaller discharge voltage. Consequently, the discharge voltage reaches the cut-off voltage earlier; resultantly, the actually discharged capacity is smaller. In short, a given cell has a maximum (latent) cell capacity (practical specific capacity), but the discharged capacity is always smaller than this value. The remnant electrode materials remain as the charged state in the cell (un-discharged capacity). Thus, it is said that the practical specific capacity (a fixed value for a given cell) = discharged capacity + un-discharged capacity. With an increase of discharge current, the un-discharged capacity increases at the expense of the discharged one.

Self learning 12-4

The theoretical specific energy of cells is calculated by multiplying the theoretical specific capacity by the average discharge voltage. For lead-acid cell, the theoretical specific energy is calculated by 83.4 A h/kg × 2.0 V = 167 W h/kg. However, the practical specific energy is about 30~50 W h/kg because of the passive components loaded in the cells. The practical specific energy of lead-acid cells is different according to their application; ~ 50 W h/kg for golf carts and ~30 W h/kg for UPS. Consider the following points to explain the reason for the difference: (1) Lead-acid cells for golf carts are used as the power source, but (2) the lead-acid cells for UPS are float-charged, under which condition the cells remain at the fully charged state waiting for use. In the latter case, the Pb grid used as the current collector for the positive electrode is corroded. Therefore, an excess of Pb grid, which is a passive component, must be loaded in the cells.

The ***nominal capacity*** is frequently used to represent the cell capacity. As mentioned above, the practical specific capacity can not represent the cell capacity in daily discharge. Rather, the discharged capacity is the actually obtained cell capacity. A problem here is that the discharged capacity depends on the discharge condition (discharge current, cut-off voltage, temperature, etc.); it is ambiguous to define the discharged capacity because the discharge condition differs according to the usage. One way to straighten this is to use the nominal capacity, which is defined as the discharged capacity obtained in the typical discharge condition in a given use. For example, in the case of a lead-acid battery for automotive SLI, the typical discharge condition is; discharge current = $c/20$, cut-off voltage = 1.75 V, and working temperature = 25°C. The discharged capacity obtained at this condition is defined as the nominal capacity for SLI application. Meanwhile, the typical discharge condition for motive power is; discharge current = $c/5$, cut-off voltage = 1.7 V, and working temperature = 30°C. Obviously, the nominal capacity for this application differs from that for SLI even if the practical (maximum, or latent) specific capacity is identical.

Lead-acid battery shows a long cell life, but is not free from ageing with repeated charge and discharge. One origin for ageing can be found by inspecting its electrode reaction mechanism; the dissolution/precipitation is involved in its charge/discharge reactions for both electrodes. Namely, the discharge reaction at the negative electrode proceeds through Pb(s) → Pb^{2+} (electrolyte solution) → $PbSO_4$(s), and the charging through the opposite path. The first step is a dissolution; Pb(s) → Pb^{2+} (electrolyte solution). The second step is the precipitation; Pb^{2+} (electrolyte solution) → $PbSO_4$(s). The first step in the charging is the dissolution; $PbSO_4$(s) → Pb^{2+} (electrolyte solution), and the second step is the precipitation; Pb^{2+} (electrolyte solution) → Pb(s). A similar dissolution/precipitation prevails for both charge and discharge at the positive electrode; PbO_2(s) → Pb^{4+} (electrolyte solution) → Pb^{2+} (electrolyte solution) → $PbSO_4$(s). In general, electrode failure is frequently observed when the electrode reaction proceeds by the dissolution/precipitation pathway. For example, let us assume that the Pb(s) particles in the charged state are spherical and 5 μm in size. When the cell is discharged and charged, the Pb(s) particles are converted into $PbSO_4$(s), and then converted back into Pb(s). It is a common observation that the Pb(s) particles precipitated after one discharge/charge cycle are not identical with the initial ones in the shape and size; rather, the shape is not spherical and the size increases. Some Pb(s) particles may fall off the electrode. If this is the case, the contact area between the Pb(s) particles and electrolyte solution may be reduced due to the particle growth. These unfavorable features lead to an increase in cell polarization and eventual cell degradation such as a decrease in cell capacity.

Self-discharge of lead-acid batteries is not severe but still cannot be ignored. When cells are not in use, two half-cells are electrically disconnected (open-circuit condition). Even under the open-circuit condition, the following two reactions proceed spontaneously as a pair as shown in Fig. 12-6; Pb(s) + HSO_4^- → $PbSO_4$(s) + 2*e* + H^+ and $2H^+$ + 2*e* → H_2(g). That is, the equilibrium potential for the former half-cell (Pb/$PbSO_4$) is more negative than that for the latter (H^+/H_2); the former has a higher oxidizing tendency, but the latter a higher reduction tendency. A galvanic cell forms at the negative electrode by itself while being isolated from the positive electrode. Four components needed for galvanic cell formation are provided from the negative electrode by itself: anode, cathode, electrolyte and the external circuit for electron transport (refer to Fig. 2-10). However, since k^0 for H_2 generation ($2H^+$ + 2*e* → H_2(g)) on the negative electrode (Pb/$PbSO_4$)

is small (Table 11-2), H_2 generation is not very severe, but cannot be ignored because the reactant for H_2 generation (H^+) is plentiful as sulfuric acid is used as the electrolyte. If H_2 generation cannot be ignored, the discharge reaction of negative electrode (Pb(s) + HSO_4^- $\rightarrow$ $PbSO_4$(s) + 2*e* + H^+) cannot be ignored; this reaction takes place at the same rate with the H_2 evolution rate. In short, the negative electrode self-discharges and Pb is consumed even under the open-circuit condition. It is a common practice that the capacity of negative electrode is intentionally adjusted to be larger than that of the positive electrode (N/P ratio = 1.2) to compensate the Pb loss due to this self-discharge. The N/P ratio refers to the capacity ratio between the negative and positive electrode.

12-1-6 Nickel-cadmium batteries

The discharge reactions of ***Ni-Cd cells*** are as follows. When manufacturing batteries, it is common to load the electrode materials in their discharged state. Ni-Cd cells are also initially loaded with the discharged materials; $Cd(OH)_2$(s) and $Ni(OH)_2$(s).

Negative electrode: $Cd(s) + 2OH^- \rightarrow Cd(OH)_2(s) + 2e$ $\quad E^0 = -0.81$ V(*vs.* NHE)

Positive electrode: $2NiOOH(s) + 2H_2O + 2e \rightarrow 2Ni(OH)_2(s) + 2OH^-$ $\quad E^0 = 0.47$ V(*vs.* NHE)

Overall: $Cd(s) + 2NiOOH(s) + 2H_2O \rightarrow Cd(OH)_2(s) + 2Ni(OH)_2(s)$ $\quad E^0_{cell} = 1.28$ V

The performance characteristics of Ni-Cd cells are as follows. Firstly, KOH does not participate in the electrode reaction, so there is no need to overdose it. Secondly, the concentration change of KOH electrolyte solution is not appreciable because water is excessively added in the electrolyte even if it is consumed during discharge. Thirdly, the equilibrium potential for hydrogen half-cell (-0.83 V *vs.* NHE) is close to the equilibrium potential of $Cd/Cd(OH)_2$ half-cell (−0.81 V *vs.* NHE), and H_2 evolution is thermodynamically possible during charge (Fig. 12-7). However, the standard rate constant (k^0) for H_2 evolution is so small on Cd electrode (kinetically slow) that H_2 generation is not severe (see Table 11-2). Note that the standard state with a unit activity and fugacity is assumed for the above discussion. Fourthly, O_2 evolution is thermodynamically possible on the

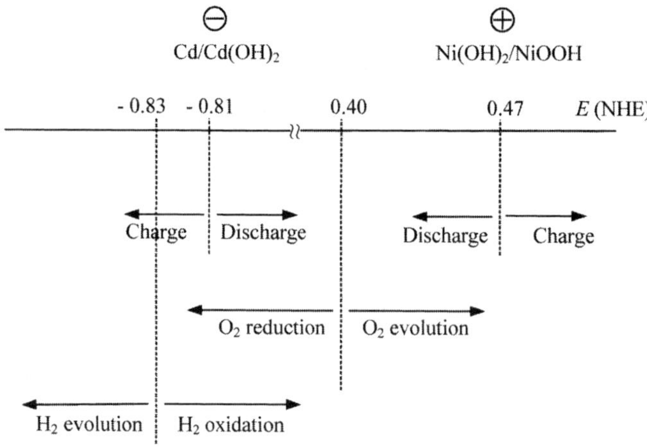

Fig. 12-7 Thermodynamically predicted potential ranges for the electrode reactions and side reactions under standard state. −0.83 V, −0.81 V, 0.4 V and 0.47 V are the equilibrium potentials (E_{eq}) of the four half-cells. The arrows indicate the potential range for each reaction.

positive electrode during charge and kinetically appreciable. Hence, the large-sized Ni-Cd cells are designed to release oxygen gas (called a vented cell). However, since O_2 can be reduced at the negative electrode ($O_2(g) + 2H_2O + 4e \rightarrow 4OH^-$) by ***internal oxygen cycle*** like the lead-acid cell, a completely sealed Ni-Cd cell is also commercialized. As shown in Fig. 12-7, if gaseous O_2 moves to the negative electrode through the separator, its reduction is thermodynamically possible and kinetically appreciable.

As is seen in Fig. 12-7, H_2 evolution on the negative electrode is thermodynamically feasible even if H_2 evolution rate is not appreciable with a very small k^0 for H_2 evolution on Cd electrode. However, for the sealed Ni-Cd cells, the pressure build-up by H_2 evolution should be completely prevented. One way for such a complete prevention is to control the charging voltage of negative electrode to be more positive than the equilibrium potential of hydrogen half-cell (−0.83 V vs. NHE) (Fig. 12-7). Such a potential control is made by two routes. Firstly, when the internal oxygen cycle proceeds, the negative electrode potential is pinned at −0.81 V ~ 0.4 V because a mixed potential forms as a result of the paired reactions of the discharge reaction of the negative electrode and O_2 reduction reaction. Secondly, an excess of $Cd(OH)_2$ is loaded in the negative electrode, which is called a ***charge reserve***. In normal charging condition, the charging reaction ($Cd(OH)_2(s)$

$+2e \rightarrow Cd(s) + 2OH^-$) proceeds on the negative electrode. The overcharge condition is introduced when $Cd(OH)_2(s)$ is exhausted after a prolonged charging. Under the overcharge condition, electrochemical reduction reactions should take place on the negative electrode. Now, $Cd(OH)_2(s)$ is depleted. Then, the next alternative reduction reaction is the H_2 evolution; $2H_2O + 2e \rightarrow H_2(g) + 2OH^-$. Note that there is a plentiful amount of reducible molecules (H_2O) in the electrolyte solution. The occurrence of H_2 evolution indicates that the negative electrode potential now moves into the H_2 evolution region (more negative than -0.83 V *vs.* NHE). The presence of charge reserve can prevent the potential to cross over -0.83 V since the charging reaction still goes on; the negative electrode potential remains at more positive value than -0.83 V.

The positive electrode, $Ni(OH)_2$/NiOOH, can be degraded due to a large volume change with repeated charge/discharge cycling. In order to mitigate this problem, $Ni(OH)_2$ is prepared as a sphere form and loaded inside the pores of Ni foam that is used as a current collector. The pores provide the enough spaces for volume expansion/contraction.

Ni-Cd cells have been used as the power sources for power tools and toys because of their relatively long cycle life and robustness. Their market share has, however, decreased significantly due to the memory effect, the pollution problem of Cd and the competitiveness of lithium secondary batteries, The ***memory effect*** refers to a phenomenon in which, when a secondary battery is charged/discharged to a certain degree, the battery discharges only to that degree as if it memories the previously discharged capacity.

12-1-7 Heat effects

The discharge reaction of Ni-Cd cell is as follows; $\Delta H = -282$ kJ/mol and $\Delta G = -256$ kJ/mol for the reaction.

$$Cd_{(S)} + 2NiOOH_{(S)} + 2H_2O \rightarrow Cd(OH)_{2(S)} + 2Ni(OH)_{2(S)}$$

$T\Delta S = -26$ kJ/mol is calculated from <12-5>. On the basis of these thermodynamic values, the cause and amount of heat evolution in Ni-Cd cell are described.

$$\Delta H = \Delta G + T\Delta S \quad \text{<12-5>}$$

$$\frac{\Delta H}{-nF} = \frac{\Delta G}{-nF} + \frac{T\Delta S}{-nF} \quad \text{<12-6>}$$

$$E_{cal} = E_{cell} - \frac{T\Delta S}{nF} \quad \text{<12-7>}$$

$$\Delta G = -nFE_{cell} \quad \text{<12-8>}$$

E_{cal} and E_{cell} are defined from <12-6> that is obtained by dividing <12-5> by $-nF$. E_{cell} is the electromotive force as shown in <12-8>. E_{cal} is called the ***calorific voltage***. As E_{cell} is derived from ΔG, E_{cal} is derived from ΔH; $E_{cal} = \Delta H/(-nF)$. Hence, E_{cal} is an electromotive force expected under the assumption that all energy corresponding to ΔH is converted into electrical energy, but it is an imaginary value because all of ΔH cannot be converted into electrical energy. In fact, only ΔG (−256 kJ/mol) can be converted into electrical energy. Then, the remnant ($T\Delta S$ = −26 kJ/mol) is evolved as a heat during discharge. From the ΔH and ΔG values of the discharge reaction, E_{cal} = 1.44 V and E_{cell} = 1.30 V are calculated.

Since the charging reaction proceeds in the opposite direction, all the thermodynamic values have the opposite sign; ΔH = 282 kJ/mol, ΔG = 256 kJ/mol, $T\Delta S$ = 26 kJ/mol, E_{cal} = −1.44 V and E_{cell} = −1.30 V. The positive (+) value of ΔH means that energy should be supplied from outside for the charging reaction. Of the total energy required (ΔH = 282 kJ/mol), ΔG = 256 kJ/mol is provided as an electrical energy by charging. The remnant ($T\Delta S$ = 26 kJ/mol) should be supplied from outside, implying that the charging reaction is endothermic ($T\Delta S$ = 26 kJ/mol). $T\Delta S$ is called the ***reversible heat effect*** (Q_{rev}) since the heat evolution during discharge (−26 kJ/mol) is identical with the heat absorption in the charge (26 kJ/mol). Obviously, it is a thermodynamic value.

Heat can also be generated by kinetic origin, which is called an ***irreversible heat effect*** (Q_{irrev} or Q_{Joule}). The total heat effect is the sum of Q_{rev}(= $T\Delta S$) and Q_{irrev} (Q_{Joule}) as

$$Q_{total} = Q_{rev} + Q_{irrev}\,(Q_{Joule})$$

Q_{rev} for discharge is defined as <12-9>. Here, i and t denote the discharge current and discharge time, respectively. It is noted that, if E_{cal} and E_{cell} are constants, which is the case for Ni-Cd cell, the value obtained by integrating the current with respect to time as in <12-10> corresponds to the discharge capacity and was defined to have a negative (−) value.

$$Q_{rev} = \int_0^t (E_{cal} - E_{cell})\, i\, dt \qquad \text{<12-9>}$$

$$\int_0^t i\, dt = -nF \qquad \text{<12-10>}$$

Q_{irrev} during discharge is defined as <12-11>. Here, $E_{discharge}$ is the time-dependent discharge voltage. $E_{discharge}$ has the relation of <12-12>, so <12-11> can be rewritten as <12-13>.

$$Q_{irrev}(Q_{Joule}) = \int_0^t (E_{cell} - E_{discharge})\, i\, dt \qquad \text{<12-11>}$$

$$E_{discharge} = E_{cell} - cell\ polarization \qquad \text{<12-12>}$$

$$Q_{irrev}(Q_{Joule}) = \int_0^t (\eta_a + \eta_c + iR_{total})\, i\, dt \qquad \text{<12-13>}$$

<12-13> implies the following. Firstly, Q_{irrev} is determined by the cell polarization. As η_a, η_c and iR_{total} are determined by the discharge current, Q_{irrev} is also decided by the discharge current; a larger discharge current causes a larger cell polarization and eventually a larger Q_{irrev}. Secondly, since Q_{irrev} has a negative (−) value according to <12-10>, it means a heat evolution (exothermic).

The total heat effect in charging is also the sum of Q_{rev} and Q_{irrev}. Q_{rev} is given as <12-9>. In this case, i and t denote the charging current and charging time, respectively. Q_{irrev} in charging is given by <12-14>. Here, E_{charge} refers to the time-dependent charging voltage. Since E_{charge} is defined as <12-15>, <12-14> can be rewritten as <12-16>.

$$Q_{irrev}(Q_{Joule}) = \int_0^t (E_{charge} - E_{cell})\, i\, dt \qquad \text{<12-14>}$$

$$E_{charge} = E_{cell} + cell\ polarization \qquad \text{<12-15>}$$

$$Q_{irrev}(Q_{Joule}) = \int_0^t (\eta_a + \eta_c + iR_{total})\, i\, dt \qquad \text{<12-16>}$$

Two features can be identified from <12-16>. Firstly, Q_{irrev} is determined by the cell polarization. <12-16> looks the same as <12-13>, but the meaning is different. In

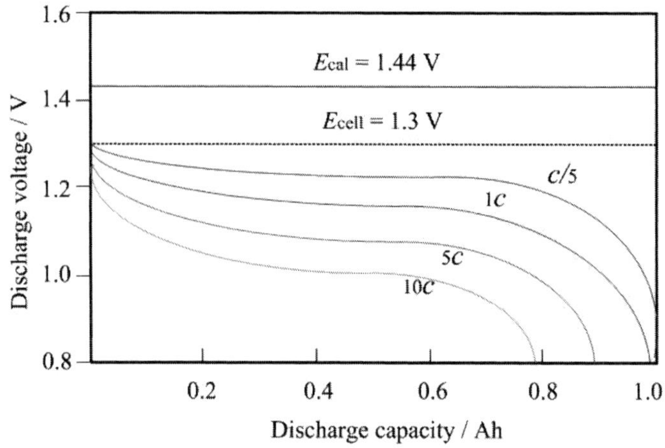

Fig. 12-8 The galvanostatic (constant-current) discharge voltage profiles ($E_{discharge}$) of a Ni-Cd cell according to the discharge current. Note that E_{cal} and E_{cell} are constant (flat) throughout the discharge. Nominal cell capacity = 1.0 A h. Cut-off voltage = 0.8 V.

<12-13>, the cell polarization is the sum of η_a, η_c and iR_{total} that change with time during discharge, but in <12-16>, the cell polarization is relevant to the charging reaction. Secondly, since Q_{irrev} has a negative (−) value according to <12-10>, it means a heat evolution (exothermic).

Fig. 12-8 shows the galvanostatic (constant-current) discharge voltage profiles ($E_{discharge}$) of a Ni-Cd cell according to the discharge current. The discharge cut-off voltage is 0.8 V, and the discharge rate is controlled in the range of *c*/5 to 10*c*. The nominal cell capacity defined at *c*/5 rate, 0.8 V cut-off and 25℃ is 1.0 A h. As shown, E_{cal} and E_{cell} are constant (flat) throughout the discharge. Meanwhile, note here that both E_{cal} and E_{cell} gradually decrease with discharge for lead-acid cells and lithium secondary cells.

At the discharge rate of *c*/5, Q_{rev} and Q_{irrev} can be estimated on Fig. 12-8. Q_{rev} can be estimated by using <12-9>, wherein the integral value of the current is the discharge capacity since E_{cal} and E_{cell} are constant. Therefore, the area of the rectangle made by (E_{cal} − E_{cell}) and discharged capacity (1.0 A h) corresponds to Q_{rev} in Fig. 12-8. The details are explained on Fig. 12-9. Since $\Delta H = E_{cal} \times (-nF)$ representing the total energy change of the discharge reaction, it is shown as a rectangular area in Fig. 12-9-a. Here, the discharged capacity is $-nF = -1.0$ A h at *c*/5 rate. Further, as $\Delta G = E_{cell} \times (-nF)$,

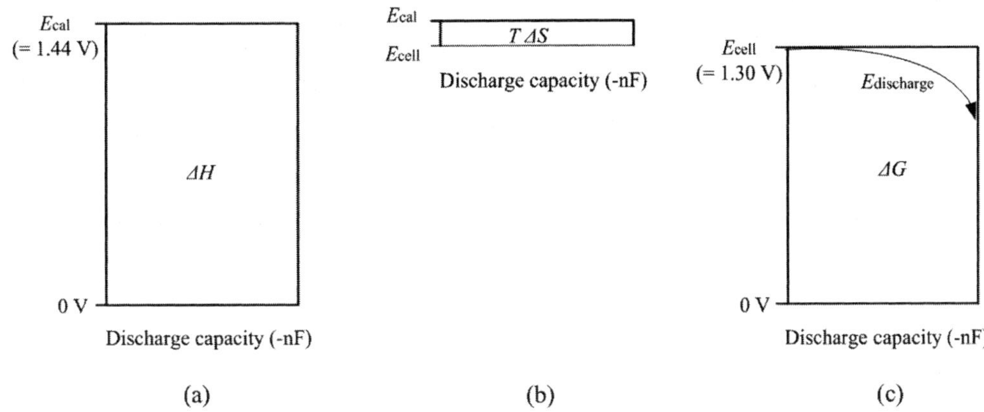

Fig. 12-9 A schematic diagram of ΔH, $T\Delta S$, ΔG, Q_{irrev}, and discharged energy from Fig. 12-8. In (c), the arrow indicates the direction of discharge.

it is shown as a rectangular area in Fig. 12-9-c. Consequently, Q_{rev} corresponds to the area of the rectangle shown in Fig. 12-9-b. In Fig. 12-9-c, ΔG corresponding to the area of the rectangle can be converted into electrical energy by discharge. However, all of it cannot be converted because of Q_{irrev}, which is defined as <12-11> and thus corresponds to the area between E_{cell} and $E_{discharge}$. Q_{irrev} has a negative (−) value; it is exothermic; it cannot be converted into electric (discharged) energy. Only the remnant energy (ΔG − Q_{irrev}) can be converted into the electric (discharged) energy. This discharged energy is expressed as the area under the $E_{discharge}$ in Fig. 12-9-c; it corresponds to the area subtracting the wedge-shaped one on the top-right from the total rectangle. In summary, the energy as much as ΔH is generated through the discharge reaction at a given *c* rate. Q_{rev} is released as heat (exothermic). The energy as much as the remaining ΔG may be converted into electrical energy, but only the energy excluding Q_{irrev} can be converted into the discharged energy; the energy corresponding to Q_{irrev} is wasted as a heat (exothermic). Note here that Q_{irrev} is decided by cell polarization, which increases with an increase of discharge current, explaining that the discharged energy becomes smaller when discharged at higher current.

Self learning 12-5

Fig. 12-9 shows the schematic drawing of ΔH, $T\Delta S$, ΔG, Q_{irrev}, and discharged energy that are derived from the data obtained at $c/5$ in Fig. 12-8. Draw the picture from the data obtained at $10c$ in Fig. 12-8, and confirm the following. (1) The discharged capacity at $10c$ (0.78 A h) is smaller than that at $c/5$ (1.0 A h), such that the areas corresponding to ΔH, $T\Delta S$, and ΔG are smaller. (2) The discharged energy ($\Delta H - Q_{total} = \Delta G - Q_{irrev}$) is smaller for the $10c$ discharge.

Fig. 12-10 shows the energy balance for the charging process. ΔH, representing the total energy required for the charging reaction, is expressed as the area of a rectangle consisting of $E_{cal} \times (-nF)$ (Fig. 12-10-a). Of the required ΔH, $T\Delta S$ (Q_{rev}) shown as a rectangle in Fig. 12-10-b is provided from outside (endothermic). The remnant energy required is ΔG, which corresponds to the area of a rectangle consisting of $E_{cell} \times (-nF)$ in Fig. 12-10-c. ΔG is provided as an electrical energy by the charging reaction. However, during the charge, Q_{irrev} is additionally needed due to cell polarization. Here, Q_{irrev} is not used for charging, rather it is wasted at a heat (exothermic). Note that Q_{irrev} has a negative (−) value for both discharge and charge; this is wasted as a heat.

As described in Fig. 12-2, the energy efficiency of a secondary battery is decided by both the Coulombic efficiency and voltage efficiency (= discharge voltage/charge voltage). The discharge voltage is always smaller than the charging voltage, such that the voltage

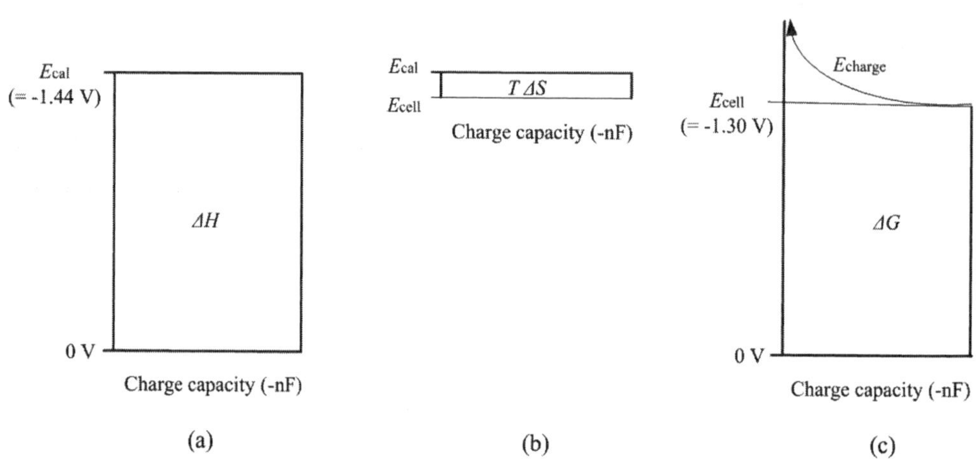

Fig. 12-10 Schematic diagram of ΔH, $T\Delta S$, ΔG, Q_{irrev}, and charged energy during the charging process. In (c), the arrow indicates the direction of charging.

efficiency is < 1.0. This manifests itself that the energy efficiency must be < 1.0; the electrical energy recovered by discharge is smaller than the energy stored by charging. The energy not recovered is dissipated to outside as a heat. Here, the loss of energy is encountered in both charge and discharge. The Q_{irrev} evolved in both charge and discharge corresponds to the heat loss for each process. As shown in <12-13> and <12-16>, the Q_{irrev} is determined by cell polarization. Hence, the Q_{irrev} can be minimized by minimizing the cell polarization encountered in both charge and discharge process.

Self learning 12-6

Frequently, battery becomes hot when discharged due to heat generation. Find the causes of heat generation in terms of thermodynamics and kinetics. Is it possible for cell temperature to drop when discharged? Answer the same question for charging.

Self learning 12-7

Describe the effect of cell polarization on: (1) discharge voltage, (2) charging voltage, (3) discharged capacity under a given cut-off voltage range, (4) discharged energy, (5) Q_{irrev} during charging, (6) Q_{irrev} during discharge, and (7) energy efficiency.

12-1-8 Nickel-metal hydride batteries

The ***Ni-metal hydride battery*** is identical in configuration to the Ni-Cd battery except that a hydrogen storage alloy is used as the negative electrode. The discharge reaction at the negative electrode is as follows.

$$MH(s) + OH^- \rightarrow M(s) + H_2O + e \qquad E^0 = -0.83 \text{ V}(vs. \text{ NHE})$$

Here, MH(s) and M(s) refer to the hydrogen-stored and hydrogen-released alloy, respectively. Since the same positive electrode is used as for the Ni-Cd battery, E^0_{cell} = 1.30 V.

The detailed discharge reaction at the negative electrode is as follows.

$$MH(s) \rightarrow M(s) + 1/2\ H_2(g)$$
$$1/2\ H_2(g) + OH^- \rightarrow H_2O + e$$

The hydrogen molecules stored in the alloy are released, and the released hydrogen is electrochemically oxidized. Since $E^0 = -0.83$ V (*vs.* NHE) for hydrogen half-cell ($H_2O + e = 1/2\ H_2(g) + OH^-$), the standard electrode potential including the H_2 storage/release reaction is also $E^0 = -0.83$ V (*vs.* NHE).

The characteristics of Ni-metal hydride batteries are as follows. Firstly, like Ni-Cd cells, KOH does not participate in the electrode reaction. Secondly, since the generation and storage of hydrogen are involved in the negative electrode, the cells are manufactured in a sealed-type only. Thirdly, as in the case of Ni-Cd cells, O_2 evolution at the positive electrode is thermodynamically feasible, and the evolved oxygen is converted into water on the negative electrode by ***internal oxygen cycle***.

Several kinds of ***hydrogen storage alloys*** have been developed. A representative example is $MmNi_5$, wherein Mm is a misch (mixture in German) metal with La as the main component. In nature, La, Ce, Pr, and Nd coexist but the separation into each element is very difficult due to the similarity in atomic size (called Lanthanide contraction); the pure La is expensive, but Mm is cheaper, thus used as a mixture. Recently, a hydrogen storage alloy, in which a part of Ni in $MmNi_5$ is replaced by Mn, Al and Co has been developed.

For charging of this battery, hydrogen storage is required, so that it was difficult to charge at a high rate, and the charging characteristics were somewhat insufficient as compared to the Ni-Cd battery. These drawbacks have, however, been solved; a high-rate charge/discharge becomes possible. In addition, the energy density has been greatly improved to be closer to that of lithium secondary batteries. Although this cell is used as a power source for small electronic devices, the market share has been greatly reduced due to the emergence of lithium secondary batteries. At present, this cell is used as a power source for some hybrid electric vehicles.

12-1-9 Lithium secondary batteries

The ***lithium secondary batteries*** have been developed with the aim of converting the lithium primary cells into the rechargeable ones. Lithium metal was used as the negative electrode in the earlier development. For rechargeable batteries, Li plating reaction is required on the negative electrode. Here, two problems were encountered. Firstly, lithium metal is electrodeposited as a form of ***dendrite***, which may lead to an ignition or explosion due to internal short; the dendritic lithium metal penetrates through thin polymer separator. Secondly, since ***dead lithium*** is produced, an excessive amount of lithium metal should be loaded. That is, when the lithium metal is electrodeposited, the newly formed lithium metal surface is exposed to the electrolyte solution, and the electrolyte is decomposed to form SEI. When the newly deposited lithium metal particles are surrounded by SEI layer, which has ionic conductivity but no electronic conductivity, they are electronically isolated from current collector; they are not active anymore (dead) for the electrode reaction.

In order to prevent the dendritic growth of Li metal and suppress the formation of dead lithium, several attempts are being made, such as developing new electrolytes, additives, and using lithium alloys. Also, many efforts to develop lithium metal negative electrode are in progress to take advantages of the very large theoretical capacity (3861 mAh/g), which is much larger than that of graphite (372 mAh/g).

Graphite has been developed as the replacement of lithium metal, and is still in wide use today. Fig. 12-11 shows the ranges of equilibrium potential (E^{c}_{eq} and E^{a}_{eq}) of some electrode materials. The equilibrium potential of graphite half-cell ($C_6 + Li^+ + e = LiC_6$, C_6 means the six-membered carbon unit in graphite) varies within 0.2 ~ 0.3 V (*vs.* Li/Li^+) upon charge/discharge, which is very close to that of Li/Li^+ half-cell; therefore, when used as the negative electrode, the working voltage of lithium secondary cells is not significantly smaller compared to that adopted lithium metal.

Graphite has a layered structure (Fig. 12-12), and ***intercalation/de-intercalation*** of Li^+ is possible through the interlayer gap. The intercalation refers to the co-injection of Li^+ and electrons into the Li^+ storage sites and electron storage sites, respectively, while the structural frameworks being maintained. The de-intercalation is the opposite process. Here,

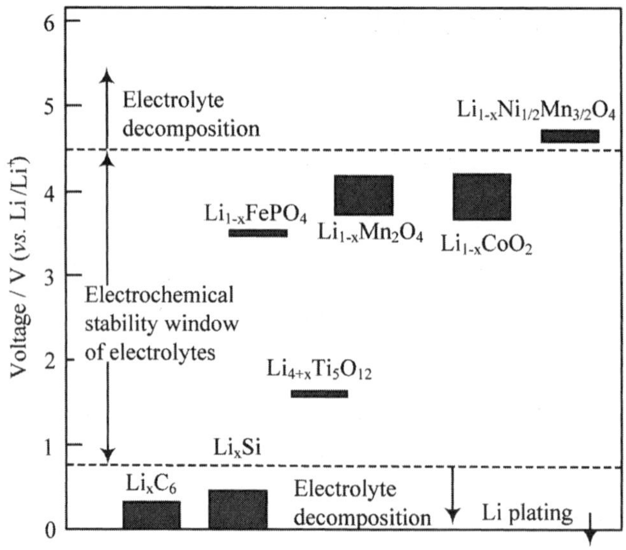

Fig. 12–11 E^{c}_{eq} and E^{a}_{eq} of some electrode materials for lithium secondary batteries, and the electrochemical stability window of common electrolytes.

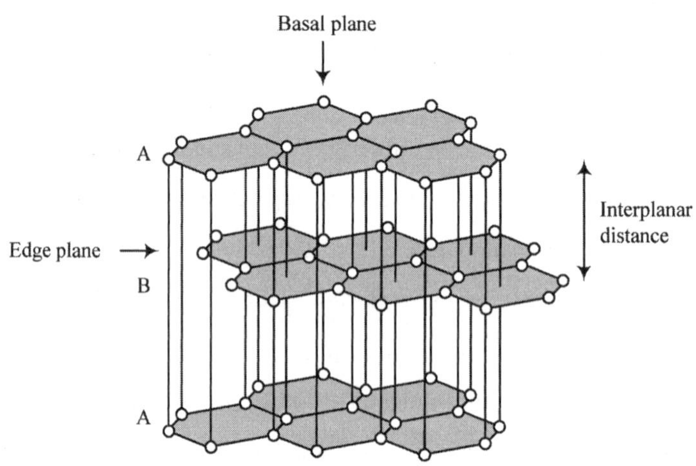

Fig. 12–12 Structure of graphite.

it is noted that Li^+ and electrons are co-injected in intercalation, and co-removed upon de-intercalation. Namely, when Li^+ having a positive charge enters the structure of an electrode material, electrons having a negative charge must accompany it to maintain electroneutrality. The injected Li^+ is stored at the Li^+ storage sites that are populated at the interlayer gap, while the injected electrons are stored at the graphene layers to be de-localized. In the de-intercalation process, both Li^+ and electrons are released from their sites.

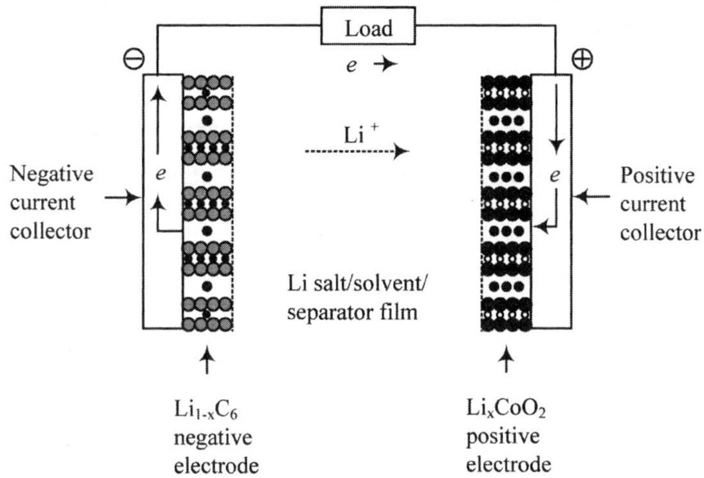

Fig. 12-13 Configuration of lithium-ion battery.

As a positive electrode, some metal oxides having a layered structure (e.g. Li_xCoO_2) or a spinel structure (e.g. $Li_xMn_2O_4$) are employed. Some metal phosphate (e.g. Li_xFePO_4) is also used. The intercalation/de-intercalation (co-injection/co-removal of Li^+ and electrons) is also involved in these metal oxides and phosphates. Since the E_{eq} of metal oxide half-cells is about 4 V (*vs.* Li/Li^+), the discharge voltage becomes $3.5 \sim 4.0$ V when combined with the graphite negative electrode. The full-cell fabricated with graphite and Li_xCoO_2 (or $Li_xMn_2O_4$) with a non-aqueous electrolyte is called a rocking-chair battery because Li^+ are transferred between two electrodes by intercalation and de-intercalation upon charge/discharge. Nowadays, ***lithium-ion batteries*** are more common name (Fig. 12-13). The highlighted 'ion' implies that lithium component always exists as an ion (Li^+) but not as a metal (Li); they can thus be distinguished from the troublesome lithium secondary cells fabricated with lithium metal negative electrode.

Self learning 12-8

Even if it is claimed that lithium always maintains an ionic state in lithium-ion battery, lithium metal is electrodeposited when the cell is charged either at high current or at low temperature, which may cause safety problems. It is noted that secondary batteries are electrolytic cell when charged. From <11-1> and Fig. 12-11, explain that lithium electrodeposition is possible in these cases.

Since the E_{eq} of graphite half-cell ($C_6 + Li^+ + e = LiC_6$, $E^0 = 0.2 \sim 0.3$ V *vs.* Li/Li^+) is close to that of Li/Li^+ half-cell, the electrolyte is subjected to a strong reduction condition of graphite electrode. In Fig. 12-11, the electrochemically stable potential range, which is called the ***electrochemical stability window***, of common organic electrolytes is presented. At below ca. 0.8 V (*vs.* Li/Li^+), the electrolyte is decomposed by reduction to form a SEI layer on negative electrode, and by oxidation at above ca. 4.3 V (*vs.* Li/Li^+) on positive electrode; the electrochemical stability window of common organic electrolytes is thus $0.8 \sim 4.3$ V (*vs.* Li/Li^+).

Fig. 12-14 shows the constant-current charge/discharge voltage profiles of a Li/graphite cell, in which ethylene carbonate (EC) is used as the solvent. When the charging current is applied in the first cycle, the voltage of graphite electrode decreases from ca. 3.0 V (*vs.* Li/Li^+) to ca. 0.1 V. In the forthcoming discharge period, the graphite electrode potential restores back to 3.0 V. A noticeable feature in the first charging is the appearance of a plateau-like voltage profile at ca. 0.6 V, at which the organic solvent (EC) in the electrolyte is decomposed by reduction to form a SEI layer. With the charge going on, the graphite electrode potential is further reduced with the evolution of several voltage plateaus in the range of $0.1 \sim 0.3$ V (*vs.* Li/Li^+). Note that the voltage shows a flat profile at the plateau regions. These are associated with Li^+ intercalation into the graphene layers along with electron injection, which is called a ***stage phenomenon*** (Fig. 12-15). The Li^+ intercalation is a systematic process, in which Li^+ ions are inserted into the interlayers given by two adjacent graphene layers. In the initial period of intercalation, the stage-4 forms, in which only one out of four interlayers is filled by Li^+ ions. With the intercalation

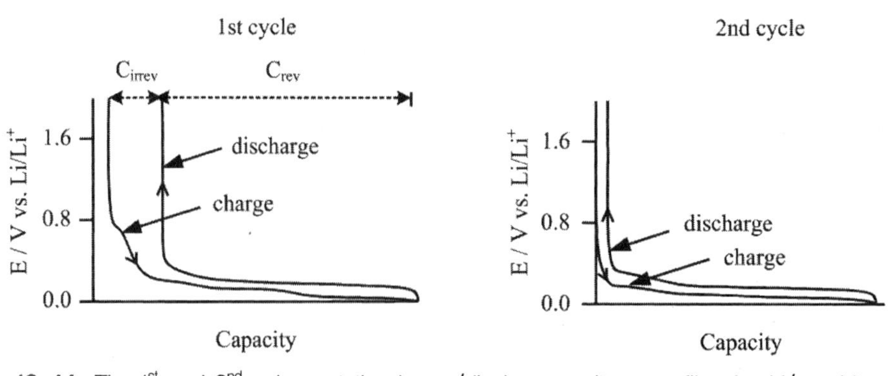

Fig. 12-14 The 1st and 2nd galvanostatic charge/discharge voltage profile of a Li/graphite cell.

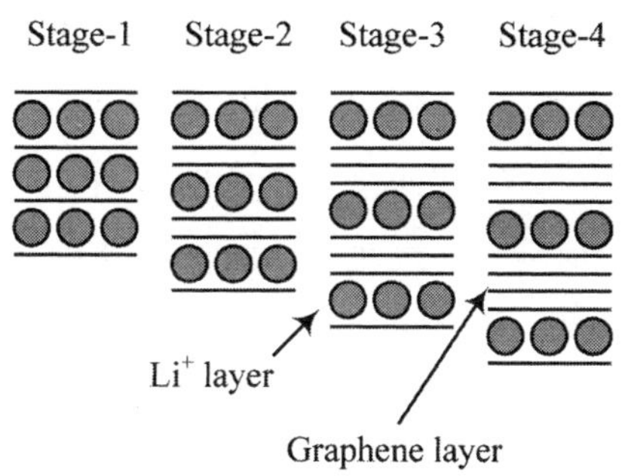

Fig. 12-15 Stage phenomenon by intercalation and de-intercalation of Li^+ in graphite electrode.

going on, the stage-3 (Li^+ ions fill one out of three interlayers) and the stage-2 (one out of two interlayers are filled by Li^+ ions) and eventually stage-1 forms, in which phase all the Li^+ storage sites are filled by Li^+ ions. Along with Li^+ insertion, the co-injected electrons are stored in the graphene layers. The Li^+ de-intercalation proceeds in the opposite path; from the stage-1 to stage-2, stage-3, stage-4, and finally the graphite with no Li^+ in. The Li^+ intercalation proceeds by consecutive two-phase reactions between the stages (e.g. stage 1/stage 2), such that the charging voltage profile shows the consecutive voltage plateaus. The similar consecutive voltage plateaus are observed during discharge since the discharge also proceeds by consecutive two-phase reactions. It is distinguished from the sloped voltage profiles that are observed with the single-phase charge/discharge reaction occurring in $LiCoO_2$ positive electrode (refer to Example 9-1).

As is seen in Fig. 12-14, the discharged capacity in the first cycle is smaller than the charged capacity. The difference is called an ***irreversible capacity*** (C_{irrev}), which corresponds to the electric charges consumed for some parasitic reactions; e.g. the SEI formation (e.g., EC + Li^+ + e → Li_2CO_3(s) + oligomers/polymers). Here, as the electric charges (electrons) and Li^+ are irrevocably consumed, they are not retrieved in the forthcoming discharge period. On the other hand, the electric charges (electrons) and Li^+ consumed for the consecutive two-phase intercalation reactions can be restored by the consecutive two-phase de-intercalation reaction upon discharge. Hence, it is called a ***reversible capacity*** (C_{rev}). For graphite, C_{rev} is theoretically 372 mAh/g, which means that 372 mAh/g of electric charges are required to complete the charging reaction ($C_6 + Li^+ + e \rightarrow LiC_6$), and the same amount of electric charges are recovered by discharge.

Also seen in Fig. 12-14 is that the plateau-like profile at 0.6 V observed in the first charging disappears in the second charging. This illustrates that the electrolyte decomposition and concomitant SEI formation is not required in the second charging. This is due to the passivating role of SEI layer that is generated in the previous (first) charging to prevent further electrolyte decomposition. Due to the absence of C_{irrev}, electric charges are consumed only for intercalation, and the same amount of charges are restored upon de-intercalation; the charged capacity and discharged capacity are identical.

Since graphite exists in nature, it is mined and used as the negative electrode. Most of the ***natural graphites***, however, have the shapes that are not suitable for electrodes, such as plates or flakes, such that they are converted into a spherical shape with a size of about 10 μm by mechanical rounding. Here, two advantages exist. Firstly, the graphite loading in the electrode layers can be increased as the packing density of rounded particles is higher than that of irregularly-shaped ones. Secondly, Li^+ intercalation rate can be enhanced by randomizing the orientation of edge planes. As is seen in Fig. 12-12, Li^+ ions move in through the edge plane rather than the basal plane. For the plate- or flake-type graphites, the edge planes tend to orient in the direction parallel to the surface of current collector upon spread onto a metal foil. It is called a preferred orientation that is frequently observed with plate-like materials. Unfortunately, the direction of current flow is perpendicular to the orientation of edge planes, such that Li^+ intercalation is sluggish. The orientation of edge planes can be randomized by mechanical rounding to facilitate Li^+ intercalation.

Artificial graphites are also used as the negative electrode. The typical example is ***g-MCMB*** (graphitized mesocarbon microbeads) having a spherical shape. For their preparation, petroleum or coal pitches are used as the starting material. The pitches are heat-treated to obtain spherical liquid crystalline (anisotropic) pitch beads. After partial cross-linking and carbonization (at about 1000 ℃ in inert atmosphere), the pitch beads are graphitized by heating at 2500 ~ 3000 ℃. The natural graphites have the advantage of slightly higher capacity and lower price, whereas the artificial ones have the advantage of higher initial Coulombic efficiency (smaller irreversible capacity) due to less serious side reactions.

The ***Li alloying materials*** such as Si and Sn that can make alloys with lithium have been proposed as the negative electrode. They show a very large theoretical capacity, but have a serious problem of volume expansion/compression upon charge and discharge. The practical use is thus hindered due to this volume-change problem. In order to take advantage of their large electrode capacity, silicon monoxide (SiO) instead of Si is added (about 10 %) to graphite electrode; the resulting SiO/graphite electrode gives a slightly enlarged electrode capacity. In the first charging period, Li_xSi, Li_2O and Li_4SiO_4 are generated from the SiO component. The Li_xSi participates in the alloy/de-alloy reaction ($Li_xSi \leftrightarrow Si + xLi^+ + xe$) to give additional capacity to the graphite component. The oxides (Li_2O and Li_4SiO_4) play as a buffer for volume change encountered during the charge/discharge of Si component.

The materials that react with lithium through a ***conversion reaction***, such as CoO, are also considered as the negative electrode: $CoO + 2Li^+ + 2e \leftrightarrow Co + Li_2O$. They show a large theoretical capacity, but two drawbacks; a large volume change and a serious voltage hysteresis. The latter means a large difference between the charge and discharge voltage, which is caused by a large cell polarization, which is in turn inherited from a large activation energy (overpotential) for break/formation of metal-oxygen bonds.

Lithium titanium oxide ($Li_4Ti_5O_{12}$), which is charged and discharged at about 1.5 V (*vs.* Li/Li^+) is also considered as the negative electrode. The working electrode potential is higher than that of graphite, such that the working voltage of the full-cells fabricated with this negative electrode is smaller than those fabricated with graphite. This oxide electrode shows a very stable cycle performance and considered to be safer because the chance for Li plating and the electrolyte decomposition/SEI formation is low at such a high working electrode potential (1.5 V *vs.* Li/Li^+) (Fig. 12-11).

Lithium cobalt oxide ($LiCoO_2$) and ***lithium nickel oxide*** ($LiNiO_2$) of a layered structure, ***lithium manganese oxide*** ($LiMn_2O_4$) of a spinel structure, and ***lithium iron phosphate*** ($LiFePO_4$, LFP) of an olivine structure have been developed and commercialized as the positive electrode. The charging reaction of $LiCoO_2$ is $LiCoO_2 \rightarrow Li_{1-x}CoO_2 + xLi^+ + xe$. Here, the charging is limited at x = 0.5; the charge and discharge are performed in the range of $LiCoO_2 \leftrightarrow Li_{0.5}CoO_2$, which gives the electrode capacity of 140 mAh/g.

This is because several unfavorable features are encountered at $x > 0.5$. Firstly, the phase transitions to inactive phases occur. Secondly, O_2 evolution from the lattice oxygen is more prone with an increase of x; O_2 evolves at lower temperature when cell temperature increases under some abuse conditions, which can induce a fire or explosion. Thirdly, the charging voltage increases to be higher than 4.3 V (*vs.* Li/Li^+), from which the oxidative decomposition of electrolyte become severer. Several attempts are being made to increase x; to increase the electrode capacity. For example, the phase transitions to inactive phases have been suppressed by a doping (replacing a part of Co with other metals); the doped one can be charged at $x > 0.5$. The oxidative decomposition of the electrolyte, which is severe at $x > 0.5$, can be suppressed by a coating of inert oxide on the surface of $LiCoO_2$ particles.

$LiNiO_2$ having the layered structure shows a larger reversible capacity than $LiCoO_2$, but it is difficult to synthesize a pure $LiNiO_2$. In addition, there are many problems such as O_2 evolution and oxidative decomposition of electrolyte, which are also prevailing in $LiCoO_2$. These problems are being solved by substituting a part of Ni with other metals. $LiNi_{0.8}Co_{0.15}Al_{0.05}O_2$ (NCA, 15% of Ni is replaced by Co and 5% by Al) and $LiNi_{1/3}Co_{1/3}Mn_{1/3}O_2$ (NCM) were developed and commercialized. In the latter, only the Ni component is active for the electrode reaction. Hence, many efforts are being made to increase the Ni content in $LiNi_xCo_yMn_zO_2$ by minimizing the adverse effects brought by high Ni content. Along this line, $LiNi_{0.6}Co_{0.2}Mn_{0.2}O_2$ and $LiNi_{0.8}Co_{0.1}Mn_{0.1}O_2$ have been commercialized, and new materials with even higher Ni content are being developed.

$LiMn_2O_4$ having a spinel structure is charged around 4.0 V through $LiMn_2O_4 \rightarrow 2(\lambda\text{-}MnO_2) + Li^+ + e$. In this spinel lattice, Li^+ diffusion pathway is developed in three-dimensional fashion, which enables a high-rate charge/discharge. Also, thermal stability is superior to other oxides; the temperature at which O_2 evolves from the lattice oxygen is higher than that for the above-mentioned layered oxides. This manganese oxide, however, shows the following demerits that are associated with the generation of Mn^{3+}, which is frequently observed when the spinel oxide is over-discharged. Note that the oxidation state of Mn in the charged sate (λ-MnO_2) is 4+, but decreases with discharge. The normal discharge reaction is $2(\lambda\text{-}MnO_2) + Li^+ + e \rightarrow LiMn_2O_4$: the oxidation state of Mn decreases down to 3.5+ with the normal discharge. When over-discharged, however, more Li^+ and electrons (e) are injected into the spinel lattice through $LiMn_2O_4 + Li^+ + e \rightarrow Li_2Mn_2O_4$; the oxidation state of Mn is 3+ for $Li_2Mn_2O_4$; the Mn valence can be reduced down to 3+ when over-discharged. Three unfavorable features are expected with the formation of Mn^{3+} in the spinel lattice. First, the Li^+ diffusion pathway is blocked by Li^+ in $Li_2Mn_2O_4$. Second, the cubic lattice converts into tetragonal one by Jahn-Teller distortion, causing a lattice strain. Third, Mn^{2+} is generated through the disproportionation reaction; $2Mn^{3+} \rightarrow Mn^{4+} + Mn^{2+}$. The as-generated Mn^{2+} is easily dissolved into the electrolyte solution. To avoid Mn^{3+} formation, the charge and discharge are performed only in the range of λ-MnO_2 (Mn^{4+}) $\leftrightarrow$ $LiMn_2O_4$ ($Mn^{3.5+}$). As shown in Example 12-1, however, Mn^{3+} can be generated even under the normal discharge condition by some local over-discharges.

Example 12-1

The electrode performance can be deteriorated due to local over-discharge in $LiMn_2O_4$. Explain that the local over-discharge is encountered when $LiMn_2O_4$ particles are not uniformly distributed in the electrode plate. Also explain that the surface region of $LiMn_2O_4$ particles is prone to over-discharge when discharged at a high rate.

Solution

In the electrode manufacturing process, active material particles may be unevenly distributed on the two-dimensional plane of metal foil. If λ-MnO_2 (the charged state of $LiMn_2O_4$) particles are scarcely distributed in a certain part of electrode plane. The λ-MnO_2 particles located in that part are discharged faster than the densely distributed particles. Under this circumstance, the scarcely distributed λ-MnO_2 particles can be in an over-discharge state, while the other particles are still being discharged. Hence, mixing and coating should be strictly managed so that the active material particles are uniformly distributed on the electrode plate.

During discharge, Li^+ in electrolyte solution must reach the λ-MnO_2 particle surface through diffusion in solution and then move to the inside of particles through solid-state diffusion. In general, the solid-state diffusion is much slower than that in the solution; Li^+ reaching the λ-MnO_2 surface can be accumulated on the surface because of the slower diffusion into the bulk. Since Li^+ concentration is larger at the surface region as compared to the bulk of λ-MnO_2, the surface region is over-lithiated (over-discharged) while the normal discharge proceeds at the bulk. The faster the discharge rate, the arrival rate of Li^+ is higher to cause a more Li^+ accumulation on the surface; the over-discharge there is more serious.

Another troublesome of the spinel Mn oxide is the Mn dissolution into electrolyte solution. The Mn dissolution proceeds by the attack of acids that are generated either through the decomposition of electrolyte solution or by the reaction between $LiPF_6$ and H_2O. Note that the former is the most popularly used Li salt for lithium cells and the latter is present in electrolyte solution as an impurity. The dissolution of Mn from $LiMn_2O_4$ means a decrease in the electrode capacity, but the capacity decrease due to Mn dissolution is not substantial because the Mn dissolution itself is not significant. Actually, Mn dissolution causes a greater damage to the negative electrode. Namely, the dissolved Mn^{2+} diffuses toward the negative electrode to be electrodeposited. The electro-deposited metallic Mn not only destroys the SEI layer on the negative electrode (graphite) surface, but also causes a continued electrolyte decomposition and concomitant SEI formation on its (metallic Mn) surface. Note that the electrodeposited metallic Mn particles can convey electrons to their surface where the solvent molecules in the electrolyte are reduced/decomposed. The continued SEI formation leads to an increase of irreversible capacity at the expense of reversible one. Also, cell polarization increases due to an increase of surface film resistance (R_{SEI}).

$LiFePO_4$ (LFP) having an olivine structure is charged/discharged with a flat voltage profile at ~3.5 V. The voltage plateau is caused by the two-phase reaction; $LiFePO_4 \leftrightarrow FePO_4 + Li^+ + e$. $LiFePO_4$ is converted into a completely different phase, $FePO_4$, by charging. It is distinguished from $LiCoO_2$, which maintains a single phase (x = 1.0 to 0.5 in Li_xCoO_2) during the charge; $LiCoO_2 \rightarrow Li_{0.5}CoO_2 + 0.5Li^+ + 0.5e$. As the positive electrode, $LiFePO_4$ has the disadvantage of a low working electrode voltage, but has superior thermal and chemical stability to the other oxide electrodes ($LiCoO_2$, $LiNiO_2$, and $LiMn_2O_4$). The strength of P-O bond in PO_4^{3-} is so strong that O_2 is not easily evolved even when cell temperature rises; the metal-oxygen bond strength in the oxides is weaker to be readily broken to release O_2. $LiFePO_4/FePO_4$ shows a poor electronic conductivity as an electrode, but this problem has been solved by a coating of carbon.

Example 12-2

The electrode reactions in lithium secondary batteries can be roughly divided into four types: (1) intercalation/de-intercalation, (2) alloying/de-alloying, (3) conversion, and (4) dissolution/precipitation. Compare their characteristics.

Solution

(1) ***Intercalation/de-intercalation***; Li^+ and electrons are co-injected into the crystallographic sites while structural frameworks being maintained. For this reaction, the sites for storing Li^+ and electrons should be provided in the crystal lattice of active materials. The layered oxides and graphites have two-dimensionally distributed Li^+ storage sites, and the spinel oxides have three-dimensionally distributed Li^+ storage sites. The graphene layers are the electron storage sites for graphites, while the metal component is the electron storage sites for metal oxides. The active materials involved with intercalation/de-intercalation show a very stable charge/discharge cycling behavior because the lattice structures are intact even after a repeated charge/discharge cycling, but the electrode capacity is limited, which is determined by the smaller one out of Li^+ storage sites and electron storage sites. For example, CoO_2 (the fully charged state of $LiCoO_2$) can take one Li^+ per CoO_2 unit according to the discharge reaction: $CoO_2 + Li^+ + e \rightarrow LiCoO_2$; there is one Li^+ storage site per CoO_2 unit. However, the Li^+ storage sites are not fully utilized; only one-half is used in the practical charge/discharge: $Li_{0.5}CoO_2 + 0.5Li^+ + 0.5e \rightarrow LiCoO_2$. This is due to the fact that Co^{3+} in the discharged state, which is the electron storage sites, is oxidized only to $Co^{3.5+}$ to avoid electrode degradation. This manifests itself that the practically usable electron storage sites is 0.5 per CoO_2 unit. Since the electron storage site (one-half for CoO_2 unit) is smaller than the Li^+ storage site (one for CoO_2 unit), the electrode capacity is decided by the smaller one (electron storage site).

(2) ***Alloying/de-alloying***; Si and Sn can make alloy phases with lithium. They give a very large theoretical capacity. Their practical use is, however, limited due to massive volume change upon charge and discharge.

(3) ***Conversion reaction***; some metal oxides are converted into elemental metals by a reaction with lithium; $SnO_2 + 4Li^+ + 4e \rightarrow Sn + 2Li_2O$. After the reaction, nano-sized Sn particles are dispersed within nano-sized Li_2O matrix. The rate for reverse reaction is fast enough due to large contact area between two nano-sized reactants; metal and Li_2O. Note that when Sn and Li_2O, both of which are micron-sized powder, are reacted in laboratory, the rate is extremely slow due

to insufficient contact between two particles. They show a large theoretical capacity (710 mAh/g for SnO_2), but suffer from a large volume change and a serious voltage hysteresis. Since the metal-oxygen bonds should be broken and formed, the oxides of late-transition metals, which have a relatively weaker metal-oxygen bonds, are lithiated by the conversion reaction. Meanwhile, the oxides of early-transition metals with stronger metal-oxygen bonds react more often with lithium by intercalation/de-intercalation rather than the conversion reaction.

(4) ***Dissolution/precipitation***; lithium metal is charged/discharged according to dissolution/precipitation when used as the negative electrode. The cycling property is poor due to undesired shape changes.

Fig. 12-16 presents the manufacturing process of a cylindrical lithium secondary battery. For the positive electrode plate, a slurry is first prepared by dispersing the mixture of $LiCoO_2$ powder, carbon powder, and poly(vinylidene fluoride) binder in a solvent (*N*-methyl pyrrolidone). The resulting slurry is coated on aluminum foil. For the negative electrode plate, the mixture of graphite powder and polymer binder is dispersed in the solvent to prepare a slurry, and then the slurry is coated on copper foil. Thereafter, the electrode plate is dried and cut to a desired size. A separator film is placed between

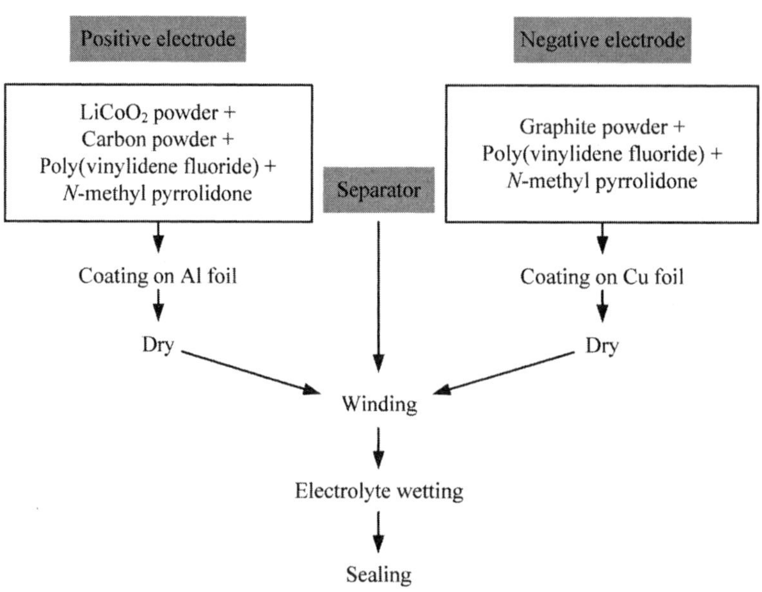

Fig. 12-16 Manufacturing process of a cylindrical-type lithium-ion battery.

the positive and negative plates, and the three layers are wound and inserted into a can. After that, the electrolyte solution is introduced into the can to fill the pores of separator and two electrode layers, which is followed by sealing the cells. Up to this stage, both electrode materials remain in each discharged state. Now, the cells are charged/discharged several times before being come into the market (this is called a formation). Fig. 12-17 shows the structure of a cylindrical lithium secondary battery. This is also called a ***jelly-roll-type cell***. One of this type is 18650 cell, which is named because the cylindrical cell has a diameter of 18 mm and a height of 65 mm.

In the jelly-roll-type cell, the volume occupied by the separator is larger compared to the bobbin-type ones since the separator film is inserted between two electrode plates. Therefore, the active material loading into the cell is smaller than that for the bobbin-type one; the cell capacity is also smaller. However, since the electrode area is larger and the electrode gap is narrower for the jelly-roll cell, both $R_{separator}$ and $R_{solution}$ are smaller; the cell polarization is smaller for the jelly-roll cell. As a result, both the discharge voltage and discharged capacity do not decrease significantly even at a large discharge current. Also, the charging voltage is not so high even under a high-current charging. In short, the jelly-roll cell can be charged/discharged at high currents. In opposite, the bobbin cells

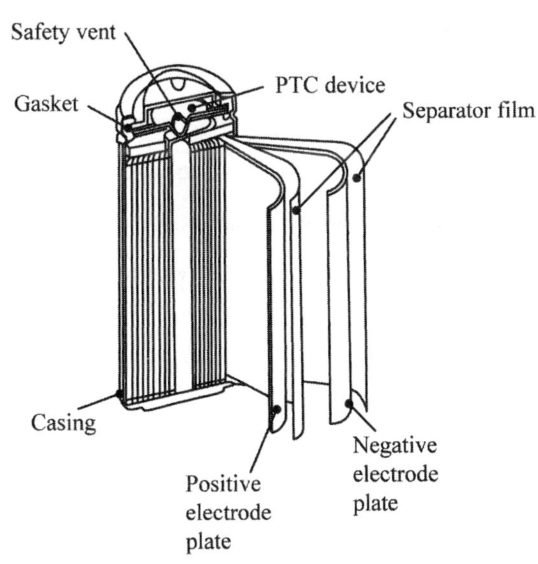

Fig. 12–17 Structure of a jelly–roll–type cylindrical lithium–ion battery.

show a higher cell capacity, but their power characteristics are poorer; the allowed charge/discharge current is limited due to a relatively large cell polarization.

In order to be used in lithium secondary batteries, organic solvents must be stable against decomposition under both strong reducing condition (negative electrode) and strong oxidizing condition (positive electrode). Unfortunately, organic solvents carrying these properties are very rare. Only some organic linear carbonates and cyclic carbonates meet this condition. Table 3-4 lists the organic carbonates that are used in lithium-ion batteries. Other demanding properties of organic solvents include a low freezing point, high boiling point, high dielectric constant, and low viscosity. Since any solvents can not satisfy all these demands, it is more common to use a mixed solvent. It should be noted that PC (propylene carbonate) can not be used for graphite negative electrode since it is inserted into the graphite interlayers to break the layered structure (this is called an exfoliation). EC (ethylene carbonate) is free from exfoliation, but its viscosity is quite high. Hence, EC is used for graphite negative electrode but as a mixture with a less-viscous linear carbonates (DMC or DEC shown in Table 3-4).

A separator is used in lithium secondary batteries to prevent the electrical short between two electrodes and to carry the electrolyte solution inside its pores to impart ionic conductivity. If it is too thick, a large $R_{separator}$ causes a substantial cell polarization. If it is too thin, poor mechanical strength causes a tear and eventual internal short circuit. The inexpensive non-woven fabrics are used in lead-acid batteries, but they are not employed in lithium secondary batteries because their pore size is large; the needle-shaped dendritic lithium metal that may be generated in some conditions can penetrate the large-sized pores to cause an internal short circuit. A porous polyethylene (PE) or PP (polypropylene) film is widely used, and a multilayer film composed of PP/PE/PP is also used in some cells.

Self learning 12-9

The porous polyethylene (PE) film is used as a separator for lithium-ion batteries. The melting point of PE film is 135 °C. If the cell temperature approaches this due to heat generation in some abuse conditions, the PE film melts and the pores are clogged to destroy the ion conduction channels; $R_{separator}$ becomes extremely high due to a very poor ion conductivity. Using the closed-loop concept, explain that the current flow in the cell is shut down under this circumstance to prevent additional safety accidents. Here, it is also possible for two electrodes to be short-circuited since the PE film intervening between two electrodes shrinks after being melted. Propose how to prevent the short-circuit even after the melt-down of PE film. For a reference, the melting point of porous polypropylene (PP) film is 210 °C.

In a lithium polymer battery, lithium metal is used as the negative electrode. As a separator, some polar polymers (e.g. PEO, polyethylene oxide) are used, in which lithium salt (e.g. $LiPF_6$) is dissolved. Here, the polymer (PEO) is a solid. The lithium salt ($LiPF_6$) is dissociated to generate Li^+ and PF_6^- that are coordinated by the oxygen atoms in PEO: the ions are solvated by the oxygen moiety of PEO. Further, the ions can migrate within an electric field. Hence, the PEO/$LiPF_6$ is called a solid polymer electrolyte (SPE) as it shows an ionic conductivity as a solid polymer. In lithium-ion batteries, an organic solvent (e.g. EC) is responsible for dissolving lithium salts. The polymer membrane (e.g. porous polyethylene film) separator serves as a support for soaking an electrolyte solution and as a physical barrier to prevent electric short between two electrodes. Somewhat differently, however, the polymer (PEO) in the solid polymer electrolyte plays two roles; the physical barrier and dissolving lithium salts. Obviously, the former is true as the separator is a solid. The latter is possible because the polymer carries a polar functional group. The ionic conductivity of solid polymer electrolytes is, however, not high because ion conduction is carried by the segmental motions of polymer chains. To facilitate the segmental motions, some organic solvents serving as a plasticizer are added. The addition of plasticizer solvents leads to two features: an increase of ionic conductivity and a gelation of solid polymer electrolyte. The gel-shaped combination (PEO + Li salts + plasticizer) is thus called as a gel polymer electrolyte (GPE). Practically, the GPE can not be used alone as a separator in lithium secondary cells because of its poor mechanical strength; it is used together with a robuster polymer film (e.g. porous PE). The resulting film (GPE + porous PE) is employed as a separator in a lithium-ion polymer battery, which is fabricated as a pouch-type cell. This cell is known to be safer than lithium-ion cells because of the absence of liquid electrolyte and no leakage problem.

The active materials in lithium secondary cells (e.g. graphites and $LiCoO_2$) are solid powder and so as the carbons used to decrease the electrode resistance. These solid powders should be dispersed onto a metal foil current collector. One way to make a strong binding between the powders and metal foil is to use a polymeric binder. In practice, poly(vinylidene fluoride) (PVdF) is widely used because this polymer satisfies the performance requirements for the binder: good retention of powders, strong adhesion to metal foils, electrochemical stability over a wide potential range, low swelling by solvents, high solubility in solvents, and easy evaporation. In the electrode preparation step, the binder (PVdF) is dispersed in a solvent (N-methyl pyrrolidone, NMP) along with the solid components (active materials and carbon). The polymer binder is dissolved into the solvent while the solid components are intact. This slurry is then applied on the current collector. After that, the solvent (NMP) is removed by evaporation, during which the binder (PVdF) returns back to the solid form and is dispersed within the electrode layer to act as a binder. The unfavorable feature of NMP is the toxicity and high boiling point (81~82°C at 10 mm Hg), which requires a large-scale facility and cost to evaporate and recover/regenerate it. When water (H_2O) is used as a solvent, these problems are alleviated. In fact, SBR (styrene butadiene rubber) binder is used along with water solvent in the manufacturing process of the negative electrode plates.

The metal oxide positive electrode materials have poor electrical conductivity, such that the contribution of electrode resistance ($R_{electrode}$) to cell polarization can not be ignored. As a way to reduce the electrode resistance, carbons with high electrical conductivity are added into the electrode layers. When the conductive carbon loading is increased in the electrode preparation step, the electrode resistance shows a sharp decrease at some point but remains without further decrease even with additional carbon loading (this is

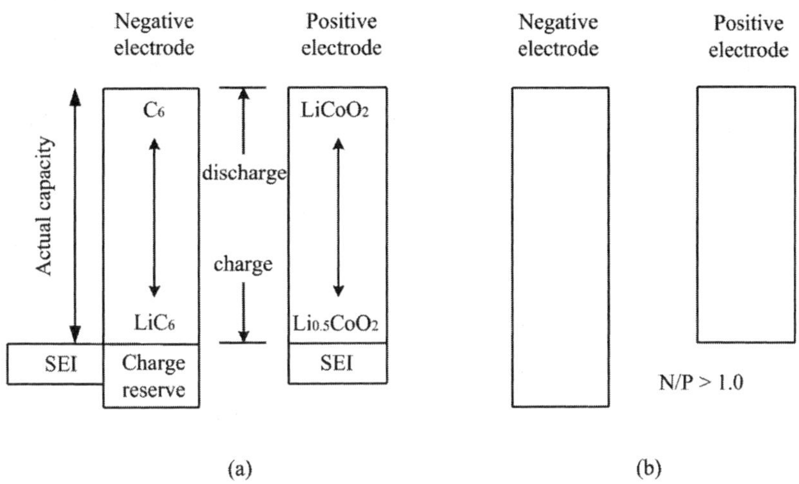

Fig. 12-18 Balance of electrode capacities (cell balance) in a lithium-ion battery.

called a percolation curve). Therefore, the carbon loading is made to be slightly larger than this critical point. The conductive carbons can be broadly classified into two types; carbon blacks and graphites. The former is composed of nanometer-sized carbon particles, and when it is loaded in electrode plates, it intervenes between the active material particles and improves the conductivity. The graphite-type ones are plate-shaped with a larger size, such that they are spread in a wide range within the electrode plate. Recently, carbon nanotubes are utilized as an alternative to the graphite-type conducting agent.

Fig. 12-18 illustrates the ***cell balance*** in a lithium secondary battery composed of a graphite (C_6) and $LiCoO_2$ electrode. With the graphite negative electrode, the formation of SEI and thus irreversible capacity encountered in the earlier charging period is unavoidable. The SEI formation also prevails with lithium metal negative electrode. However, the SEI formation process differs for the two negative electrodes. In the case of lithium metal, SEI forms by the reductive decomposition of electrolyte when lithium metal and electrolyte (organic solvent) come into contact. It is contrasted by the graphite negative electrode, in which the electrolyte decomposition and concomitant SEI formation does not take place upon a simple contact, but when the negative electrode potential is lowered (Fig. 12-14) in the charging process. For the SEI formation, Li^+ and electrons are required: EC (ethylene carbonate) + Li^+ + e → Li_2CO_3(s) + oligomers/polymers. For lithium metal negative electrode, Li^+ and electrons required for SEI formation are provided by the Li metal itself: $Li \rightarrow Li^+ + e$. Note that Li^+ and electrons (e) are readily generated because Li metal has a very high oxidizing tendency. However, the graphite can not provides Li^+ and electrons by itself. Li^+ is supplied from the electrolyte solution, but the only way to supply electrons is from the positive electrode. The positive electrode produces electrons and Li^+ (e.g. $LiCoO_2 \rightarrow Li_{1-x}CoO_2 + xLi^+ + xe$) in the charging period of graphite (C_6)/$LiCoO_2$ cell, and the as-generated electrons are provided to the graphite to enable SEI formation. Here, Li^+ and electrons consumed for SEI formation can not be restored for the discharge of the positive electrode ($Li_{1-x}CoO_2 + xLi^+ + xe \rightarrow LiCoO_2$). That is, a part of positive electrode capacity is irreversibly consumed for SEI formation. The actual capacity corresponds to the area indicated by the rectangle of positive electrode in Fig. 12-18-b.

In the practical lithium secondary cells (e.g. graphite/$LiCoO_2$), the negative electrode capacity is adjusted to be larger than that of positive one; ***N/P ratio*** > 1.0. That is, an excess amount of graphite is loaded in the negative electrode as a ***charge reserve***. The reason for this is to remove the risk factors that may be encountered upon overcharging. Namely, when the graphite electrode is overcharged, the electrode potential approaches 0.0 V and lithium is electrodeposited, which leads to a risk of an internal short circuit due to the dendritic growth of Li metal. When the $LiCoO_2$ positive electrode is overcharged, however, the potential of positive electrode goes up to 4.3 V or more, which may lead to a phase change, O_2 evolution, and oxidative decomposition of electrolyte. Here, the risk factors encountered at the negative electrode are more serious than those on the positive one, such that the prevention of overcharge at the negative electrode is more concerned. When the negative electrode capacity is larger than the positive one (with the charge reserve) as shown in Fig. 12-18-b, in some overcharging condition, the negative electrode is still under normal charging while the positive electrode being overcharged: the charging reaction ($C_6 + Li^+ + e \rightarrow LiC_6$) still continues in the negative electrode since there is an excess amount of graphite (C_6) as a charge reserve; lithium electrodeposition ($Li^+ + e \rightarrow Li$) can be avoided.

Self learning 12-10

Even if a lithium-ion battery is designed with an N/P ratio of > 1.0, if graphite particles are not evenly distributed on an electrode plate because the mixing, slurry preparation, and coating are not carefully controlled in the electrode manufacturing process, the lithium metal deposition is highly probable at the point where the graphite particles are scarcely populated. Explain this.

Since lithium-ion cells are manufactured with an N/P ratio of > 1.0 as mentioned above, there is a possibility of overcharging on the positive electrode, which leads to a failure of the positive electrodes. To prevent this, some ***overcharge protection additives*** for the positive electrode are added into the electrolyte solution. A typical example is 2,4-di-fluoroanisole (DFA), which is soluble in a typical electrolyte solution and is oxidized at near 4.3 V. When the positive electrode ($LiCoO_2$) starts to be overcharged ($Li_{0.5}CoO_2 \rightarrow Li_{0.5-x}CoO_2 + xLi^+ + xe$), the positive electrode potential increases to approach 4.3 V. At this time, since DFA is oxidized at 4.3 V, the overcharge of $Li_{0.5}CoO_2$ can be prevented. That is, instead of overcharging (an oxidation reaction) of the positive electrode, DFA oxidation proceeds. Then, the positive electrode potential is fixed (pinned) at 4.3 V; no more overcharging thereafter. Furthermore, the oxidation product of DFA diffuses to the negative electrode and is reduced back to DFA, which diffuses back to the positive electrode to play the overcharge protection role.

Nowadays, a variety of electrolyte additives has been utilized in lithium secondary cells. A small amount of additives is added into electrolyte solution with a specific purpose; they are called ***functional electrolyte additives***. For example, the above-mentioned DFA is added to protect the overcharge of the positive electrode. Other examples are; the SEI former on negative electrode, the inhibitor for metal dissolution from positive electrodes, the fire retardant, and the wetting ability improver for electrode plates. The role of ***SEI former*** on negative electrode is to generate a thermally/chemically stable SEI layer on negative electrode. EC (ethylene carbonate) in the commonly used electrolyte solution is decomposed to generate an SEI layer on graphite negative electrode. The EC-derived

SEI is, however, thermally unstable and readily dissolved or decomposed when cell temperature increases. The result is the exposition of graphite surface to the electrolyte solution, which causes an additional decomposition of EC and SEI formation. At least, two unfavorable impacts are expected from this. First, the cell capacity decreases because the electric charges are consumed to repair the damaged SEI layer; the as-consumed electric charges are not recovered. Second, the SEI layer thickens, which leads to an increase of R_{SEI}, and eventually an increase of cell polarization. Vinylene carbonate (VC) undergoes a reductive decomposition at a higher potential than EC to form SEI. The VC-derived SEI is thermally more stable than the EC-derived one. When a small amount of VC is added into the electrolyte solution containing EC, VC is decomposed to produce an SEI layer prior to EC in the first charging period. Note that the negative electrode potential decreases from ca. 3.0 V to 0.1 V (*vs.* Li/Li^+) during the charging. When the VC-derived SEI layer has a sufficient thickness, electron tunneling rate becomes negligibly small. Then, the reductive decomposition of EC, which is thermodynamically allowed when the electrode potential further decreases in the continued charging, is negligible due to the negligible tunneling rate. In short, the graphite surface is covered only by the VC-derived SEI layer, which maintains the SEI function without dissolution or decomposition even when cell temperature increases.

When a lithium-ion battery is exposed to high temperature or the cell temperature rises to about 100 °C or higher in some abuse conditions (internal short circuit, external short circuit, and overcharging, and etc), the SEI layer is damaged and then exothermic reactions such as electrolyte decomposition are followed. When the cell temperature rises to above 200 °C, O_2 can be evolved from the positive electrode. Then, the combustible ingredients (organic solvent, separator, carbon, etc.) in the cell are burned by the evolved O_2, which may lead to an explosion. This is called ***thermal runaway***. In order to prevent this disaster, safety must be taken into consideration in the material and cell design. For example, the surface area/volume ratio of batteries should be increased to facilitate heat dissipation. In addition, if necessary, a cooling device should be added.

Lithium secondary batteries were first commercialized in 1992, and their uses are expanding as a power source for mobile phones, notebook computers, electric vehicles, and energy storage systems (ESS). To explore additional markets, the safety must be secured, and a continuous improvement in energy, power, charging rate, and price is demanding.

Self learning 12-11

The jelly-roll-type cell shown in Fig. 12-17 is manufactured by inserting a separator film between two electrode plates. Assuming that the wound positive plate/separator/negative plate is unwound, check that the electrode area facing to each other is much larger than that of the bobbin-type cell shown in Fig. 12-3. In general, the cells working in aqueous electrolytes are manufactured as a bobbin-type, whereas those working in non-aqueous electrolytes (lithium primary and lithium secondary cells) are mostly manufactured as a jelly-roll type. Note that the ion conductivity is poorer for the non-aqueous electrolytes. Using <12-3>, explain why lithium-ion cells are fabricated as a jelly-roll type. Confirm that the bobbin-type cells give a larger cell capacity compared to the jelly-roll one because a larger amount of active electrode material can be loaded due to the smaller volume occupied by the separator.

12-1-10 Sodium-sulfur battery

The electrode material of ***Na-S battery*** is the metallic Na and sulfur (S); both electrode materials are liquid at the working temperature of 300 ~ 400 °C (Fig. 12-19).

As the electrolyte, β-alumina having Na^+ conductivity is used; to ensure high Na^+ conductivity, the cell works at 300 ~ 400 °C. The discharge reaction is $2Na(l) + 3S(l) \rightarrow Na_2S_3(l)$, and $E^0_{cell} = 2.08$ V at 350 °C. The theoretical specific energy is 755 Wh/kg, which is much larger than that of lead-acid batteries. Initially, it was developed as a power source for automobiles, but recently it has been developed as a high-capacity energy storage device by taking advantage of its high specific energy.

12-1-11 Redox-flow battery

In ***redox-flow battery***, the redox couples in their charged state that are dissolved in electro-

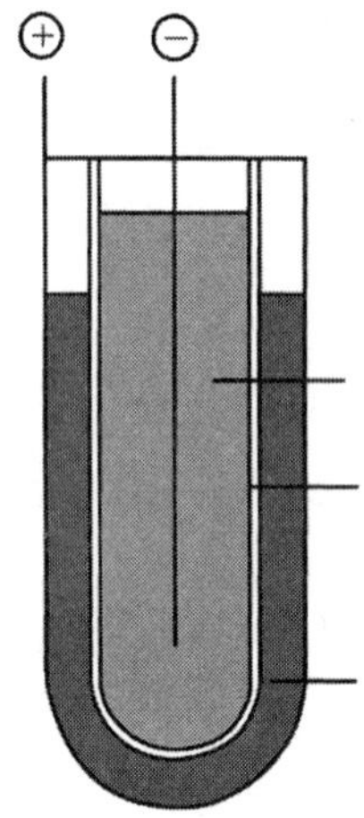

Fig. 12-19 Structure of sodium-sulfur (Na-S) secondary battery.

lyte solution are pumped into the cell stack and discharged on inert electrodes (Fig. 12-20). For charging, electric energy is supplied to proceed the opposite reactions. For example, upon discharge in a vanadium redox-flow cell, V^{2+} dissolved in sulfuric acid is oxidized to V^{3+}, while VO_2^+ in sulfuric acid is reduced to VO^{2+}. In the charging process, V^{3+} is reduced back to V^{2+} and VO^{2+} is oxidized to VO_2^+. The H^+ exchange membrane is used as the separator, and the operating voltage is about 1.26 V at room temperature.

Unlike other batteries, redox-flow cells have an unique feature that can control the energy (capacity) and power (current) separately; it is called a decoupling of energy and

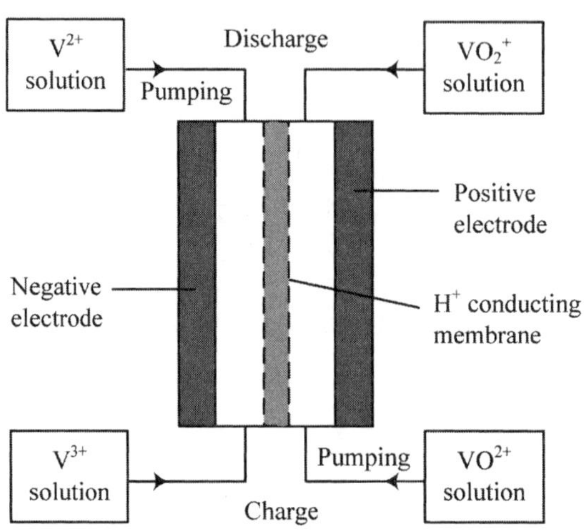

Fig. 12-20 Configuration of vanadium redox-flow battery.

power. In other cells including lithium-ion, when the power (the discharge current) is increased, the discharged capacity (energy) decreases inevitably (e.g. lead-acid cell, Fig. 12-5 and SL 12-3); the energy and power are coupled. In redox-flow cells, the capacity (energy) is determined by the tank size storing the redox couples, and the power (current) is determined by the size of cell stack (total area of inert electrode). The two variables (tank size and stack size) can be independently controlled.

When using an aqueous electrolyte, the working voltage is theoretically < 1.23 V due to water decomposition (H_2 and O_2 evolution); practically, however, it is larger than 1.23 V because the inert electrodes used show a very small standard rate constant (k^0) for H_2 and O_2 generation reaction. In aqueous electrolytes, the solubility of common redox pairs is not sufficient, which limits the energy (capacity) of the cells. These drawbacks may be mitigated by using non-aqueous electrolytes. Since the electrochemical stability window is wider for non-aqueous electrolytes, the working voltage can be enlarged up to > 2.0 V. Redox pairs that are highly soluble in non-aqueous electrolytes have also been developed.

12-2 Supercapacitors

Supercapacitors are divided into two types: ***electric double-layer capacitors*** (EDLCs) and ***pseudo-capacitors***. The former utilizes ideally polarizable electrodes to store charges on electric double-layer. As the electric double-layer forms on electrode surface, the electrode materials of a large surface area with high electronic conductivity are employed; the typical example is activated carbons.

Pseudo-capacitors store electric charges through a faradaic reaction. When electrode potential changes, two kinds of currents flow; faradaic and non-faradaic. The former involves the electrochemical reaction on electrode (faradaic reactions), whereas the latter is associated with electric double-layer charging. In a linear sweep voltammogram, the electric double-layer charging current shows a linear relation with scan rate ($i_C \propto v^1$) as shown in <8-12>. The same relation ($i_p \propto v^1$) is observed when electroactive materials are adsorbed or immobilized on electrode surface, and participating in faradaic reactions as shown in <8-37>. The electrodes with adsorbed or immobilized electroactive species are categorized into pseudo-capacitors because they show the capacitor-like behavior ($i_p \propto v^1$); "pseudo" comes from the fact that they are not capacitors intrinsically, but appear to be a capacitor. Meanwhile, when faradaic reactions take place only near electrode surface, the current shows $i_p \propto v^1$ behavior because mass transfer is not needed for the faradaic reaction. This can be put into the category of pseudo-capacitors because of the $i_p \propto v^1$ behavior. The typical example is small-sized RuO_2, in which the faradaic reaction is $RuO_2 + xH^+ + xe = H_xRuO_2$. Here, the diffusion rate of H^+ in solution is high, such that H^+ concentration near electrode surface is high enough; it appears that the mass transfer of H^+ from the bulk solution is not needed. Furthermore, H^+ diffusion length into the bulk of RuO_2 particles is very short since the particle size is small. As a result, the mass transfer of H^+ in both H^+-containing solution and bulk of RuO_2 is high enough; the faradaic reaction ($RuO_2 + xH^+ + xe = H_xRuO_2$) is limited by the charge transfer step just like the electroactive species that are adsorbed or immobilized on electrode surface.

Fig. 12-21 illustrates the charge storage mechanism in an EDLC that is composed of

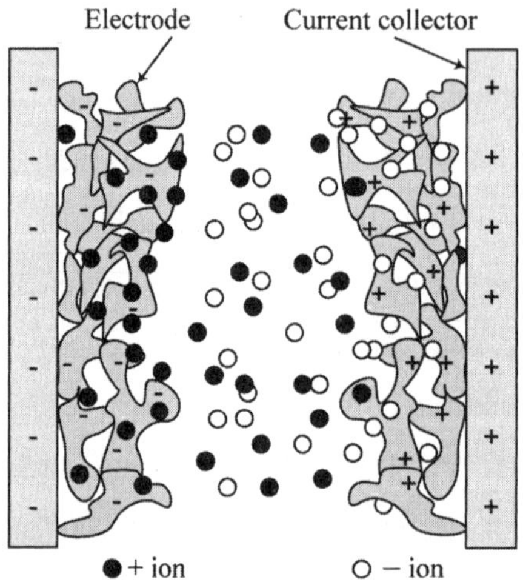

Fig. 12-21 Configuration of a symmetrical electric double-layer capacitor (EDLC).

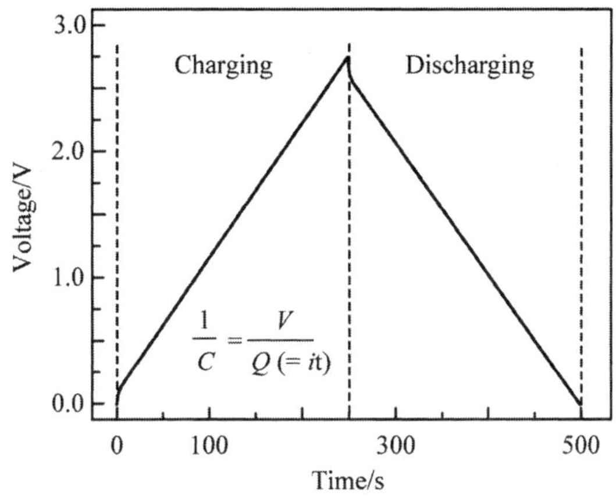

Fig. 12-22 The constant-current charge/discharge voltage profile of a symmetric EDLC.

two identical electrodes (called a symmetric EDLC); one for negative and the other for positive electrode. When charged, the negative electrode is polarized to have negative charges; positive charges on the positive electrode. Then, positive ions in the electrolyte solution move to the negative electrode to form an electric double-layer there, and negative ions to the positive electrode to form an electric double-layer. The stored charges are discharged through the opposite path. Meanwhile, in the charging process, both positive and negative ions are depleted in the solution; this principle can be utilized to remove ions from solutions, thus called a ***capacitive de-ionization***.

Fig. 12-22 shows a charge/discharge voltage profile of an EDLC obtained under a constant-current condition. Since it is a symmetric cell, the cell voltage is 0.0 V when fully discharged. Upon charging, the cell voltage increases. Here, since $V = (1/C) \times Q = (1/C) \times (it)$ (C is the cell capacitance), there appears a linear relationship between the charging voltage and time. The same is true for discharging. The cell capacitance can be obtained from the inverse of the slope.

Self learning 12-12

Sketch the equivalent circuit for a symmetrical EDLC and explain why the discharge voltage drops vertically at the beginning of discharge and then linearly decreases as is seen in Fig. 12-22.

Self learning 12-13

Prove that the stored electrical energy in a symmetric EDLC is given by $1/4CV^2$, where C is the capacitance of the half-cell and V is the cell voltage.

Fig. 12-23 compares the specific energy and specific power of electrical energy storage or generation devices; this plot is called a ***Ragone plot***. As is seen, supercapacitors have a higher power than secondary batteries or fuel cells, but energy is lower. The high specific power for supercapacitors is available because they can be charged/discharged with a high current density. This is because electric charges are stored at the electric double-layer without any faradaic reactions (EDLC), or the faradaic reaction proceeds only on electrode surface (pseudo-capacitor). For charge/discharge of batteries, ion diffusion into the bulk of electrode materials is required, which is such a slow process that high-current charge/discharge is not allowed; the power characteristics of batteries are inferior to supercapacitors. Due to the absence of bulk ion-diffusion, the deterioration of supercapacitor electrodes upon charge/discharge is not so severe, which is advantageous in that it can be charged/discharged semi-permanently. However, since electric charges are stored only on electrode surface, the specific energy is much smaller due to the smaller storage capacity. On the

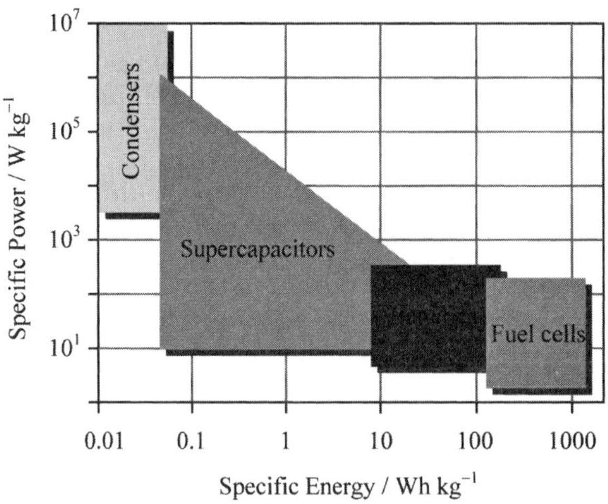

Fig. 12-23 Ragone plot comparing the specific power and specific energy of condensers, supercapacitors, batteries, and fuel cells.

contrary, since faradaic reactions proceed inside the bulk of electrode materials in batteries, the electric storage capacity is much larger to give a much larger specific energy.

12-3 Fuel cells

Thermal power plants burn fuels (petroleum, coal or natural gas) to obtain heat, and use the heat to turn turbines to obtain electricity. Fuel cells are similar to thermal power plants in that they use fuel (hydrogen, carbon monoxide, or hydrogen-containing compounds) to obtain electricity. However, fuel cells have a higher power generation efficiency in theory because they do not require heat conversion or turbine rotation. Moreover, there is no emission of NO_x and SO_x that are generated during fuel burning, and there is no noise generated from the turbine. In addition, CO_2 emission due to the use of fossil fuels can be eliminated when hydrogen is used as a fuel, which is produced by water electrolysis using renewable energy such as solar power or wind power generation.

Table 12-2 compares the performance characteristics of fuel cells, which are divided into several types according to the used electrolyte. The direct methanol fuel cell (DMFC) is similar to the polymer electrolyte membrane fuel cell (PEMFC), but differs because methanol instead of hydrogen is used as the fuel. Table 12-3 lists the oxidation and reduction reactions involved in the fuel cells using hydrogen as a fuel.

Alkaline fuel cells (AFC) use 35 ~ 50% KOH as the electrolyte. It is necessary to purify hydrogen and oxygen because $KHCO_3(s)$ is precipitated; $KOH + CO_2 \rightarrow KHCO_3(s)$; CO_2 impurity should be removed from H_2 and air (O_2). The precipitate with low solubility deposits on the membrane to inhibit ion conduction. There are examples already used as a power source for space shuttles.

Phosphoric acid fuel cells (PAFCs) use phosphoric acid as the electrolyte, which is thermally stable up to 150 ~ 220 °C. The cell operates at this temperature. There exist two merits by working at this temperature. First, phosphoric acid shows a high enough

Table 12-2 Performance characteristics of fuel cells

	AFC	PAFC	MCFC	SOFC	PCFC	PEMFC	AEMFC	DMFC
Electrolyte	KOH	H_3PO_4	Li_2CO_3/K_2CO_3 Li_2CO_3/Na_2CO_3	Y_2O_3/ZrO_2	$Ba(Zr,Ce)O_3$, CsH_2PO_4	H^+-exchange polymer membrane (PEM)	OH^--exchange polymer membrane (AEM)	PEM
Ion carrier	OH^-	H^+	CO_3^{2-}	O^{2-}	H^+	H^+	OH^-	H^+
Fuel electrode	Pt/Pd	Pt/carbon	Ni/Al	Ni/ZrO_2	$Ni/Ba(Zr,Ce)O_3$	Pt/carbon	Pt or PtRu//carbon	Pt-Ru/ carbon
Air electrode	Au/Pt	Pt alloy/ carbon	Li-doped NiO	Sr-doped $LaMnO_3$	(Pr,Ba,Sr) $(Co,Fe)O_5$	Pt/carbon	Non-precious catalysts	Pt/carbon
Working Temp.(°C)	<100	150-200	600-700	500-1000	300-500	<120	<100	<100
Fuel	H_2	H_2	H_2, CO	H_2, CO	H_2	H_2	H_2	Methanol
Efficiency (%)	60	40	50	60	60	60	60	30
Stack Power (kW)	1~100	5~400	300~3,000	1~2,000	N.A.	1~100	1~100	1~100
Application	Space shuttle	Distributed generation	Large scale generation	Auxiliary power, Distributed power generation	Auxiliary power, Distributed power generation	Transportation Distributed power generation, backup power	Distributed power generation, backup power	Mobile devices

AFC; Alkaline fuel cell, PAFC; Phosphoric acid fuel cell, MCFC; Molten carbonate fuel cell
SOFC; Solid oxide fuel cell, PCFC: Protonic ceramic fuel cell, PEMFC; Polymer electrolyte membrane fuel cell or proton-exchange membrane fuel cell, AEMFC: Alkaline membrane fuel cell, DMFC; Direct methanol fuel cell

Table 12-3 Electrode reactions of fuel cells that use hydrogen as a fuel

Fuel cell	Anodic reaction	Cathodic reaction
AFC	$H_2 + 2OH^- \rightarrow 2H_2O + 2e$	$1/2O_2 + H_2O + 2e \rightarrow 2OH^-$
AEMFC	$H_2 + 2OH^- \rightarrow 2H_2O + 2e$	$1/2O_2 + H_2O + 2e \rightarrow 2OH^-$
PEMFC	$H_2 \rightarrow 2H^+ + 2e$	$1/2O_2 + 2H^+ + 2e \rightarrow H_2O$
PAFC	$H_2 \rightarrow 2H^+ + 2e$	$1/2O_2 + 2H^+ + 2e \rightarrow H_2O$
PCFC	$H_2 \rightarrow 2H^+ + 2e$	$1/2O_2 + 2H^+ + 2e \rightarrow H_2O$
MCFC	$H_2 + CO_3^{2-} \rightarrow H_2O + CO_2 + 2e$	$1/2O_2 + CO_2 + 2e \rightarrow CO_3^{2-}$
SOFC	$H_2 + O^{2-} \rightarrow H_2O + 2e$	$1/2O_2 + 2e \rightarrow O^{2-}$

ionic conductivity at this temperature. Second, the catalyst shows a stability against poisoning at this temperature, such that low-purity hydrogen can be used as a fuel. PAFCs are evaluated to be suitable for the distributed power generation.

Molten carbonate fuel cells (MCFCs) use an eutectic mixture of 62 mol% Li_2CO_3/38 mol% K_2CO_3 or 52 mol% Li_2CO_3/48 mol% Na_2CO_3 as the electrolyte, both of which melt at above 500 °C to be molten salts. The cell operates at 600 ~ 700 °C. Since the operating temperature is high, the direct reforming of fossil fuels (e.g. $CH_4 + H_2O \rightarrow CO + H_2 + CO_2$) is possible, and thus it is evaluated as suitable for large-scale power generation or cogeneration using waste heat. In the direct reforming, natural gas is introduced into the cell to obtain hydrogen or carbon monoxide, which are then utilized as a fuel. The energy required for the steam reforming (e.g. $CH_4 + H_2O \rightarrow CO + H_2 + CO_2$) is supplied by the heat generated by the fuel cell operation. Note that heat evolves during the operation of galvanic cells due to thermodynamic and kinetic origins as described in 12-1-7. For MCFCs, since heat is utilized for the steam reforming, the stack can be effectively cooled to allow a stable operation for a prolonged period. Some Ni-based electrodes are used without CO or CO_2 poisoning problem. The power density is relatively low (~ 0.1 W/cm^2). When operated for more than 30,000 h, the loss of molten salts is appreciable due to corrosion and evaporation, which is the most critical problem to be solved.

Solid oxide fuel cells (SOFCs) use the metal oxide (yttria-stabilized zirconia, YSZ, Y_2O_3/ZrO_2) as an electrolyte, which shows an oxide ion (O^{2-}) conductivity (Fig. 12-24). This cell operates at 500 ~ 1,000 °C to ensure a high enough O^{2-} conductivity for the solid electrolyte (Y_2O_3/ZrO_2). Due to the high working temperature, fossil fuels can be directly reformed inside of the cells to produce H_2 and CO, which are then used as a

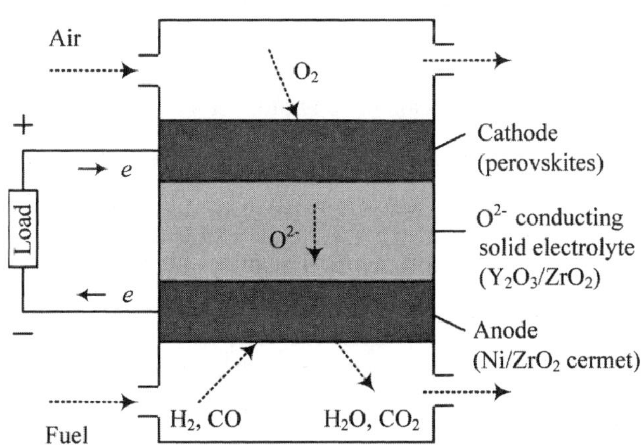

Fig. 12-24 Ion and electron flow in solid-oxide fuel cell (SOFC).

fuel as in MCFCs. Since both the electrodes and electrolyte are metal oxides rather than noble metals, they are inexpensive, but the difficulty in sealing between solid components and the material degradations due to thermal conductivity difference are of problem. In order to solve these problems, many efforts have been made; to develop a thin electrolyte (several to several tens of μm) or an alternative solid electrolyte with high ionic conductivity. To take advantages of high power density (several W/cm^2), high power generation efficiency (> 60%), and the versatility of fuels (methane, gasoline, diesel, etc.), continuous development are in progress.

Self learning 12-14

YSZ (yttria-stabilized zirconia, Y_2O_3/ZrO_2), which is used as the electrolyte for SOFC, is also used as an electrolyte for a potentiometric oxygen sensor that measures the oxygen partial pressure inside an automobile engine. The structure of this sensor is:

$$O_2 \text{ (outside air, 0.21 atm), Pt|YSZ|Pt, } O_2 \text{ (engine)}$$

This sensor measures the potential difference between the Pt electrode in contact with oxygen in engine (O_2, engine) and another Pt electrode exposed to ambient air (oxygen partial pressure = 0.21 atm). The potential-determining equilibrium for both half-cells are $O_2 + 4e = 2O^{2-}$. Derive how the potential difference between two Pt electrodes changes according to O_2 (engine).

Proton-conductive ceramic fuel cells (PCFCs) use a ceramic electrolyte having H^+ conductivity. Since the activation energy for H^+ conduction is smaller than that for oxide ion (O^{2-}) conduction, the PCFCs can be operated at lower temperature (300~500 ℃) than SOFCs, which has the advantage of suppressing high-temperature material degradation and material costs. However, there is a difficulty in the synthesis of proton-conductive ceramics. Moreover, they show a poor chemical stability against water and carbon dioxide.

Polymer electrolyte membrane fuel cells (PEMFCs) use a proton-exchange membrane as an electrolyte. The unit cell is composed of a fuel electrode, a polymer electrolyte, and an air electrode. These units are repeatedly stacked to form a fuel cell. They have been evaluated as suitable for residential power generation or as a power source for fuel

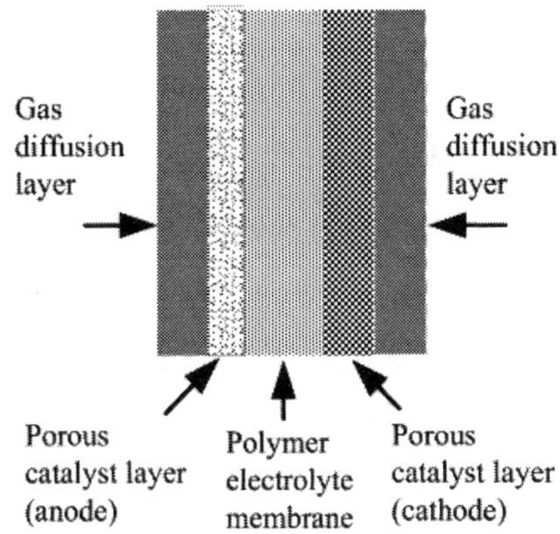

Fig. 12-25 Unit cell structure of PEMFC.

cell vehicles. Nafion is currently the most widely used proton-exchange membrane, whose chemical composition is given in Fig, 11-7. In this polymer, the hydrophilic part ($-SO_3^-$ H^+) and the hydrophobic moiety of F-containing carbon chains are phase-separated. In the hydrophilic part, ions form clusters and are hydrated by water, providing a path for H^+ movement.

The unit cell structure is shown in Fig. 12-25. The fuel electrode and air electrode are located on both sides of a polymer electrolyte membrane, and a gas diffusion layer is attached to the outside of each electrode for gas diffusion. In the porous catalyst layer where the electrochemical reactions take place, Pt-based catalyst particles are dispersed on carbon particles. This structure is called a ***membrane electrode assembly*** (MEA) and the cell performance is greatly influenced by the structure of MEA (Fig. 12-26) as dis-

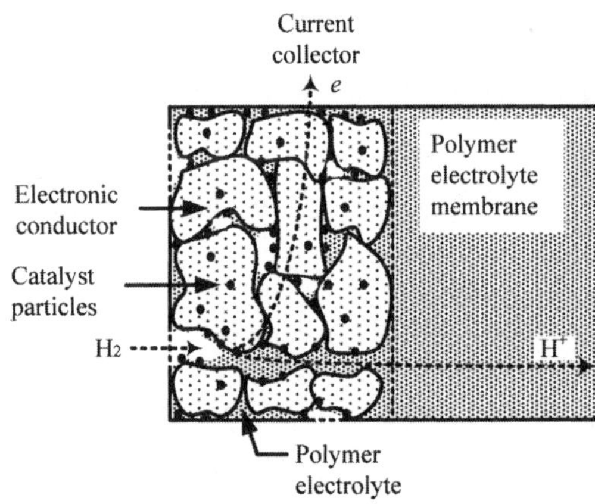

Fig. 12-26 Structure of membrane electrode assembly (MEA) in PEMFC.

cussed below.

H_2 diffuses in through the gas diffusion layer, and $H_2 \rightarrow 2H^+ + 2e$ reaction proceeds in the porous catalyst layer. Here, the generated electrons move toward the current collector through the conductive carbon, whereas the generated H^+ moves toward the air electrode through the polymer electrolyte membrane to form a closed-loop and current flow. Here, in order to transfer H^+ from the catalyst layer to the polymer electrolyte membrane, the catalyst layer should possess the H^+ conductive polymer component (e.g. Nafion); then, H^+ can move through the channels provided by the H^+ conductive polymer from the catalyst layer to the air electrode. Meanwhile, as discussed in 12-1-3, the solubility of gases is very poor in aqueous solution. Moreover, the diffusion rate of gaseous species in solution is about 4-6 orders of magnitude smaller than that of gas-phase diffusion. Here, if the catalyst surface is fully wetted by water, which is added to dissociate the acid moiety (e.g. $-SO_3H$ in Nafion) and to hydrate the resulting ions, H_2 must reach the catalyst surface by solution-phase diffusion. Hence, to secure the gas-phase diffusion of H_2, the catalyst layers must provide void spaces. Also, for the movement of generated electrons and H^+, the contact area made by the Pt particles, H^+ conductive polymer and carbon surface must be large within the porous catalyst layer. The contact points are called ***three-phase boundary sites***. Particularly, since the expensive Pt is used, it is necessary to minimize the size of Pt particles, which can maximize the three-phase boundary sites while minimizing Pt loading.

Example 12-3

The importance of three-phase boundary sites in PEMFC has been demonstrated. Similarly, it is necessary to maximize the contact area between the electrolyte and electrode components even in lithium secondary batteries. Explain this in lithium-ion batteries and lithium-polymer batteries.

Solution

In lithium-ion cells, the contact area between the electrolyte solution and electrode material must be large enough for the electrode reaction. For example, for the charging of graphite ($C_6 + Li^+ + e \rightarrow LiC_6$), graphite ($C_6$), Li^+ and e must meet together. The electrons are easily supplied to the reaction sites because graphite particles are highly electron conductive. The supply of Li^+ is also easy because the electrolyte solution containing Li^+ penetrates well into the electrode layer and comes into contact with the surface of graphite particles.

In lithium polymer battery, where the solid polymer electrolyte (e.g. Li salts/PEO) is used as the separator, it is difficult for the solid polymer electrolyte to penetrate into the electrode layer because it is a solid; thereby, it is difficult to secure the contact area between graphite and Li^+ dissolved in the solid polymer electrolyte. As a measure, the electrolyte component (solid polymer electrolyte) is also loaded into the electrode layer as in PEMFC.

A troublesome feature for Pt catalyst in PEMFCs is the poisoning by carbon monoxide; this impurity should be removed from hydrogen fuel. The poisoning is less serious when operated at 120°C or higher. However, this high temperature is not friendly to the conventional electrolyte membrane (e.g. Nafion); the water required to impart H^+ conductivity is evaporated to decrease H^+ conductivity. In order for PEMFCs to operate at 120°C or higher with less serious Pt catalyst poisoning, phosphoric acid is tested as an alternative to water as a conductive medium. Toward this end, a polymer capable of doping phosphoric acid, such as polybenzimidazole, is being tested.

Several causes for degrading PEMFCs are known. Firstly, the fuel electrode catalyst dissolves under the strong oxidizing condition. Secondly, small holes or cracks in the polymer electrolyte membrane can cause hydrogen and oxygen to move to the opposite electrode (it is called a ***gas cross-over***). Thirdly, the desulfonation in Nafion is observed. Fourthly, Nafion is decomposed by peroxides or free radicals generated as a by-product of oxygen reduction reaction.

When the gas cross-over prevails, E_{cell} ($= E^c_{eq} - E^a_{eq}$) decreases due to the mixed-potential effect. Namely, the standard electrode potential (E^0) of hydrogen half-cell and oxygen half-cell is 0.0 V (*vs.* NHE) and 1.23 V, respectively. E_{cell} = 1.23 V under the standard state. If oxygen moves into the fuel electrode from air electrode, oxygen reduction ($O_2 + 4H^+ + 4e \rightarrow 2H_2O$) takes place on the fuel electrode as a pair with hydrogen oxidation ($H_2 \rightarrow 2H^+ + 2e$); the equilibrium potential of hydrogen half-cell (E^a_{eq}) becomes more positive (>0.0 V) by the mixed-potential effect (refer to Fig. 2-11). Similarly, when hydrogen moves toward the oxygen electrode, hydrogen oxidation ($H_2 \rightarrow 2H^+ + 2e$) and oxygen

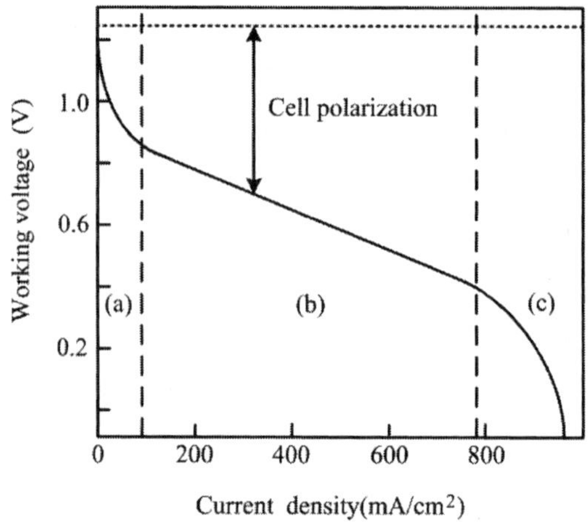

Fig. 12-27 The working voltage of a fuel cell according to current density. The evolution of cell polarization according to current density is also illustrated.

reduction (O_2 + $4H^+$ + $4e$ → $2H_2O$) are paired on the oxygen electrode; the equilibrium potential of oxygen half-cell (E^c_{eq}) becomes more negative (<1.23 V) due to the mixed-potential effect. Both results in a decrease of E_{cell} (= $E^c_{eq} - E^a_{eq}$).

The working voltage of fuel cell (E_{wk}) is decided by two factors; E_{cell} and cell polarization because E_{wk} = E_{cell} − cell polarization (= $\eta_a + \eta_c + iR_{total}$). A possible decrease of E_{cell} with gas cross-over is mentioned above. Three factors determine the size of ***cell polarization***, and the factor contributing the most changes according to current density. This feature is depicted in Fig. 12-27. As is seen, at low current density (a), the activation overpotential for both electrode reactions is dominant over the other two factors in deciding the cell polarization, so the working voltage-current density gives the relation shown in Fig. 6-13-a; current increases by the exponential function of voltage. At medium currents, the contribution of iR_{total} is dominant, such that the working voltage shows a linear relation with the current density as shown in (b) (ohmic behavior). At high current density, the contribution of concentration overpotential that is associated with the mass transfer of hydrogen, oxygen, or generated water is the largest, showing the current-voltage profile (c) (refer to Fig. 6-13-b). Overall, the working voltage decreases with an increase of current density. As a result, the power density (= current density × operating voltage) of fuel cells gradually increases as the current density increases, then decreases after a peak.

Alkaline membrane fuel cells (AEMFCs) are the modified form of the traditional alkaline fuel cells (AFCs) that use an aqueous alkaline electrolyte. In the modified version, the OH^- exchange polymer membrane (AEM) is used as the electrolyte, which shows a high OH^- conductivity under neutral or weakly basic condition. The intrinsic problems of AFCs can be avoided in the modified version: the corrosiveness of the alkaline liquid electrolyte, $KHCO_3$ deposition due to CO_2 impurity in air, sealing and pressure management. In addition, since the activation overpotential for oxygen reduction reaction (ORR) is less significant in the weakly basic condition, the inexpensive metals such as Co, Fe, Ni, Mn, and Ag can be used as the electrode catalysts. This is contrasted by the PEMFCs, which uses the noble metal catalysts (Pt) due to the acidic working condition. Also, the inexpensive hydrocarbon-based polymer electrolyte is used instead of the expensive Nafion. To take advantage of these, studies are undergoing to increase the performance and durability of AEMFCs.

Direct methanol fuel cells (DMFCs) are being evaluated as suitable for powering small portable electronic devices because they use portable and replaceable methanol as a fuel. The electrode reaction is as follows.

Anode: $CH_3OH + H_2O \rightarrow CO_2 + 6H^+ + 6e$ $\quad E^0 = 0.043$ V(*vs.* NHE)

Cathode: $3/2\ O_2 + 6H^+ + 6e \rightarrow 3H_2O$ $\quad E^0 = 1.23$ V(*vs.* NHE)

Overall: $CH_3OH + 3/2\ O_2 \rightarrow CO_2 + 2H_2O$ $\quad E^0_{cell} = 1.18$ V

The standard electromotive force $E^0_{cell} = 1.18$ V, but η_a and η_c in <12-3> are relatively large because six and four electrons are involved in the oxidation and reduction reaction, respectively. Therefore, the working voltage is low when operated under high current. At present, Pt-Ru catalyst is used as the fuel electrode. Meanwhile, when methanol moves by cross-over to the cathode, the cathode can be poisoned; an methanol-tolerant air electrode has been developed. Meanwhile, carbon monoxide and formaldehyde are generated on the anode as byproducts. Higher operating temperatures will alleviate this problem; a polymer electrolyte capable of working at > 100 °C is needed. Also, the anodes tolerant to carbon monoxide have been developed.

Problems

01 The standard electrode potentials of Ni/Ni^{2+}, Zn/Zn^{2+} and Mn/Mn^{2+} half-cells are -0.26 V, -0.76 V and -1.18 V (*vs*. NHE), respectively. When Zn metal is contacted with an alkaline KOH solution, Zn dissolution is negligible. However, when a small amount of Ni^{2+} exists in the solution as an impurity, Zn dissolution becomes serious. Identify two reduction reactions that are paired with Zn dissolution (oxidation). Note that the exchange current of hydrogen evolution on Ni electrode is much larger than that on Zn electrode (refer to Table 11-2). What happens when Mn^{2+} is present instead of Ni^{2+}? Suggest which impurity metal ions should be avoided to suppress the self-discharge of Zn electrode in Zn/MnO_2 alkaline batteries.

02 Answer the following questions on lead-acid and zinc/silver oxide batteries.

(a) E_{cell} ($= E^{c}_{eq} - E^{a}_{eq}$) for a zinc/silver oxide cell is expressed by <12-4>. Derive E_{cell} for lead-acid cell.

(b) The discharge voltage profile of zinc/silver oxide cell is very flat, but it is sloping for lead-acid cell as shown in Fig. 12-5. Explain these features.

03 Since metal corrosion is an oxidation reaction, a galvanic cell forms by pairing with reduction reactions. Assume that the following two reduction reactions in neutral solutions (pH = 7) can be paired with the corrosion of metals (Fe and Cu), and answer the following. When a corrosion is coupled with H_2 evolution (I), it is called acid corrosion, whereas the metal corrosion coupled with O_2 reduction is called oxygen corrosion (II).

$$(I): 2H_2O + 2e = H_2(g) + 2OH^- \quad E^0 = -0.83 \text{ V}(vs. \text{ NHE})$$

$$(II): O_2 + 2H_2O + 4e = 4OH^- \quad E^0 = 0.4 \text{ V}(vs. \text{ NHE})$$

(a) Calculate the equilibrium potential (E_{eq}) of the two half-cells at pH = 7.

(b) Using the E_{eq} calculated above, sketch the reduction current *vs.* potential profiles. Here, assume that $H_2(g)$ evolution current is determined by the charge transfer rate in the given potential range, whereas O_2 reduction by the mass transfer rate. Also assume the presence of limiting current in the latter. Also, sketch the metal oxidation (corrosion) current *vs.* potential profiles using the following E^0 values. Assume that the activity coefficient is unity for all metal ions and the metal oxidation current is determined by the charge transfer rate.

$$Fe^{2+} + 2e = Fe \quad E^0 = -0.44 \text{ V}(vs. \text{ NHE})$$

$$Cu^{2+} + 2e = Cu \quad E^0 = 0.34 \text{ V}(vs. \text{ NHE})$$

(c) Answer the following from the above sketches.

(i) At pH = 7, the oxygen corrosion is dominant over the acid corrosion for both Fe and Cu. Explain this.

(ii) Explain why the acid corrosion becomes more probable with a decrease in pH.

(iii) Why is the corrosion rate of Cu slower than Fe at pH = 7?

(iv) Rationalize the assumption given in (b); mass transfer determines the overall rate in the oxygen reduction reaction and limiting current appears.

(v) How does the Fe corrosion rate and mixed-potential change as the O_2 concentration in the solution increases at pH = 7? From this, suggest a method to suppress Fe corrosion.

04 The overall discharge reactions for two primary cells (alkaline and Li/SO_2 cell) are as follows.

$$2Zn + 2MnO_2 + H_2O \rightarrow 2MnOOH + 2ZnO$$

$$2Li + 2SO_2 \rightarrow Li_2S_2O_4$$

Calculate the theoretical specific energy (in Wh/kg) assuming that the average discharge voltage of alkaline and Li/SO_2 cell is 1.2 V and 2.8 V, respectively.

05 The discharge reaction of a lithium-ion battery is as follows. Calculate the theoretical specific energy assuming that the discharge range $x = 0.5$ and the average discharge voltage is 3.7 V.

Negative electrode: $LiC_6 \rightarrow Li_{1-x}C_6 + xLi^+ + xe$

Positive electrode: $Li_{1-x}CoO_2 + xLi^+ + xe \rightarrow LiCoO_2$

06 $\Delta G^0 = -2108$ kJ/mol for the propane combustion reaction; $C_3H_8 + 5O_2 \rightarrow 3CO_2 + 4H_2O$. A fuel cell is fabricated with propane and oxygen, and the cathodic reaction is $O_2 + 4H^+ + 4e \rightarrow 2H_2O$ ($E^0 = 1.23$ V *vs.* NHE).

(a) Write the anodic reaction for propane.

(b) Calculate the standard electromotive force (E^0_{cell}) for this fuel cell.

07 A redox-flow cell can be fabricated using the following two half-cells.

$Pb^{2+} + 2e = Pb$ $\qquad E^0 = -0.125$ V(*vs.* NHE)

$PbO_2(s) + 4H^+ + 2e = Pb^{2+} + 2H_2O$ $\qquad E^0 = 1.47$ V(*vs.* NHE)

(a) Write the charge/discharge reactions occurring on the negative and positive electrode, respectively.

(b) An undivided cell is available. Explain why separator is not necessary.

(c) In theory, the working voltage of aqueous redox flow-cell is less than 1.23 V due to water decomposition. Is there the water decomposition problem in this redox-flow cell? If there is, how can you overcome this problem?

08 It is a common practice in lithium-ion batteries that carbon is added to enhance the electrical conductivity of the electrodes. Some carbons contain metal impurities such as Fe and Ni, which can cause problems in battery life and safety for the following reasons. The metal impurities contained in the positive electrode plate may be dissolved into the electrolyte solution and electrodeposited on the negative electrode to destroy the SEI layer and cause an electrical short between two electrodes.

(a) Show that the processes of Fe/Ni dissolution and electrodeposition are thermodynamically allowed under the operating conditions of lithium-ion batteries.

(b) Describe the process in which Fe/Ni are electrodeposited on the negative electrode to induce the self-discharge of graphite electrode, to destroy the SEI layer, and to cause an internal short circuit.

(c) One wants to add some substances that can inhibit the dissolution and electrodeposition of Fe/Ni. What additives are available to inhibit the dissolution and electrodeposition, respectively?

09 Considering the concept of closed-loop, the battery performance depends on the electrolyte resistance (R) rather than on the ionic conductivity of electrolyte. This is expressed in <12-3>. In fact, solid polymer electrolytes having lower ionic conductivity than ordinary electrolyte solutions are used as a separator/electrolyte. A solid polymer electrolyte film (10 μm in thickness) is prepared by dissolving 1.0 M $LiPF_6$ into poly(ethylene oxide), of which ionic conductivity is 10^{-5} S/cm. The ionic conductivity of a porous polyethylene film (1.0 mm thick) soaked with 1.0 M $LiPF_6$ in propylene carbonate is 10^{-2} S/cm.

(a) Calculate $R_{separator}$ for both.

(b) Rationalize why the solid polymer electrolyte can be used as a separator even if its ionic conductivity is poorer.

CHAPTER 13

Photoelectrochemistry and solar cells

In photoelectrochemistry, semiconductor electrodes are used because they can absorb light and convert absorbed light into electrical energy or chemical energy. Solar cells (photovoltaic cells) convert light energy into electrical energy, photoelectrolysis cells decompose water to obtain oxygen and hydrogen, and photocatalytic cells generate hydroxyl radicals (OH •), which are used in detoxification of contaminants.

The electrical properties of solids are explained on the basis of band theory. Unlike metals, there are band gaps in semiconductors, which determine the electrical and optical properties of semiconductors. Meanwhile, the density of charge carriers in semiconductors is relatively lower than free electrons in metals. Due to the difference in charge carrier density, the interface characteristics of semiconductor/solution are totally different from those of metal/solution interface. Such unique properties of semiconductor/solution interface enable the operation of various types of photoelectrochemical cells.

13-1 Semiconductor theory

Solids comprise a large number of atoms to form a specific crystal structure. As shown in Fig. 13-1, electrons in atoms fill atomic orbitals (*AO*) from the lowest level, forming the filled and vacant *AO*. If two atoms form a molecule, molecular orbitals (*MO*) are generated by a combination of *AO* and electrons are filled from the lowest *MO*. If a large number of atoms are combined like solids, *AO* will form a large number of *MO*. Here, as the number of atoms to be bonded increases, the energy gap between *MO*s gradually decreases, and consequently energy levels can be considered to be connected to each other. This is called an energy band in contrast to the discrete energy level. The electron-filled band is called as a valence band, and the empty one as a conduction band. The highest energy level of the valence band is called the ***valence band edge***, and the lowest energy level of the conduction band is called the ***conduction band edge***. The conduction band and valence band are separated by a ***band gap*** ($\boldsymbol{E_g}$), which determines the

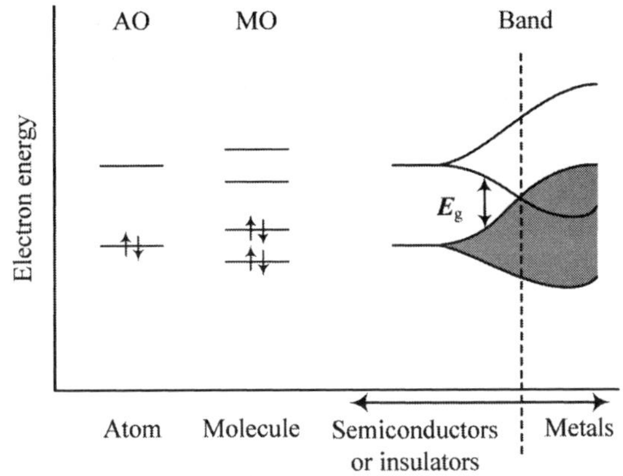

Fig. 13–1 Formation of molecules and solids by atomic bonds, and formation of energy bands in solids.

electrical and optical properties of solids. Since the band gap has a unit of energy, eV is used as a unit; the electrode potential is denoted by E, so the energy $\boldsymbol{E}$ is distinguished by a bold face. Normally, if $\boldsymbol{E}_g < 3.0$ eV, it is categorized into a semiconductor, while it is called an insulator if $\boldsymbol{E}_g > 3.0$ eV. This will be explained from the group IV element. Diamond (C), silicon (Si), germanium (Ge) and α-tin (α-Sn) all have the same crystal structure and bonding properties. However, diamond ($\boldsymbol{E}_g$ = 5.4 eV) is an insulator, Si ($\boldsymbol{E}_g$ = 1.1 eV) and Ge ($\boldsymbol{E}_g$ = 0.7 eV) belong to semiconductor, and α-Sn ($\boldsymbol{E}_g$ = 0.09 eV) has a metallic characteristic.

Semiconductors show electric conductivity due to an excessive amount of electrons existing in conduction band or an excessive holes forming in valence band. Both electrons and holes participate in electrical conduction so as to be called ***charge carriers***. There are three main ways for the generation of charge carriers in semiconductors. Firstly, electrons existing in valence band are excited to conduction band by thermal energy. Since the average thermal energy kT = 0.026 eV at room temperature, electron excitation by heat is possible only when the band gap of semiconductors is small enough. Since electrons are excited from valence band, the same number of holes are generated in valence band. This is called an ***intrinsic semiconductor***. Secondly, electrons in valence band can also be excited to conduction band by absorption of photons if the photon energy is larger than the band gap; $h\nu > \boldsymbol{E}_g$. Hence, the minimum energy (maximum wavelength, λ_{bg})

Table 13–1 Band gap and the maximum wavelength of absorbable light (λ_{bg}) for semiconductors

Semiconductor	Band gap (eV)	λ_{bg} (nm)
Si	1.1	1130
GaAs	1.4	890
CdSe	1.7	730
Fe_2O_3	2.1	590
TiO_2	3.0	410
ZnO	3.2	390
SnO_2	3.5	350

of light that a semiconductor absorbs is determined by <13-1>.

$$\lambda_{bg}\ (\text{nm}) = 1240/\boldsymbol{E}_g\ (\text{eV}) \qquad \text{<13-1>}$$

Table 13-1 compares the band gap ($\boldsymbol{E}_g$) and λ_{bg} for some semiconductors. From <13-1>, it is expected that the semiconductors of $\boldsymbol{E}_g$ = 1.5 ± 0.5 eV are the choice for solar cells since the photovoltaic efficiency of a solar cell is maximized when the light of 600 ~ 1000 nm is used in sunlight.

The third way to create charge carriers is by doping, which introduces a new energy level inside the band gap by either changing the stoichiometry of solids or introducing foreign elements into solid lattice. This is called an ***extrinsic semiconductor*** and distinguishes it from intrinsic semiconductor (e.g. pure Si, Ge, etc.). A typical example of introducing foreign element into solid lattice is to introduce a group V element (phosphorus, P) or a group III element (boron, B) into the group IV solid matrix (Si). The doping of P introduces a donor level in band gap near the conduction band edge as shown in Fig. 13-2-a, which is filled by electrons. Since the energy gap between the conduction

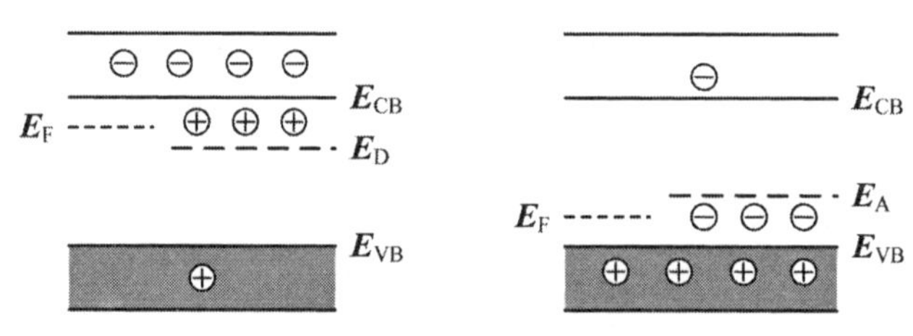

Fig. 13–2 Formation of extrinsic semiconductors: (a) *n*–type by doping with P in Si and (b) *p*–type by doping with B.

band edge (E_{CB}) and donor level (E_D) is very small, the electrons in the donors can be easily excited to the conduction band even at ambient temperature. As explained above, the number of electrons in conduction band and holes in valence band is the same in the intrinsic Si semiconductor that is not doped with P, but the number of electrons in the conduction band is far exceeded the number of holes in the valence band due to P doping. In fact, when 1 ppm of P is doped into Si, the density of electrons in the conduction band at room temperature is $5\times10^{16}/cm^3$ but the density of holes in the valence band is $4000/cm^3$; the electrons in the conduction band are the major charge carrier, hence it is called an *n*-type extrinsic semiconductor (or abbreviated as an ***n-type semiconductor***). When B is doped in the Si lattice, vacant acceptor levels are created within band gap near the valence band edge as shown in Fig. 13-2-b. The electrons in the valence band are readily excited to the acceptor level by heat; the holes of the same number as the acceptor are generated in the valence band. The density of holes in the valence band exceeds the density of electrons in the conduction band, so that the major charge carriers are the holes; this is called a ***p-type semiconductor***.

Self learning 13-1

TiO_2 belongs to an *n*-type semiconductor, but any foreign elements are not doped. How does it become an *n*-type extrinsic semiconductor?

The ***Fermi level*** (E_F) is shown in Fig. 13-2, which is defined as the energy level at which the probability of electron filling is 50%. In metals, the probability of electron filling at the energy level immediately above E_F is 0 %, but it is 100% at the energy level immediately below E_F; hence, the probability of electron filling at E_F is 50 %. As discussed in Chap. 1-1-1, the electron energy in metal electrodes can be controlled by electrode potential. Here, the control of electrode potential means a control of Fermi level.

In the case of intrinsic semiconductors, E_F is located at the middle of band gap. However, since *n*-type extrinsic semiconductors have more electrons in the conduction band as compared to intrinsic semiconductors, the energy level for 50 % probability for electron filling (E_F) is shifted toward the conduction band. It is located between the conduction band edge ($\boldsymbol{E}_{\mathrm{CB}}$) and donor level ($\boldsymbol{E}_{\mathrm{D}}$) as shown in Fig. 13-2. For *p*-type semiconductors, $\boldsymbol{E}_{\mathrm{F}}$ is located between the valence band edge ($\boldsymbol{E}_{\mathrm{VB}}$) and acceptor level ($\boldsymbol{E}_{\mathrm{A}}$).

13-2 Semiconductor/electrolyte interface

In solid-state physics, the Fermi level and energy band are defined with respect to a vacuum, so that the electrical potential is also defined respect to a vacuum. In electrochemistry, however, electrode potential is defined based on NHE (normal hydrogen electrode). As shown in Fig. 13-3, when the electron energy in vacuum is defined as 0.0 eV, the energy of NHE corresponds to −4.5 eV. The Fermi level corresponding to the standard electrode potential (E^0) of half-cells can be converted using <13-2>. For example, since E^0 = 0.77 V (*vs*. NHE) for the Fe^{2+}/Fe^{3+} half-cell, the energy against the electron in vacuum corresponds to −5.27 eV. Further, as shown in Fig. 13-3, the Fermi level of Au is −5.1 eV against the electron in vacuum.

$$\boldsymbol{E}_{\mathrm{F}} = -4.5\ \mathrm{eV} - e_0 E^0 \qquad \text{<13-2>}$$

Assume that the Fermi level ($\boldsymbol{E}_{\mathrm{F,\ O/R}}$) of a redox pair (e.g. Fe^{2+}/Fe^{3+}) is lower than the Fermi level ($\boldsymbol{E}_{\mathrm{F}}$) of an *n*-type semiconductor as shown in Fig. 13-4. When the *n*-type semiconductor is contacted with the redox pair in solution, electrons that are the major charge carriers of *n*-type semiconductor move to the redox pair of lower energy level.

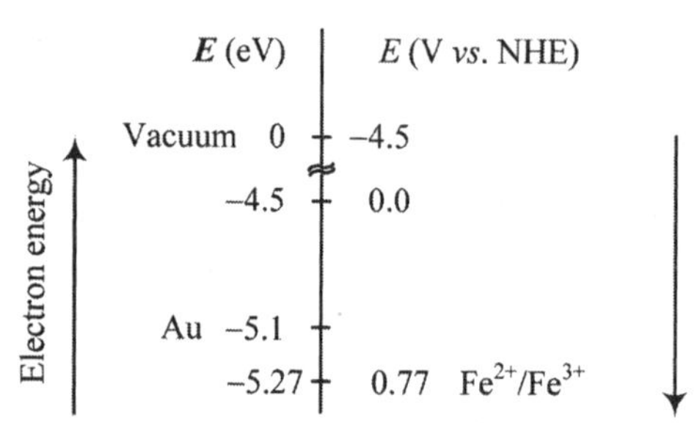

Fig. 13-3 The scale of electron energy (*E* in eV) *vs.* vacuum and electrode potential (*E*) *vs.* NHE.

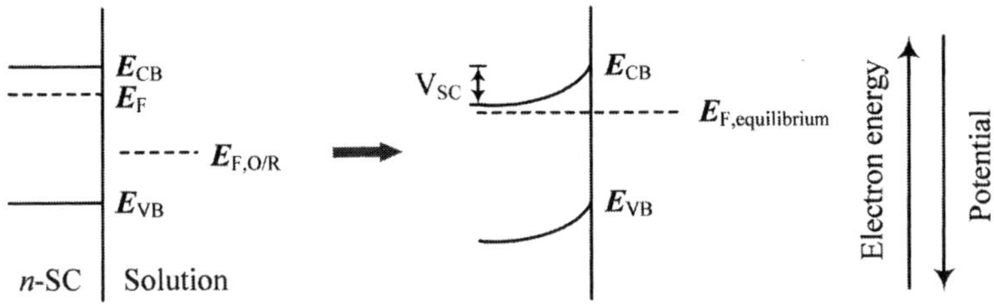

Fig. 13-4 Band bending phenomenon at the *n*-type extrinsic semiconductor contacting with a solution containing a redox pair.

As a result, the Fermi level of the redox pair increases because electrons are filled in the empty MO. Conversely, the *n*-type semiconductor loses electrons, so the Fermi level is lowered. This electron transfer proceeds until the Fermi level of the semiconductor and redox pair is equal to reach an equilibrium state ($E_{F,\ equilibrium}$). Through this process, an excess of positive charge in the semiconductor and an excess of negative charge (negative ions) in the solution are generated and distributed in each space charge region.

Here, a question arises as to how thick the space charge regions in the semiconductor and solution side are. In the case of metal electrode, even if an excessive charge is generated, the space charge region is extremely shallow because the excessive charge is negligibly small compared to the free electrons in metal. In addition, the amount of ions in electrolyte is much smaller than free electrons, such that the space charge region in the electrolyte side is much wider (ca. 10-1000 Å). This is why we ignore the space charge region in metal but consider only the space charge region in the solution side (Fig. 3-23, see Section 3-9). When the charge carrier density is compared for an *n*-type semiconductor and electrolyte solution, the electron density in the former is much smaller than the ion concentration in electrolytes (see SL 3-6), Therefore, we consider only the space charge region in semiconductor side ignoring the space charge region in electrolyte solution. In the space charge region of semiconductor, there appears a band bending as shown in Fig. 13-4. Since the excess negative charge is populated in the solution side while excess positive charge is arranged in the semiconductor, electron energy on the solution side is higher than that on the semiconductor side; the electrical potential is lower in the solution side. The electrical potential within the space charge region of semiconductor increases toward the inside according to the Poisson equation <3-47>. It is also noted in Fig. 13-4

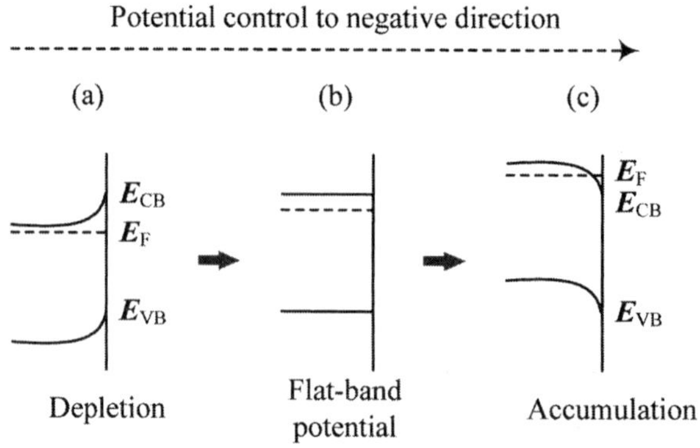

Fig. 13-5 Change of Fermi level by controlling the electrode potential of *n*-type semiconductor to the negative direction.

that a potential difference (V_{SC}) evolves within the space charge region of semiconductor electrode.

Like metal electrodes, the potential of semiconductor electrodes can be controlled to change the electron energy (i.e., the Fermi level). In Fig. 13-5, the potential of semiconductor electrode is controlled to be more negative from (a) to (c). In (a), electrons, which are the major charge carrier of *n*-type semiconductor, migrate to the inside of semiconductor because the electron energy is lower at the inside; electrons are depleted in the space charge region; thus it is referred to as a ***depletion state***. When the electrode potential changes to a more negative value, the Fermi level increases and also the degree of band bending gradually decreases to reach a state where there is no band bending. It is referred to as a ***flat-band potential*** (b). When the potential moves to even more negative direction, the Fermi level (E_F) becomes higher than the conduction band edge (E_{CB}) and the band is bent in the opposite direction, which is called an ***accumulation state*** (c). At this point, electrons are injected into the conduction band, such that the semiconductor electrode shows the metal-like conduction behavior.

13-3 Photoelectrochemistry in semiconductor

As shown in Fig. 13-6, when an n-type semiconductor electrode in the depletion state is irradiated by light having a higher energy than the band gap ($h\nu > E_g$), photons are absorbed to excite the electrons in the valence band. As a result, holes in the valence band and electrons in the conduction band are generated. Under the depletion state, the electrons migrate to the inside of the semiconductor due to the given electric field in the space charge region, and the holes move to the surface. Here, if there is any electroactive chemical species (R) in the solution near the surface, the holes (h^+) react with the chemical species; the electroactive chemical species is oxidized; $R + h^+ \rightarrow O$. In normal condition without irradiation, the concentration of holes is negligible for an n-type semiconductor electrode, such that the oxidation reaction is impossible even if electroactive species exists in the solution near the semiconductor surface. However, when the light ($h\nu > E_g$) is irradiated, holes are generated and migrate to the surface to participate in the oxidation reaction. This is called a ***photo-anodic current***; the oxidation current is induced by photons (irradiation).

Fig. 13-7 shows the potential-dependent photo-anodic current. When the potential of n-semiconductor electrode moves to the positive direction without irradiation, no oxidation current flows because the holes concentration is negligibly small. Even under irradiation, if the applied potential is more negative than the flat-band potential, an accumulation state is reached, where the holes generated by irradiation move to the semiconductor inside; oxi-

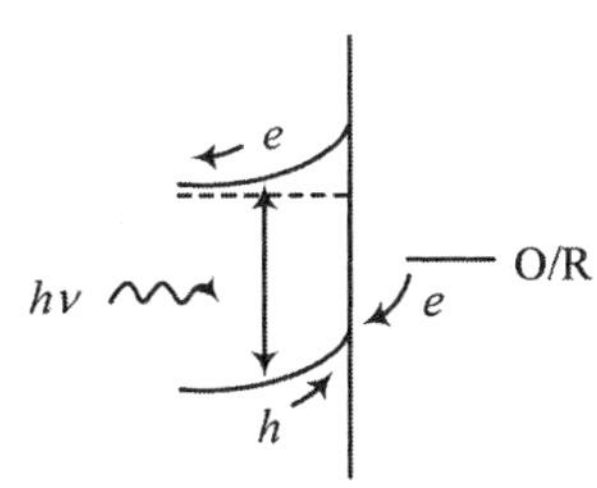

Fig. 13-6 Generation of electrons and holes under irradiation in the depletion state of n-type semiconductor, and flow of photo-anodic current.

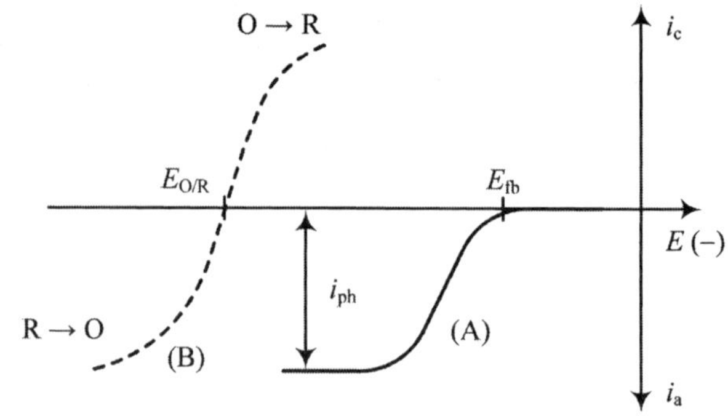

Fig. 13-7 Photo-anodic current under irradiation on *n*-type semiconductor contacting with an electroactive chemical species in solution.

dation reaction is impossible. However, if the applied potential moves more positive than the flat-band potential with irradiation, a depletion state is reached and electrons/holes are generated. Now, the holes (h^+) move to the surface to participate in the oxidation reaction ($R + h^+ \rightarrow O$); a photo-anodic current flows (curve *A*). Fig. 13-7 illustrates that, when a metal electrode is used, the oxidation current ($R \rightarrow O + ne$) is observed when the applied potential becomes more positive than the equilibrium potential of *O/R* half-cell ($E_{O/R}$) (curve *B*). When an *n*-type semiconductor electrode is used with irradiation, however, the oxidation reaction takes place from E_{fb}; the oxidation reaction is allowed at less positive potential (curve *A*) as compared to the metal electrode (curve *B*). Hence, this phenomenon is called a ***photoassisted electrode reaction*** because light energy assists the oxidation reaction to occur at a lower potential.

When an electrochemical cell is fabricated with a semiconductor working electrode, a metal counter electrode, and an electrolyte, the electrons and holes generated by irradiation can be utilized for production of chemicals or electricity. Such cells are called a ***photoelectrochemical cell*** (PEC), which are further divided into three types.

Self learning 13-2

Consider the electrons in the conduction band in Fig. 13-6 and in Fig. 13-2-a. Differentiate the generation path for the two.

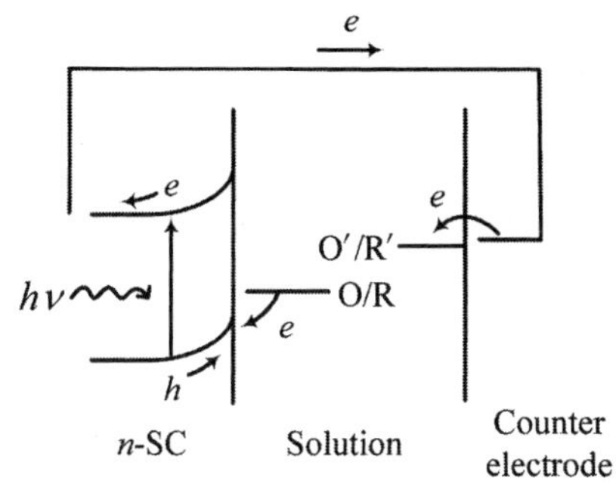

Fig. 13-8 Configuration of photosynthetic cell.

Fig. 13-8 shows the configuration of a ***photoelectrosynthetic cell*** (or photoelectrolytic cell). Since the equilibrium potential of two half-cells shows $E_{O/R} > E_{O'/R'}$, the *O/R* pair has a higher reduction tendency while the other (*O'/R'*) higher oxidation tendency. However, since the overall reaction consists of oxidation of *R* and reduction of *O'*, the overall reaction is not spontaneous ($\Delta G > 0$); light energy is needed to proceed the combined reaction. It is similar to the electrolytic cell (e.g. brine electrolysis) in that the non-spontaneous reaction is promoted by the assist of light energy. A typical example is the photoelectrolysis of water to produce hydrogen and oxygen. The half-cell reactions and the overall reaction proceed in solution as follows.

Anode: $2H_2O \rightarrow O_2 + 4H^+ + 4e$ $\quad E^0 = 1.23$ V(*vs.* NHE)

Cathode: $4H^+ + 4e \rightarrow 2H_2$ $\quad E^0 = 0.0$ V(*vs.* NHE)

Overall: $2H_2O \rightarrow 2H_2 + O_2$ $\quad (\Delta G^0 = 475$ kJ/mol)

Note that the equilibrium potential of oxygen half-cell under standard state is more positive than that of hydrogen half-cell, such that the paired reaction of O_2 evolution/H_2 evolution is not spontaneous ($\Delta G^0 > 0$); the assist of light energy is needed. Under irradiation, holes are generated and move to the surface of semiconductor electrode under a depletion state, water molecules give electrons to the electrode and are oxidized to O_2 and H^+. The electrons in the conduction band, which are generated by light absorption, move through the external circuit to the counter electrode to reduce H^+ to produce H_2.

Self learning 13-3

In Fig. 13-8, the equilibrium potential of two half-cells ($E_{O/R}$ and $E_{O'/R'}$) must have a certain range of values to enable the cell operation. What range of values should they have?

In the ***photocatalytic cell*** shown in Fig. 13-9, the combined reaction of R oxidation and O' reduction can take place spontaneously because $E_{O/R} < E_{O'/R'}$, which represent the equilibrium potentials for two half-cells. Thus, light energy is used only to overcome the activation barrier for the reaction. A typical example is the photocatalytic cell, in which acetic acid is oxidized and oxygen is reduced on n-TiO_2 semiconductor electrode under irradiation. The holes generated by light absorption under depletion state move to the electrode surface and oxidize OH^- adsorbed on the surface into hydroxyl radicals.

$$h^+(s) + OH^- \rightarrow OH\bullet$$

The as-generated hydroxyl radicals oxidize acetic acid. Meanwhile, the electrons generated by light absorption reduce Ti^{4+} to Ti^{3+} in the n-TiO_2 lattice; Ti^{3+} then reacts with oxygen. In the detoxification process for removing water contaminants, oxidizing agents are used to decompose the contaminated organic substances into CO_2, H_2O, and HCl. Hypochlorite (ClO^-) is a typical oxidizing agent, but many organic contaminants react with it to generate toxic substances. The generation of toxic substances is not serious when the hydroxyl radicals ($OH\bullet$) are used as the oxidizing agent, which are conventionally produced by either reacting hydrogen peroxide (H_2O_2) with Fe^{3+} (called the Fenton

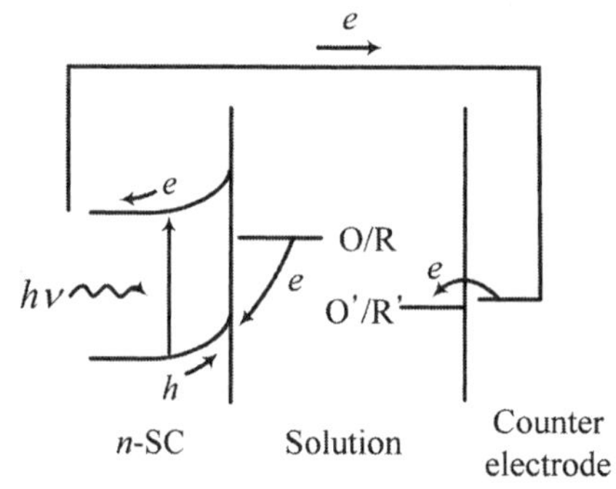

Fig. 13–9 Configuration of photocatalytic cell.

reaction), or by irradiating ultraviolet lights on ozone (O_3) or hydrogen peroxide. As described above, the hydroxyl radicals can also be produced by irradiating lights on n-TiO_2; irradiation on n-TiO_2 powder that is dispersed in a contaminated water with oxygen injection, hydroxyl radicals (OH•) are generated, which decompose the organic contaminants. This photoelectrochemical cell is similar to the cell shown in Fig. 13-9, but different in that it uses a part of n-TiO_2 particles as an anode while another part as a cathode without use of any counter electrode (see Section 2-4). Here, the electrons generated by light absorption move to the n-TiO_2 surface to reduce oxygen.

Self learning 13-4

In the photocatalytic cell of Fig. 13-9, the equilibrium potential of two half-cells ($E_{O/R}$ and $E_{O'/R'}$) must have a certain range of values to enable the cell operation. What range of values should they have?

13-4 Solar cells or photovoltaic cells

Fig. 13-10 shows a schematic diagram of a solar cell (or a ***photovoltaic cell***). The photo-anodic reaction proceeds on the n-type semiconductor electrode, and the oxidized compound (O) is reduced back to the initial compound (R) on the counter electrode (see Fig. 13-6). The combined reactions proceed spontaneously without providing electrical energy from outside, and the electric current flows through the external load to provide electric energy; it is a device that converts light energy into electric energy.

In order to induce the photoassisted electrode reaction by using semiconductor electrodes, electrons and holes should be generated by light absorption, which is possible when the band gap of semiconductors is smaller than the photon energy; $E_g < h\nu$. n-TiO_2 semiconductor can not utilize more than 95% of light because the band gap is too large (E_g

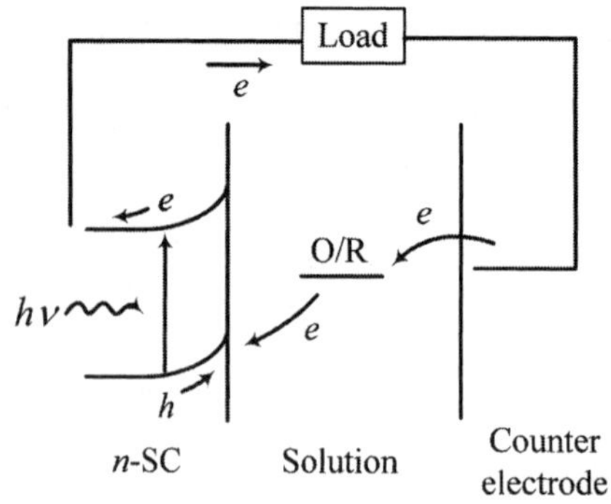

Fig. 13–10 Configuration of photovoltaic cell.

= 3.0 eV, see Table 13-1); it absorbs only the lights of shorter wavelength. A dye sensitization method has been attempted to enlarge the absorption range to the longer wavelength. Since dyes absorb the lights in wide wavelength range, a ***dye-sensitized solar cell*** using the dyes has been developed. Fig. 13-11 shows the schematic diagram of this. The dye (D) adsorbed on a semiconductor (e.g. n-TiO_2) absorbs the visible light to be converted into the excited dye (D^*); electrons in the *HOMO* are excited to the *LUMO*. Here, if the conduction band edge of the semiconductor is lower than the *LUMO* of the excited dye (D^*), electrons move from the excited dye to the semiconductor. The excited dye is then converted into the oxidized form (D^+) after losing electrons. Now, if R (e.g.

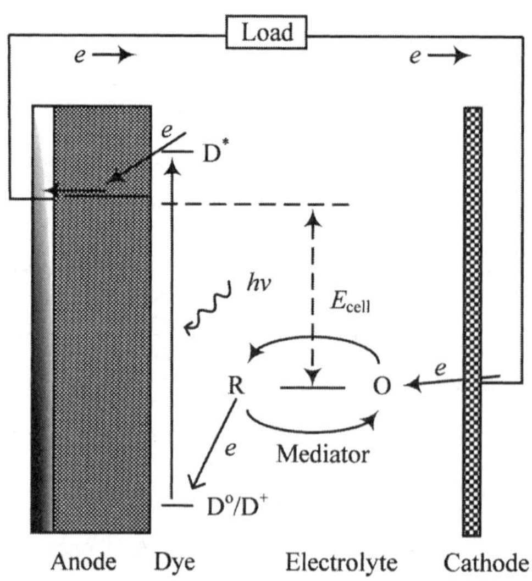

Fig. 13–11 Operation principle of dye–sensitized solar cell.

iodide, I^-) is present in the solution, the oxidized dye (D^+) can be converted into the initial state (D) by taking electrons from R; R is oxidized into O (triiodide, I_3^-). The as-regenerated O takes the electrons, which are provided from the anode through an external load, to be R. This sequential process is continued with continued irradiation. Here, a closed-loop is formed; electrons move along the clockwise direction as is seen in Fig. 13-11. The net result is the electron flow through the external load to provide electric energy to the load; it is a photovoltaic cell as it converts light energy into electric energy. The maximum voltage for this photovoltaic cell (E_{cell}, electromotive force or open-circuit voltage) is the difference between the flat-band potential (E_{fb}) and the equilibrium potential ($E_{O/R}$) of the O/R half-cell (e.g. I_3^-/I^- redox mediator). Since the practical working voltage (E_{wk}) is determined by <12-3>, it is necessary to reduce the overpotentials at two electrodes and the total resistance to obtain a high working voltage. As the semiconductor electrodes, nano-sized particles with large surface area are used in order to adsorb a large amount of dye molecules.

The dye-sensitized solar cell is a kind of electrochemical cells that are composed of two electrodes and an electrolyte. The solid-state solar cells are completely different in that they are electrolyte-free and all the components are solid. The examples are the *p-n* junction solid solar cell, donor/acceptor-type solar cell, and perovskite solar cell. The ***p-n* junction solid solar cell** is fabricated by adhering a *p*-type silicon or compound semiconductors (e.g. $CuIn(Ga)Se_2$) to an *n*-type one. Fig. 13-12 shows their working principle.

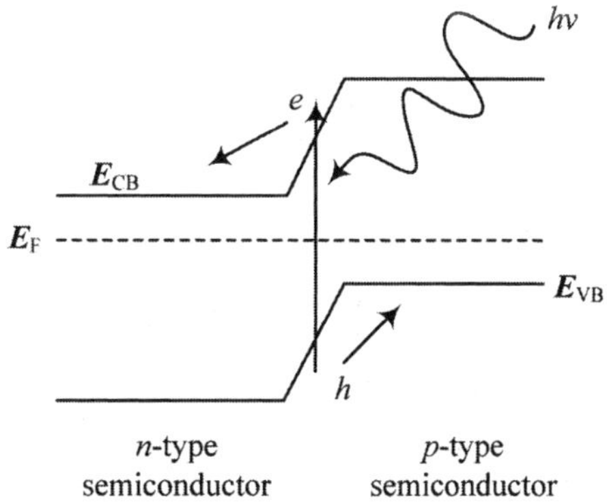

Fig. 13-12 Operation principle of *p-n* junction solid-state solar cell.

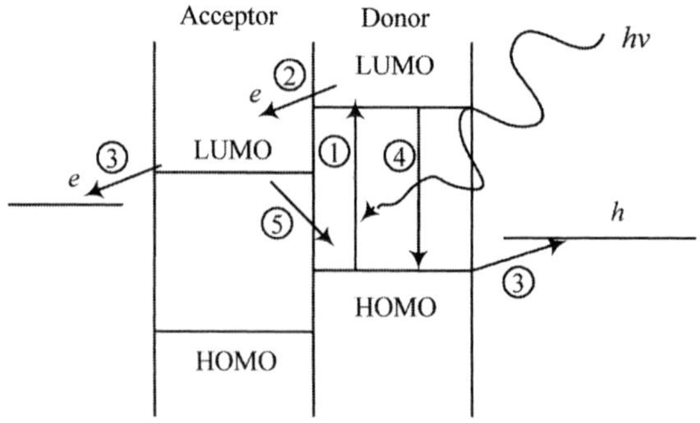

Fig. 13-13 Operation principle of donor/acceptor-type organic solar cell.

Electrons and holes are generated by light absorption, and the as-generated electrons move toward the *n*-type semiconductor while the holes are transferred toward the *p*-type one; a charge separation occurs. The as-separated electrons pass through an external load to recombine with the holes, an electrical energy is provided to the external load. The charge separation is the requisite for this solar cell. If electron/hole recombination occurs before the charge separation, the light energy is converted (wasted) into heat; the light conversion efficiency decreases.

Fig. 13-13 illustrates the working principle of the donor/acceptor type solar cell. When the donor absorbs light, the electrons in the *HOMO* level of donor are excited to the *LUMO* and a negative polaron is produced at the *LUMO*, while a positive polaron being generated at the *HOMO* (process ①). If the *LUMO* of acceptor is lower than the *LUMO* of donor, electrons move to the acceptor (process ②), which are further transferred to the anode (process ③). On the other hand, the positive polaron located at the *HOMO* of donor moves to the cathode. Through these processes, the electrons and holes generated by light absorption are separated; they pass through the external load to provide electrical energy. Since electron/hole recombination is possible inside the donor (process ④) or at the donor/acceptor interface (process ⑤), the minimization of this is a key factor for increasing the light conversion efficiency. To facilitate the charge separation, the contact area between the donor and acceptor should be maximized. Toward this end, two kinds of polymers functioning as a donor and acceptor respectively are blended, or a nano-sized substance serving as an acceptor and a polymeric donor can be mixed as a form of composite.

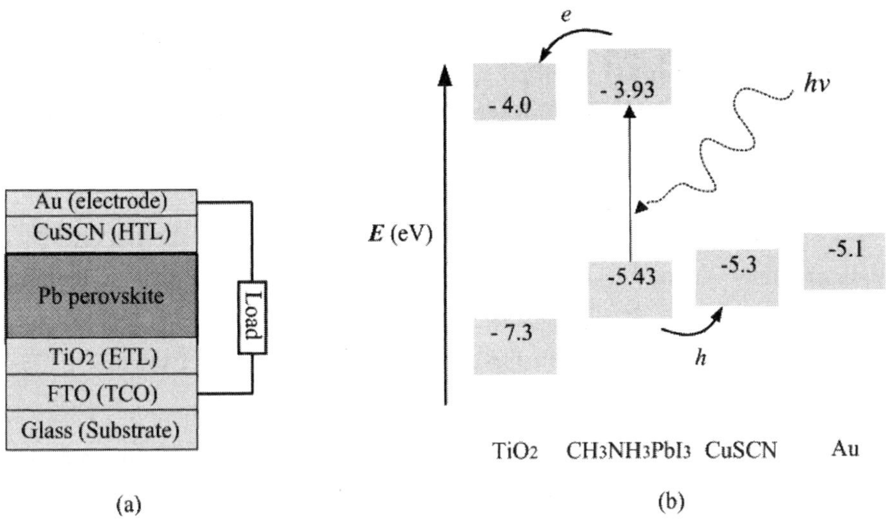

Fig. 13-14 (a); The structure of a perovskite solar cell, (b); Energy level diagram. HTL; hole transporting layer, ETL; electron transporting layer, FTO; fluorine-doped tin oxide, and TCO; transparent conductive oxide.

In the dye-sensitized solar cell shown in Fig. 13-11, $CH_3NH_3PbX_3$ (X = Br or I) that has a perovskite structure (AMX_3, A; cation, M; metal ion, and X; anion) can be used as a dye. In addition, if a hole transporting material (HTM) capable of holes transferring is utilized, a solid-state dye-sensitized solar cell can be fabricated; the liquid electrolyte in which the redox mediator is dissolved is not required. By modifying this concept, various types of ***perovskite solar cells*** have been developed. Fig. 13-14 shows a typical example. In this solar cell, $CH_3NH_3PbI_3$ absorbs light to generate electrons and holes. This material has a band gap of 1.5 eV and λ_{bg} = 827 nm; it can absorb a wide range of visible light. Electrons generated by light absorption are transferred to an electron transporting layer (ETL), which is an *n*-type semiconductor, and holes are transported to a Au electrode through the CuSCN (hole transporting) layer that is a *p*-type semiconductor. In short, electrons and holes are separated, which meet through an external load to be recombined to provide electrical energy. The Pb-containing perovskite materials are suitable for this use; their band gap is suitable for light absorption and can further be optimized by the compositional change of the compound. These materials have many advantages as a light absorbing material for solar cells; a strong light absorption, a long diffusion distance of charge carriers, and a less serious recombination of electrons and holes.

Problems

01 Answer the following from the current-potential curves shown in Fig. 13-7. In the figure, $E_{fb} = -0.2$ V (*vs*. NHE) for the semiconductor electrode, and the used *O/R* redox pair is I_3^-(0.5 *M*) and I^-(0.5 *M*), respectively. Assume that light is irradiated at a dose of 5×10^{15} photons/s, and 100 % of the irradiated light produces electrons and holes, and the electron-hole recombination is negligible (this is not the case in reality). To simplify calculations, ignore overpotentials and *iR* drops.

(a) What is the electromotive force (or open-circuit voltage) of this photovoltaic cell?

(b) What is the magnitude of current when the light is irradiated?

(c) What is the working voltage (E_{wk}) of this photovoltaic cell when light is irradiated?

02 Fig. 13-15 shows the conduction band edge ($\boldsymbol{E}_{CB}$), valence band edge ($\boldsymbol{E}_{VB}$), and the band gap ($\boldsymbol{E}_g$) of several *n*-type semiconductors; for example, for *n*-TiO_2, $\boldsymbol{E}_{CB} = -4.3$ eV, $\boldsymbol{E}_{VB} = -7.3$ eV, and $\boldsymbol{E}_g = 3.0$ eV. The equilibrium potentials (E_{eq}) of hydrogen and oxygen half-cell at pH = 0 are also shown.

(a) When a platinum electrode is used as the counter electrode, which semiconductors can be used as the anode for water photoelectrolysis in thermodynamic point of view?

(b) Suggest the problems in terms of kinetics.

(c) Assume pH = 14 and answer (a) above.

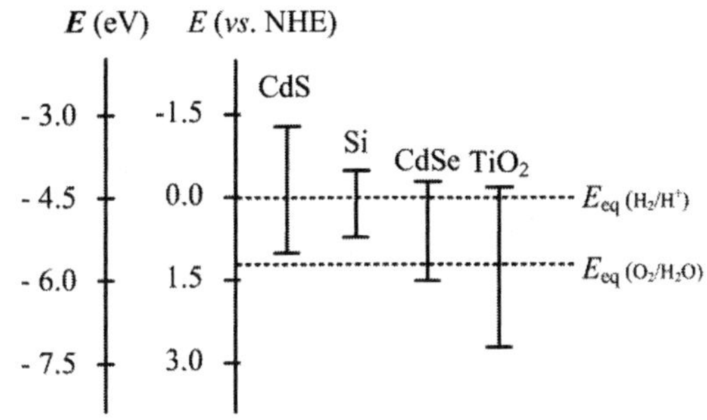

Fig. 13–15 Band structure of some n–type semiconductors and the equilibrium potential of hydrogen and oxygen half–cell at pH = 0.

03 A photocathodic current is possible in *p*-type semiconductor electrodes. Answer the following questions.

(a) Sketch the location of $\boldsymbol{E}_{\mathrm{CB}}$, $\boldsymbol{E}_{\mathrm{F}}$, and $\boldsymbol{E}_{\mathrm{VB}}$ for a *p*-type semiconductor.

(b) When light is not irradiated, which electrochemical reaction is possible?; oxidation or reduction?

(c) To induce a photocathodic current, how can you control the applied potential on the *p*-type semiconductor electrode from the flat-band potential?

(d) For a photocathodic current, what value should the $E_{\mathrm{O/R}}$ of redox pair (equilibrium potential of *O/R* half-cell) have as compared to $\boldsymbol{E}_{\mathrm{CB}}$, $\boldsymbol{E}_{\mathrm{F}}$, and $\boldsymbol{E}_{\mathrm{VB}}$?

Index

The author

Seung Mo Oh
(1988~2018) Professor in School of Chemical and Biological Engineering in Seoul National University
(2019~present) Chair professor in School of Energy and Chemical Engineering in Ulsan National Institute of Science and Technology

Electrochemistry 4th edition

저 자	오승모	저자와의 협의하에 인지생략
발행인	김지영	
발행처	자유아카데미	
주 소	경기도 파주시 회동길 37-42 파주출판도시	
전 화	031-955-1321	
팩 스	031-955-1322	
홈페이지	www.freeaca.com	
전자우편	main@freeaca.com (대표) editor@freeaca.com (편집) crm@freeaca.com (영업)	
등 록	제406-2003-017호, 1980. 7. 12	
제4판1쇄	2024년 1월 30일 발행	

ISBN 979-11-5808-566-7 93430

■ Physical constant

Term	Symbol	Value
Avogadro's number	N_A	6.02214×10^{23} mol^{-1}
elementary charge	e	1.60218×10^{-19} C
Faraday constant	F	9.64853×10^{4} C mol^{-1}
molar gas constant	R	8.31447 J mol^{-1} K^{-1}
Planck constant	h	6.62607×10^{-34} J · s
Boltzmann constant	k	1.38065×10^{-23} J K^{-1}
permittivity of free space	ε_0	8.85419×10^{-12} C^2 N^{-1} m^{-2}

■ Unit and conversion

Unit	Symbol	Value
calorie	cal	4.184 J
energy	erg	10^{-7} J
electron volt	eV	1.602×10^{-19} J
angstrom	Å	10^{-10} m = 10^{-8} cm = 10^{-4} μm = 10^{-1} nm
ampere	A	C s^{-1}
coulomb	C	F V
farad	F	C V^{-1}
joule	J	N · m
newton	N	kg · m s^{-2}
volt	V	J C^{-1}
watt	W	J s^{-1}

- **Evaluated constants for 25°C (298.15 K)**

$$kT = 4.116 \times 10^{-21}\ \text{J} = 25.69\ \text{meV}$$

$$RT = N_A kT = 2.478\ \text{kJ mol}^{-1} = 592\ \text{cal mol}^{-1}$$

$$f = F/RT = 38.92\ \text{V}^{-1}$$

$$1/f = RT/F = 0.02569\ \text{V}$$

$$2.303/f = 2.303\ RT/F = 0.05916\ \text{V}$$